Water Security

The purpose of this book is to present an overview of the latest research, policy, practitioner, academic, and international thinking on water security—an issue that, like water governance a few years ago, has developed much policy awareness and momentum with a wide range of stakeholders. As a concept it is open to multiple interpretations, and the authors here set out the various approaches to the topic from different perspectives.

Key themes addressed include:

- Water security as a foreign policy issue
- The interconnected variables of water, food, and human security
- Dimensions other than military and international relations concerns around water security
- Water security theory and methods, tools and audits

The book is loosely based on a Master's level degree plus a short professional course on water security both given at the University of East Anglia, delivered by international authorities on their subjects. It should serve as an introductory textbook as well as be of value to professionals, NGOs, and policymakers.

Bruce Lankford is Professor of Water and Irrigation Policy in the School of International Development at the University of East Anglia, UK.

Karen Bakker is a Professor in Geography, Canada Research Chair in Political Ecology, and Director of the Program on Water Governance at the University of British Columbia, Canada.

Mark Zeitoun is a Reader in the School of International Development at the University of East Anglia, UK, and Co-Director of the UEA Water Security Research Centre.

Declan Conway is Professor of Water Resources and Climate Change in the School of International Development, University of East Anglia, UK.

Water Security
Principles, Perspectives, and Practices

Edited by Bruce Lankford,
Karen Bakker, Mark Zeitoun,
and Declan Conway

First published 2013
by Routledge
2 Park Square, Milton Park, Abingdon, Oxon OX14 4RN

Simultaneously published in the USA and Canada
by Routledge
711 Third Avenue, New York, NY 10017

*Routledge is an imprint of the Taylor & Francis Group,
an informa business*

British Library Cataloguing-in-Publication Data
A catalogue record for this book is available from the British Library

Library of Congress Cataloging-in-Publication Data
Water security : principles, perspectives and practices / edited by
Bruce Lankford, Karen Bakker, Mark Zeitoun and Declan Conway.
    pages cm
  Includes bibliographical references and index.
 1. Water supply—Government policy.   2. Water security.
3. Water resources development.   I. Lankford, Bruce A.
  HD1691.W3646 2013
  333.91—dc23
  2013015504

ISBN: 978-0-415-53470-3 (hbk)
ISBN: 978-0-415-53471-0 (pbk)
ISBN: 978-0-203-11320-2 (ebk)

Typeset in Sabon
by Apex CoVantage, LLC

Contents

Preface

It is perhaps not surprising that the increasing use of the term *water security* mirrors growing concerns over how society's needs for resources will be met and who will gain relative to others. Water security is a lens through which to understand the risks of a lack of water, poor quality water, and inadequate flood protection, as well as how these are distributed. This contrasts with a vernacular understanding of the notion of security, which for water would imply that in facing these risks, it should be appropriated, annexed and secured. What this book (and literature elsewhere) shows is that a 'securitisation' interpretation of water security is unhelpful in its framing of the challenge of managing water.

Yet water security resonates with donors, countries, individuals, and organisations. Water security invokes the ideas of risk but also action—that water insecurities exist and can be managed—and in a much more acute way than the rather more neutral term IWRM (integrated water resources management) ever did. While some water managers, politicians, CEOs, or individuals might see water security in its unilateral 'securing' way, many more intuitively understand and witness the shared nature of the resource. This collective, integrated, and action-oriented view of water and water security is, arguably, common knowledge amongst water scientists, managers, and users. Thus, although this new term has arrived, their experiences tell them that the challenges of managing and sharing water remain considerable and highly complex. In short, little has changed; it would be an absurdity for water managers and scientists to be exhorted to rethink water in a new securitising way.

Nevertheless, even when the intention is better management, appropriative and securitising forces are generated continuously, sometimes subtly and unwittingly, potentially exacerbating shortages and resulting in inequitable distribution of water and water benefits. For example, placing an irrigation scheme above a small town can disrupt shares of river water during droughts and dry seasons, or in another example, the introduction of water charges for drinking water might marginalise the poorest in a community. To uncover and mitigate these appropriative forces for the benefit of public and environmental goods and services requires an understanding of many factors—entirely the ethos and concern of integrated water resources management. Therefore, perhaps what is interesting about water security is that, by implying unilateral action to securitise a resource, it acts as guard against this phenomenon reminding us that water is prone to capture, and therefore is best managed when understood collectively and governed cooperatively. A balance exists; to reflect upon possible inequities and insecurities implied by water security combined with the use of integrative adaptive frameworks (represented by IWRM) to deliver collective water solutions.

In this book, our chapter authors reflect on the idea of water security, applying their own specific lenses and experiences. As editors, we sought out authors that would cover the key themes in water and water security (e.g., law, climate change, domestic use, cities) and, without giving our contributors a strong predetermined interpretation of water security, we asked them to write from their perspectives. The collective effort in this book may not be a complete picture of the myriad dimensions of water and water security, but they do reflect some of the tensions and difficulties faced by science in accommodating the idea of water security (in terms of principles, perspectives, and practice) while navigating political, economic, environmental, and social concerns and demands for a sometimes limited and usually highly variable resource.

We have been helped through the editing process by many people. We warmly acknowledge the support of Earthscan, particularly Tim Hardwick and Ashley Wright, and also their copyediting team.

Considerable thanks are also due to all our reviewers who provided comments on drafts by our chapter authors: Maite Aldaya, Tony Allan, Henning Bjornlund, Janosz Bogardi, Vincent Casey, Anton Earle, Marie Ekstrom, Tom Franks, Dustin Garrick, Phil Hirsch, Holger Hoff, Guy Howard, William Howarth, Bryan Karney, Nicole Kranz, Tobias Krueger, Jamie Linton, Bjørn-Oliver Magsig, Michael Mason, Owen McIntyre, Katharine Meehan, Doug Merrey, Francois Molle, Jason Morrissette, Mike Muller, Peter Newborne, Micheal Norton, Thomas Perreault, Jaimie Pittock, Brian Richter, Chris Scott, Jan Selby, Afreen Siddiqi, Suvi Sojamo, Larry Swatuk, Erik Swyngedouw, Dan Tarlock, and Govindarajan Venkatesh. Your insights and diligence greatly improved this book's exploration of the idea of water security.

Bruce Lankford, Karen Bakker, Mark Zeitoun, Declan Conway
10 April 2013

Contributors

Mike Acreman is Head of Hydro-Ecology and Wetlands at the Centre for Ecology and Hydrology, Wallingford, UK, and Visiting Professor at University College London. His expertise is in environmental flows in rivers and wetland hydrology. In the UK, he is advisor to the Environment Agency and Natural England, with recent work focusing on defining environmental flows for the Water Framework Directive and impact assessment of wetlands. He was a lead author on freshwater systems in the UK National Ecosystem Assessment and he sits on the WWF-UK Programme Committee. Internationally, he is an advisor to IUCN, the Convention on Wetlands (Ramsar), and the World Bank advisory panel on environmental flows. He has recently edited a Special Issue of *Hydrological Sciences Journal* on Ecosystem Services of Wetlands. He is currently leading work for the Conventions on Biological Diversity and Wetlands (Ramsar) on the role of biodiversity in the water cycle.

J. A. (Tony) Allan heads the London Water Research Group at King's College London and SOAS. He specialises in the analysis of water resources in semi-arid regions and on the role of global systems in ameliorating local and regional water deficits. He pointed out that water-short economies achieve water and food security mainly by importing water-intensive food commodities. He coined the concept of *virtual water*. His ideas are set out in *The Middle East Water Question: Hydropolitics and the Global Economy* and in a new book, *Virtual Water*. He is currently working on why the accounting systems in the food supply chain are dangerously blind to the costs of water and of misallocating it. In 2008, he was awarded the Stockholm Water Prize in recognition of his contribution to water science. In 2011, he became Académico Correspondiente Internacional of the Academy of Sciences of Spain.

Karen Bakker is a professor in geography, Canada Research Chair in Political Ecology, and Director of the Program on Water Governance at the University of British Columbia (www.watergovernance.ca). Her research interests span political economy, political ecology, and water management, with a current focus on water supply privatization, delegated water management, transboundary water governance, and access to urban water supply in developing countries. She has published in development studies, environmental studies, geography, urban studies, and interdisciplinary environmental science (including science, global environmental change, and world development). Her two most recent book publications are: *Eau Canada* (UBC Press, 2007) and *Privatizing Water: Governance Failure and the World's Urban Water Crisis* (Cornell University Press, 2010). Fluent in French and Spanish, Dr. Bakker regularly acts as an advisor to governments and nongovernmental and international organizations, including the OECD, various UN agencies, DfID, and the Conference Board of Canada.

Rutgerd Boelens is Associate Professor in Water Management and Social Justice, Wageningen University, The Netherlands, and Visiting Professor at the Catholic University of Peru. His research focuses on water rights, legal pluralism, cultural politics, and political ecology in Latin America and Spain. He directed the Water Law and Indigenous Rights (WALIR) program; is coordinator of the international Justicia Hídrica/Water Justice alliance (comparative research on water accumulation and conflict in the Americas, Europe, Africa, Asia); and directs the programs "Struggling for Water Security" and "The Transnationalization of Local Water Struggles" of the Netherlands Organization for Scientific Research. Recent books include *Liquid Relations: Contested Water Rights and Legal Complexity* (with Roth/Zwarteveen, Rutgers University Press, 2005); *Water and Indigenous Peoples* (UNESCO, 2006); *Aguas Rebeldes* (IEP/Imprefepp, 2009); *Out of the Mainstream: Water Rights, Politics and Identity* (with Getches/Guevara, Earthscan, 2010); and *Justicia Hídrica: Accumulation, Conflict and Civil Society Action* (with Cremers/Zwarteveen, IEP, 2011).

Jonathan Chenoweth is a lecturer in natural resources management in the Centre for Environmental Strategy at the University of Surrey. His research focuses upon the institutional and policy dimensions of water management and sustainable development in developed and developing regions including in the UK and elsewhere in Europe, the Middle East, and Africa. His recent work has focused on coping with water scarcity, water resources management in the context of climate change, attitudes to water supply services, and the role of the public versus private sector in the management of water and sanitation services.

Floriane Clément worked as a postdoctoral fellow and researcher at the International Water Management Institute (IWMI) in Hyderabad, India, from 2008 to 2012. She is now posted at the IWMI Nepal Office. She has a multidisciplinary background, with an engineering degree, an MSc in environmental sciences, and a PhD in geography/political science from Newcastle University (UK). In her pre-PhD life, she gained a solid experience on industrial and public water issues while working at the Regional Chamber of Commerce and Industry Paris—Ile-de-France. Her main research interest is the analysis of the gap between discourses and practices/outcomes in government-led and donor-funded programmes in the field of natural resource management. Her research has cut across spatial scales and research perspectives, combining institutional analysis, discourse analysis, and political economy. She has work experience in France, Vietnam, India, Nepal, and Sri Lanka. She has published in international peer-reviewed geography and political science journals.

Declan Conway is Professor of Water Resources and Climate Change in the School of International Development, University of East Anglia, UK. His research concentrates on the interactions between climate, water resources, and society, with extensive experience in developing countries. He has a broad base of expertise that encompasses detailed knowledge of climate models, scenario generation, climate impacts assessment, and research and applied consultancy on policy and adaptation issues. He has long-term research interests in Ethiopia, the Nile Basin, and China. He is a founding member of UEA Water Security and is closely involved in the Tyndall Centre for Climate Change Research. He enjoys interdisciplinary research that links the biophysical and social contexts of climate and water.

Christina Cook is a post-doctoral fellow in the department of geography at the Hebrew University of Jerusalem. Her interdisciplinary training in resource management and sustainability, law, and biological sciences grounds her sociolegal research in water governance. Her current research interests include the politics of scale in water governance, intersections between water and land governance, drinking water governance and metagenomics, and water rights regimes.

Anton Earle is a geographer with an academic background in environmental management, specialising in transboundary integrated water resource management, facilitating the interaction between governments, basin organisations, and other stakeholders in international river and lake basins. He is experienced in institutional development and policy formation for water resource management at the interstate level in the Southern and East African regions, the Middle East, and internationally. In 2010, he was the lead editor for the Earthscan book *Transboundary Water Management: Principles and Practise*, aimed at practitioners and advanced students in that field. He is the Director of Capacity Development at the Stockholm International Water Institute (SIWI) and is completing a PhD in Peace and Development at the School of Global Studies at the University of Gothenburg. His thesis investigates the role of nonstate actors in transboundary water management processes.

Malin Falkenmark is Professor of Applied and International Hydrology, and tied to both the Stockholm Resilience Center at Stockholm University and Stockholm International Water Institute (SIWI). Her particular interests are interdisciplinary, with focus on similarities and differences between different regions, especially linkages between land and water and their policy implications. She has worked on water scarcity and water security issues for most of the last two decades, also studying the crucial role of the global water cycle as the bloodstream of the biosphere, deeply involved not only in human life support but also in generating environmental side effects from human efforts to harvest water, energy, and biomass from the natural landscape. She is a Global 500 Laureate and has been awarded the Swedish KTH Great Prize, the International Hydrology Prize, the 2005 Crystal Drop Award of IWRA, the Volvo Environment Prize, and the Prince Albert II of Monaco Award.

Antony Froggatt has studied energy and environmental policy at the University of Westminster and the Science Policy Research Unit at Sussex University. He is currently an independent consultant on international energy issues and a senior research fellow at Chatham House (also known as the Royal Institute for International Affairs). While working at Chatham House he has specialised on energy security and, in particular, working in emerging economies with extensive work in China on the establishment and methodologies of low carbon economic development. He was also an associate fellow at Warwick Business School from 2006 to 2007 and gave lectures at the Ecole des Mines de Nantes in France. He has worked as a consultant with environmental groups, academics, and public bodies, including the European Parliament and Commission in Europe and Asia, specialising in the development of policies, initiatives, and capacity building.

Dustin Garrick is a research fellow at the University of Oxford specialising in comparative water policy and economic analysis. His research examines the effectiveness of policy responses to water scarcity and climate risk in large transboundary rivers, particularly in semi-arid basins within or shared by federal countries. This research applies concepts and methods of institutional economics and draws heavily on the Institutional Analysis and Development Framework advanced by Elinor Ostrom and colleagues. Dr. Garrick has over a decade of experience at the intersection of water research and policy, with a geographic focus in the Western United States and Australia. Before joining Oxford in 2011, he was a Fulbright Scholar in Australia (2010–2011), where he examined water trading and river basin governance in the Murray-Darling Basin. He maintains an active water policy and economics research programme in Australia, and is a research associate with the Centre for Water Economics, Environment and Policy at Australia National University. He holds a PhD in geography (University of Arizona) and master's degree in environmental science and policy (Columbia University).

Nick Hepworth has worked in water and environment management for 20 years in Africa, Asia, Europe, and South America. As a practitioner, regulator, consultant, and researcher, he has helped governments, development agencies, communities, NGOs, and multinational corporations to identify and deliver on opportunities for a sustainable and fair future. His current research interests include action research to support citizen agency and institutional accountability to improve the effectiveness of aid and governance, 'rethinking' capacity and corporate water engagement. He provides strategic advice and analysis to governments in Europe and Africa, the World Bank and the UN, and to NGOs including IUCN, WWF, and WaterAid. He is Director of Water Witness International, an independent research and advocacy charity working for equitable water management in developing countries. He is a founding board member of the Alliance for Water Stewardship, co-convener of the Corporate Water Research Network, and is a Research Fellow at the University of East Anglia, UK.

Robert Hope is an economist at the University of Oxford whose research explores the relationship between water and development. His research has focused on understanding tradeoffs, choices, and outcomes in balancing multiple and competing water users across space and time. This has included two research strands: (1) river basin/catchment studies under varying hydrological, institutional, and sociopolitical contexts, and (2) designing and evaluating sustainability of water supply systems. He leads an interdisciplinary, cross-departmental research group on mobile water for development which seeks to design, test, and evaluate innovative applications of mobile communications technology for water security and poverty reduction in developing countries (http://oxwater.co.uk). His work has been funded by DFID, UK research councils (ESRC, NERC), World Bank, OECD, the Gates Foundation, and the Skoll Foundation in Colombia, Costa Rica, India, Kenya, Mozambique, Senegal, South Africa, Tanzania, Uganda, Zambia, and Zimbabwe.

Thoko Kaime is lecturer in law in the School of Law and Deputy Director of the Environmental Regulatory Research Group at the University of Surrey. He also serves as Senior Teaching Fellow at the School of Oriental and African Studies (SOAS), University of London. He has previously studied law at the Universities of Malawi, Pretoria, and Western Cape, and was awarded his PhD at SOAS. He maintains broad interests in the area of public international law and the social-legal critique of law and legal policy.

Bruce Lankford is Professor of Water and Irrigation Policy in the School of International Development at the University of East Anglia, UK. He has worked for more than 25 years in the fields of irrigation and water resources management, starting in Swaziland in 1983. His main research and advisory work covers water management in sub-Saharan Africa on the following themes: river basin management; irrigation policy in Africa; the use of games in democratizing discussions on water management, performance, and water allocation; system- and farmer-centred infrastructure design for water management; irrigation efficiency and productivity; and tradeoffs related to catchment and water-related ecosystem services. He is a Co-Director of the UEA Water Security Research Centre and a Fellow of the Institution of Civil Engineers.

Christina Leb is an associate member of the Platform for International Water Law at the Faculty of Law of the University of Geneva. Her publications and research focus on the governance of transboundary watercourses and aquifers, climate change, and the human right to water. She holds a doctorate in international law from the University of Geneva and works as a consultant in the field of transboundary water resources management and international water law with international governmental and nongovernmental organizations such as the World Bank, the World Conservation Union (IUCN), and the Stockholm International Water Institute (SIWI).

Rosalind Malcolm is a professor of law, Director of the Environmental Regulatory Research Group in the School of Law at the University of Surrey, and a barrister with Guildford Chambers. Between 2005 and 2010, she was Head of the School of Law at Surrey. As an environmental lawyer, she engages in a range of research areas within that field, such as regulatory frameworks for water and sanitation; regulatory approaches for ecodesign of green product development; and approaches to compliance and enforcement across the environmental arena. She writes on regulatory environmental areas and has extensive experience working within multidisciplinary research teams, in particular working within developing countries. She also researches and writes across environmental health fields, such as food safety and statutory nuisance, where she has been particularly involved with enforcement approaches and has had extensive experience in the training, development, and capacity-building of enforcement officers.

Nathaniel Mason works on a broad range of water-related issues. His interests include the politics of how information is presented and used, the intersection of socioeconomic development and the environment (especially around natural resource management), and the financial and institutional aspects of water supply, sanitation, and hygiene services. He is based at ODI's Water Policy Programme in London.

Naho Mirumachi is a lecturer in the Department of Geography, King's College London. Trained in international relations, international studies, and human geography, she focuses on the politics of water resources management. She examines issues of power, discourse, scale, and agency in water allocation and river basin development. Her research has taken her to the Orange-Senqu River basin in Southern Africa, the Ganges River basin in South Asia, and the Mekong River basin in Southeast Asia. At King's College London, she convenes the MSc Water: Science and Governance programme and is actively involved in the London Water Research Group activities, including training policymakers on water security.

Stuart Orr is Freshwater Manager at WWF International. Much of this work explores the roles and responsibilities of business in water management challenges, as well as understanding how risk manifests for multiple stakeholders. He has published mainly on water accounting, public policy, and water-related risk, recently co-drafting guidelines for the UN Global Compact on corporate engagement in water policy. Most recently, he led the development of an online risk tool for companies and investors. He holds an MSc from the University of East Anglia and worked for many years in the private sector, mainly in Asia. He is a member of the World Economic Forum's Council for Water Security, a water advisor for the Carbon Disclosure Project (CDP), and sits on a number of corporate sustainability advisory panels. He is also on the steering board of the Water Resources Group (WRG) and the Water Futures Partnership (WFP).

Steve Pedley is a reader in environmental engineering at the Robens Centre for Public and Environmental Health, University of Surrey. He is a microbiologist with over 30 years' experience of research and consultancy work. For the past 20 years he has specialised in water quality, pollution control, and public health, with extensive experience in bacteriological and virological techniques, and public health microbiology related to water quality. He has carried out research and consultancy projects on behalf of NERC, EPSRC, UK Environment Agency, USEPA, WHO, UNEP, DFID, British Council, DANIDA, FAO, Save the Children Federation, World Bank, and the International Atomic Energy Agency. He has worked on a number of water quality monitoring and assessment projects, including assessments of national and local capacity in Kenya, Ethiopia, Zimbabwe, and Qatar, and infrastructure development in Gambia, Ghana, Uganda, Costa Rica, Turkmenistan, and Mauritius.

David Tickner is Chief Adviser on freshwater issues at WWF. He provides strategic leadership to a UK-based team that supports river conservation projects around the globe and advises governments and companies on water risks, water stewardship, and water policy. He is currently also a Non-Executive Director of Water and Sanitation for the Urban Poor (WSUP) and a Research Fellow at the University of East Anglia. Previously, Dave worked in the UK government's environment ministry, led WWF's programme for the Danube River, and advised Standard Chartered Bank on water and environment issues. He holds a PhD in hydro-ecology and he has authored, edited, or contributed to a number of articles, scientific papers, and books on water-related issues.

Jeroen Warner teaches and publishes on disaster studies at Wageningen University, where he also took his PhD degree. His main research interests in the disaster studies domain are social resilience and participation, the politics of (flood) disaster risk reduction, and the role of disaster in international relations. In the water domain, he works on domestic and transboundary water conflict, participatory resource management, and governance issues. He has published over 60 peer-reviewed articles and seven books.

Patricia Wouters is Professor of International Law; Founding Director of the Dundee UNESCO Centre for Water Law, Policy and Science; Visiting Professor at Xiamen Law School, China; and Visiting Professor at IRES, University of British Columbia. She researches issues related to the rules of law (within an interdisciplinary context) that govern international watercourses. She has presented her research around the world and published extensively on topics related to transboundary waters. She has been appointed to a number of global water policy organisations and advisory boards, including: the United Nations University Institute of Water, Environment and Health; the Technical Expert Committee of the Global Water Partnership; and Global Agenda Council on Water Security, World Economic Forum. She continues her research in international water law and strives to contribute to building a new generation of 'local water leaders' with expertise in water law (see http://dundee.academia.edu/PatriciaWouters).

Benjamin Zala is Director of the Sustainable Security Programme at the Oxford Research Group. He is also a PhD candidate in International Relations at the University of Birmingham and a member of the editorial team for the journal *Civil Wars* (published by Routledge). He has previously worked at Chatham House and the La Trobe University Centre for Dialogue, as well as teaching international relations and diplomatic history at the University of Birmingham. His research focuses on approaches to world order in international relations theory, foreign policy analysis, and global security issues. He has published on nuclear deterrence, great power politics, and nontraditional security issues including in journals such as *Cooperation and Conflict, The RUSI Journal,* and the *Nonproliferation Review.*

Mark Zeitoun is a reader in the School of International Development at the University of East Anglia, and Co-Director of the UEA Water Security Research Centre. He is interested in the ways that power asymmetry and social justice interact to influence water security and perceptions of it. The interest stems from his work as a humanitarian-aid water engineer in conflict and post-conflict zones in Africa and the Middle East. He also consults regularly on water security policy, hydro-diplomacy, and international transboundary water negotiations.

Part I

Frameworks and Approaches to Water Security

1 Introduction

A Battle of Ideas for Water Security

Mark Zeitoun, Bruce Lankford, Karen Bakker, and Declan Conway

Introduction

There may be as many interpretations of 'water security' as there are interests in the global water community. The purpose of this book is not to provide yet another interpretation, but to explore the range the interpretations cover, and to move knowledge and thought forward through debate of interpretations that hold the greatest meaning to different actors.

In meeting this goal, the book thus queries the growing discourses on water security, in particular, narrow 'securitized' approaches to the concept. In its most alarmist securitized formulation, 'water security' suggests safeguarding the resource in volumetric terms from others, and is often associated with the desire to eliminate risk and variability through climate-proofing infrastructure, for instance. While such responses can appeal to political favour and public emotion in uncertain situations, they may also exacerbate existing problems or even trigger unintended consequences by reducing the number of options available in the future.

One of this book's clear messages is that water security passes not through armoury or concrete, but through the messier realm of policy and governance. From this perspective, we argue that water security cannot be achieved at the expense of the water security of others; sustainable outcomes require reconciliation of basic needs and access to water, as well as the most assured physical and social thinking of water science by practitioners, academics, students, and professionals. Within this reduced field of water security study, the ground covered is still quite broad, and the battle of ideas very acute.

Context and Scope

Having grappled with the terms of water security through scholarship, teaching, and training with a diverse constituency, we as editors sensed before we invited contributions to this book that we would not achieve consensus on any meaning of water security. Each author was thus encouraged to approach his or her contribution without a common frame of water security in mind. Now with contributions from 27 authors in 21 chapters, we are even more confident that the term has too many disciplinary, sectoral, ideological, and geographic roots to be conveniently pinned down.

There is a general political ecology disposition amongst the chapters, in the sense that the authors hold knowledge about water to be both be socially produced and generate material consequences that can be somewhat objectively measured. There is also consistency in critical thinking, neither eschewing nor espousing the rush towards 'water security' as a meaningful and possibly innovative concept. Otherwise, each chapter presents a personal perspective and interpretation of water security, and the volume might best be seen as a collection of analytical but partisan essays. Law, environmental science, international relations, hydropolitics, geography, political economy, and political ecology are all deployed here, cumulatively building on considerable scholarly work aimed at conveying the myriad dimensions of water. The path followed bypasses both the deterministic alarmist tone (captured by the phrase 'the world is running out of water') and ungrounded theorization (of risk, for instance).

As the title of the book would suggest, each chapter offers views on perspectives, principles, and practices of water security. With particular reference to 'practices', we caution against any expectation that the book records actual practice of water security—the concept is simply too novel to be (mis)understood in that light. Rather, our contributors in Part 2 have reflected on debates regarding the application of the idea of water security. We also wish to note that the order of the chapters within each section conveys no particular meaning in terms of priority or relationship, and all comprise elements of the 'battle of ideas'. All previous assertions not withstanding, a selection of these ideas are organised into a conceptual framework through which water security can be understood, and this is presented in the final chapter.

With the authors' partisan interests in mind, readers will notice that some chapters record recent thinking in the water security debate in an apparently neutral fashion, while many others weigh in to push water science to challenge concepts that are gaining undue wide acceptance, or to develop new ones. We hope the collective effort has been sufficiently contentious that water security does build some measure of identity that separates it from other terms that have lost coherent meaning in water research (such as *integrated water resources management* and *water governance*).

The Battle of Ideas for Water Security

Within the broad range of topics covered in the volume, we find that 'water security' serves both to revitalise old ideas and to promote new ones. This comes as no surprise, at this point in time when water scientists of all types face a degree of uncertainty that has questioned the very way we approach water resource futures. Some equate this with insecurity, and see opportunities in reducing the variability of river flows, meaning dams for agriculture or hydropower (e.g., Briscoe, 2009; Muller, 2012). Alternative views question the paradigms of distribution of the possibly reduced or increased water flows between countries and communities of vastly different capacities. Water security can inform and be informed by water science, in other words, through exploration of how climate change and water communities approach the same challenges (Conway, Chapter 6), how competing perspectives on water security are articulated with distinct governance practices (Cook and Bakker, Chapter 4), and how the engineering biases and fashions have guided infrastructure choices to date (Lankford, Chapter 16). Water security reaches beyond scientists, furthermore, to those concerned about food security (Allan, Chapter 20), national security (Zala, Chapter 17), environmental sustainability of businesses (Hepworth and Orr, Chapter 14), or 'international development' (be it human or economic development [Chenoweth et al., Chapter 19; Garrick and Hope, Chapter 13] in non-industrialised or urbanised/industrialised [Boelens, Chapter 15; Earle, Chapter 7] contexts).

The range of principles invoked throughout the book is more tightly delimited. Environmental sustainability and collaboration both figure so explicitly (or implicitly) throughout the chapters that they may be considered two inseparable and fundamental elements of water security. Almost as ubiquitous are references to the interdependencies water creates with everything else in the world, including energy (Froggatt, Chapter 8), the demands of cities (Earle), climate change (Conway), and—most evidently and importantly—food (Allan; Falkenmark, Chapter 5). Equity and justice are also given considerable weight, whether explicitly (Boelens; Leb and Wouters, Chapter 3; Garrick and Hope; Hepworth and Orr, Chapter 14; Zeitoun, Chapter 2) or implicitly by recognising an environmental exigency (Tickner and Acreman, Chapter 9).

The variety of topics and principles that inform this battle of ideas can be classed into the broad groups (and probable research directions) of emerging ideas and debates, interconnectedness, comprehensiveness, and harmonisation.

Emerging Ideas and Debates of Water Security

Fortunately, tensions can fuel progress. Clément enters the debate in her discussion of water productivity versus security, for instance, reminding us that the attention paid to principles and justice in this volume rarely reflects the dynamics of the real-world political economy. Equity typically takes second place to efficiency as the guiding principle in efforts to secure water for large corporations (e.g., WRG, 2010). Similarly, the role of armed forces in water provision in country recovery and stabilisation programmes is raised as a policy goal in foreign affairs and defence circles via concerns about how water security relates to state failure (see, e.g., DNI, 2012; King, 2012; Tanzler and Carius, 2012). On the other hand, Leb and Wouters here assert that 'military security and water security in this paradigm are incongruent goals'. And Warner (Chapter 18) points out the mechanism by which the two may nonetheless meet: a short (discursive) route from 'security' to 'threat', to the legitimatized 'securitization' and then militarisation of water resources (see also Cook and Bakker; Warner; Zala; Zeitoun).

Fruitful debate also occurs amongst contributors. For example, calls by Mason (Chapter 12) for water security indicators (of water availability, access, risk, ecosystem services, and institutions) sit alongside considered arguments to question our superficial understandings of the same (Falkenmark for scarcity (5), Clément for apolitical views of nature (10), and Mirumachi, (11) for the very political and commercial nature and interests of water institutions). Likewise, Garrick and Hope, propose that issues of water stress, pollution, water variability, and climate variability are best thought of and handled as risks. Yet Lankford, Chapter 16, argues that the apportionment of risk of excessive scarcity above and beyond that caused by natural distributions of rainfall and river flow can be traced to design faults in river basin architectures. And Warner's analysis of lessons of risk management from floods in Holland cautions against the approach, due to its tendency to pass on residual risks to local people.

Hepworth and Orr grapple with the contested topic of corporate engagement in water security, convincingly demonstrating that the influence of large multinationals on local and global food and water production is so great that water security practitioners cannot afford to debate that role from the sidelines. The need to reconcile the traditional interests of corporations—preferential and sustained water access, permissive water quality objectives, and laissez-faire regulation—with the water security goal of improving the wider public good at all scales is also identified. In their comprehensive tour of water law and legal frameworks, Leb and Wouters suggest another part of that way forward: the development of guidelines that can serve both to evaluate competing claims, and, crucially, to desecuritise water conflict issues. The potential for market-based tools to balance equity and efficiency (which can work only in very well-regulated contexts, as Garrick and Hope point out) thus finds its place beside more critical persuasive views that the market should be shunned for cultural-based solutions (Boelens) and warning of the pitfalls of retaining productivity—not security, much less equity—as a guiding principle for water management.

Water Security and Interconnectedness

The interdependence of material and immaterial objects at all levels of water security are also revealed throughout the breadth of the volume. Tickner and Acreman argue, for instance, that human water security can only be achieved via environmental security. Chenoweth et al. stress connections between water functions and benefits for household members, while Earle's examination of cities confronts a concentration of human and economic activity so profound that hydropower development and protection from floods are incorporated over and above domestic water issues. Similarly, commercial efforts at resource securitisation come with the dawning realisation of corporate risk being situated within wider societal insecurities (Hepworth and Orr).

The emerging theme here is of interconnectedness—of seeing one sector or user's resource security via a wider lens of collective security. But collective water security—in the sense of securing sufficient water for all users, all uses, and at all times—is not achievable. Nor is it the entire picture: Interconnectedness implies a sharing of deficits in times of flood or famine. Conway attentively draws out lessons from the climate science community in dealing with entirely uncertain and unknown contexts, for example, while Lankford asserts that water security 'seeks, and is a consequence of, the sharing of water surpluses and deficits between different users mediated by the architecture of water infrastructure designed to address the spatial, temporal and scalar complexities of demand and supply'.

In a comparable vein, Zeitoun asserts that sustainable water security policy at the national level may be achieved following a thorough understanding and balancing of the interdependencies water has with other resources, and of the equitable sharing of both water benefits and harm. Zala's application of security studies' 'sustainable security framework' to water resources proves an effective counter to the more tapered understandings of security in traditional defence circles, precisely because it 'prioritises the resolution of the interconnected and underlying drivers of insecurity and conflict, with an emphasis on preventive rather than reactive strategies'. Pushing the hydropolitics body of research into new arenas, Mirumachi asserts that any robust understanding of 'transboundary water security' must look beyond mere treaties and institutions, towards the capacity of the actors involved—and the interdependent means and justifications employed to assure their share of the resource.

Water Security and Comprehensiveness

A second theme emerging from the book is that of completeness. Water security does more than connect interdependent users and uses; it also seems to reach for an all-encompassing and global approach. The two introductory chapters emphasise the multiple social and biophysical links between water and the political economy and other natural resources, for example. Cook and Bakker call for a wide-ranging and integrative approach to water security, while Zeitoun's 'sustainable water security web' attempts to capture the breadth of related resources and users that may be of relevance. Such broad interpretations are surely less workable than highly tailored and more precise definitions, particularly if there are specific policies or analytical objectives to be reached. A rigorous reflection on the reasons for the previously mentioned loss of meaning in research and in practice of previous water paradigms demands, however, that we also not lose sight of the bigger picture of water security. Thus, water security policy practitioners and researchers are advised to consider water security as a frame to guide analysis or policy formation (e.g., see Lankford, Chapter 21).

In keeping with this point, any frame or custom definition of water security should avoid environmental determinism and naively apolitical approaches.

This is important because poorly considered and supposedly apolitical views can lead straight into the same kind of perversion of apparently good or well-intended ideas, as Boelens points out is the case with the human right to water. If water security is considered narrowly and primarily in terms of risks and threats, the only space for politics in the policy formation is through the 'politics of fear', which would lead to the previously mentioned retreat to securitisation and protection of 'our' water. Constricted thinking of this kind is based on assumptions that water security could be made independent of multiple other natural resources and socioeconomic and political forces.

Water Security and Harmonisation

Given the complex nature of the previous two themes of water security, the next step the authors follow is through consideration of how different components are articulated with one another. For example, the importance of harmonizing water resources policy at different levels and across sectors (Pegram et al., 2011) is applied to the various interpretations of water security in chapters by Allan, Froggatt, and Falkenmark. The direct link between successful innovations in policy and suitable governance (e.g., Budds and Hinojosa, 2012) is found to hold for water security policy at the communal (Garrick and Hope) and international levels (Mirumachi). We draw further attention to three cases where authors have considered how water resources must be seen as an endeavour of coordinated concepts.

Seeking agreement between the different elements of water security, for example, Lankford calls for 'share management', which he argues can be seen not only via governance but also in terms of arrays of infrastructure. He proposes analysing how existing conventions of infrastructure architecture mediate water distribution amongst varying supplies of water to disparate needs. The imperative is to rework these physical structures so that they distribute water variability to different users providing a dynamic but more evenly shared water security.

Allan and Falkenmark each build on their foundational work on the importance of 'green water', which is perhaps the single most important concept relevant to water security at a global level (and must complement at every turn the tendency to focus on 'blue water'). Falkenmark qualifies her earlier work on water scarcity thresholds, which are widely used and misused, but rarely unpicked (e.g., Bates et al., 2008; Vörösmarty et al., 2010). Her assertion that the water stored temporarily in the soil throughout the world must be considered for any measure of water scarcity is of fundamental importance to any conception of water security, and is supported through consideration of human and biophysical vulnerability. Allan makes a similar point even more strongly, coining here the terms *food water* and *non-food water*. Water and food are not only related, he insists, but they are also so fundamentally linked that any discussion or concerns for water security that do not involve farmers are ill-founded, and probably illegitimate.

A Guide to Readers

In order to make the most of the range of topics and principles covered in this volume, we suggest that readers critically engage in the same open and creative atmosphere that the authors have. Considering the debates, comprehensiveness, interconnectedness, and harmonisation we've discussed, readers may also want to bear in mind the following suggested thought-structuring guidelines:

Refrain from seeking a perfect singular definition of water security. It is not possible to adequately capture the breadth and depth of water security in a single snappy definition. Consider instead how the concept may serve as a frame that allows space for all of the relevant links water creates with other users and sectors (those downstream, food, the less powerful, climate) for the particular context sought. Readers are invited to draw, from these chapters (e.g., the frame in the concluding chapter) and elsewhere, their own arrangement of key ideas and relations between ideas.

Question 'water security for who?' Ambiguity of purpose or intended recipients could reflect naively apolitical approaches to a fundamentally political issue. Analysis and policy and might be steered away from the less powerful and more marginalised, as a result. This issue is linked directly to queries of how 'security' is understood. Is 'security' understood in terms of freedom from fear or protection from hazards; and is it understood to derive from independence or interdependence? The perspective taken on security may lead to water security for a select group, at the cost of water insecurity for others.

Accept and embrace interconnectedness and comprehensiveness. As water is and creates interdependence with just about everything, it becomes crucial to take a transdisciplinary approach that acknowledges both the limits of knowledge and the positionality of actors involved. Here, an appreciation of context is crucial to ensuring water security for those that matter, and this implies focussed investigation of the historical, social, political, biophysical, and economic fabric of the community/basin/country in question.

The biggest question left unanswered in this book is probably *'water security how?'* We thus urge readers to broaden the range of actors that must be involved (as Allan) and the guiding concepts and precepts that must be enacted (as Zala). Although each chapter touches on the question, addressing it much more explicitly is the labour demanded of all of us. We thus end this introduction to water security by pointing to the further threads of inquiry raised in the concluding chapter, and suggesting that it is up to all of us to chart the future. We look forward to your engagement in order move the ideas forward through debate and deliberation.

References

Bates, B.C., Kundzewicz, Z.W., Wu, S. and Palutikof, J.P. (eds) (2008) *Climate Change and Water, Technical Paper of the Intergovernmental Panel on Climate Change*, IPCC Secretariat, Geneva.

Briscoe, J. (2009) 'Water security: why it matters and what to do about it', *Innovations*, vol 4, no 3, pp3–28.

Budds, J. and Hinojosa, L. (2012) 'Restructuring and rescaling water governance in mining contexts: the co-production of waterscapes in Peru', *Water Alternatives*, vol 5, no 1, pp119–137.

DNI (2012) 'Global Water Security—Intelligence Community Assessment', ICC-coordinated paper, Office of the Director of National Intelligence, U.S. Department of State, Washington, DC.

King, W. (2012) 'Water security—a matter of national defense', *Water Security, Risk and Society Conference*, 16–18 April, University of Oxford, Oxford.

Muller, M. (2012) 'Africa's path to water security, rocks hard places, road blocks', *Water Security, Risk and Society Conference*, 16–18 April, University of Oxford, Oxford.

Pegram, G., Le Quesne, T., Yuanyuan, L., Speed, R. and Li, J. (2011) *River Basin Planning: Principles, Procedures and Methods for Strategic Basin Planning*, WWF and China Water, Beijing.

Tanzler, D. and Carius, A. (eds) (2012) *Climate Diplomacy in Perspective: From Early Warning to Early Action*, Berliner Wissenschafts-Verlag, Berlin.

Vörösmarty, C.J., McIntyre, P.B., Gessmer, M.O., Dudgeon, D., Prusevich, A., Green, P., Glidden, S., Bunn, S.E., Sullivan, C.A., Reidy Liermann, C. and Davies, P.M. (2010) 'Global threats to human water security and river biodiversity', *Nature*, vol 467, pp555–561.

WRG (2010) 'Charting our water future: economic frameworks to inform decision-making', Water Resources Group: The Barilla Group, The Coca-Cola Company, The International Finance Corporation, McKinsey & Company, Nestlé S.A., New Holland Agriculture, SABMiller plc, Standard Chartered Bank, and Syngenta AG (also known as 'the McKinsey Report').

2   The Web of Sustainable Water Security

Mark Zeitoun

Introduction

Whilst the authors engaged in this book offer numerous perspectives from many disciplines, there appears to be consensus on one element of water challenges: that solutions lie as much outside the watershed as within it. Working within what Allan (1998) has called the 'problemshed', there is nonetheless the likelihood that the different approaches take us in dissonant or even competing directions. We thus run the risk of collectively developing an inadequate version of the robust yet flexible conceptualisation of water security that, I think, is required. The peril that serves to guide this chapter is that our efforts serve to develop policy that leads to selective (i.e., our own) temporary water security, at the cost of water insecurity for others.

This chapter employs political ecology to explore the idea of sustainable water security as a preferred future alternative to such conceptualisations and policies. It first briefly reviews existing conceptions of water security, to point out just how shaky the biophysical science foundation of water security is, and the extent to which social science is dismissed. The research and policy communities' broadening of focus from the 'water box' to multilateral nexus is also reviewed, and the exploration of the interdependencies of water with food, climate, and energy is the catalyst for the 'web' of water security presented in the second section. The water security web is offered as a conceptual tool that keeps attention focussed on the interdependencies, on the more structural causes of water insecurity, and on the inseparability of biophysical and sociopolitical processes. Sustainable water security is then defined and elaborated on as a suggested policy goal, with the proposal that it be based on a more thorough understanding of the interdependencies within parts of the web and on equitability. The former requires better knowledge of the hydrosocial cycle; the latter requires a return to principles: a balance between human use of water resources and related natural 'security resources', and a strong sense of egalitarian justice to guide the distribution of benefits and effects of any policy.

Restrictive and Coherent Concepts of Water Security

Forsyth (2003, p38) documents a number of poorly informed but deeply held beliefs about environmental issues, which he terms 'environmental orthodoxies'.[1] He demonstrates how the idea that forest cover improves river flow quantity and quality is thought to apply universally, for instance, despite the lack of contrary evidence. Without critical questioning at this early stage, 'water security' runs the risk of becoming a similar orthodoxy, judging by the level of confidence placed in poorly understood biophysical processes and the frequency with which social processes are dismissed or ignored.

Shortcomings in the Science

One is struck when reviewing the mounting literature on water security by how poorly the physical component of the hydrological cycle is treated (e.g., Briscoe, 2009; WRG, 2010; and many others). Despite more than a century of effort devoted to the study of the water cycle, there is no agreement on methods to calculate the basic quantities of water—in all of its forms—flowing in and out of a river basin. Watershed models developed from hydrological science are just beginning to incorporate groundwater into their water balance models. Hydrologists and hydro-geologists have yet to adequately incorporate the soil water that sits above the groundwater to sustain all rain-fed vegetation, including the bulk of global food production (Taylor, 2009). Just a few researchers are beginning to explore the evaporation and evapotranspiration processes that provide soil water, sometimes far beyond the watershed in question (e.g., van der Ent et al., 2010). Much of the literature, however, neglects to mention—much less address—these fundamental gaps in our knowledge, as if they were not central to our quest for water security.

The shortcomings of basic watershed science can have substantial ramifications on policy. Negotiations over the Nile River, for example, have taken place without recognition of the provenance (where the rain feeding the river flows originally evaporated from) and even existence (soil water, e.g., in the highlands of Ethiopia) of more than half of the water in the Nile Basin (Zeitoun, Allan, and Mohieldeen, 2010). Such biophysical interdependencies of water are indeed complex, perhaps even discouraging. One might discover, for example, that the forests of Madagascar influence the southern Nile flows via the El Nino Southern Oscillation. It could rationally follow that the island state should be invited to the negotiations table of the Nile Basin Initiative. Complex political processes can become yet more complicated, so long as the basis for definition of the physical limits is not agreed upon.

Our quest for water security is further hindered by serious shortcomings with—or ignorance of—advances in social science. For example, flawed understandings of water scarcity are perpetuated by simplistic but very popular classifications (disregarding, for the moment, the lack of reconciliation between the terms 'scarcity' and 'security'; see Chapter 10, for instance).

The national 'water stress' thresholds classification asserts that countries with less than 1,000 m³ of water available annually per person are 'chronically stressed' and those with less than 500 m³ per year are 'beyond the water barrier' (see Perveen and James, 2010, as just one example). The originator of the idea herself has proposed a more sophisticated approach to scarcity (Falkenmark et al., 2007), on which she elaborates in Chapter 5. The original thresholds can mischaracterise national water problems for neglecting, for instance, the soil water within a country and the pressure-reducing feature of food ('virtual water') imports. Perhaps most importantly, they can't allow for recognition of the very political global and local economy, which can skew inordinate amounts of water towards the wealthy and politically connected, and away from the less powerful. Despite these significant conceptual short-comings, however, studies using water stress thresholds are cited by the Inter-governmental Panel on Climate Change (e.g., Bates et al., 2008, p7) and many other high-profile climate change (e.g., Walker and King, 2008) or water secu-rity (e.g., Vörösmarty et al., 2010) studies. Their widespread use risks them becoming a fully developed, if misguided, environmental orthodoxy.

The simplifying physical and social scientific assumptions that underpin some versions of 'water security' are thus undermined by the failure to con-sider how water is distributed within a country. Recognition that water scar-city for the masses does not necessarily mean water scarcity for the economic elite led to the development of the concept of 'social' water scarcity, before even the 1970s 'limits to growth' debates (Mehta, 2011a), and this has been reemphasised over the years (as 'economic' or 'second order' scarcity) by authors of multiple disciplines (Ohlsson, 1999; Bakker, 2000; Amit and Ramachandran, 2009; FAO, 2009; Johnson et al., 2011; Mehta, 2011b). Considering the social side of water scarcity means investigating politics, ethics, justice, economics, and human water and food consumption. While the validity of the concept is not debated (see UNDP, 2006; Mollinga, 2008; FAO, 2011), it is largely ignored by the previously mentioned high-profile and more-influential physical studies, and rather wilfully by policymakers who may prefer to deal with numbers and the illusion of certainty than politics. Nonetheless, privileging research of the relatively neat biophysical aspects of water security over its messy social realities cannot be expected to form a cohesive basis for policy on water security.

Shifts to Interdependencies

Such studies emphasising the environmental aspects of water security are placed in the view of the evolution of the concept just beyond the purely anthropocentric definitions (see Chapter 4). They still sit firmly within the watershed, however, and fail to consider the interdependency of water and water use with other natural resources such as food, climate, and energy—not to mention social issues. For their restricted/focused breadth and water resource–centric perspective, this body of work is referred to here as *water resources security.*

Our perspective must broaden and shift if we are to reflect the nuance and importance of the interdependencies. The clear relation between water and food production has proven the starting point for that broadening, as in the Ministerial Declaration on Water Security at the 2000 World Economic Forum. The resulting virtual water and water footprint work is increasingly taken up by water research institutes, think tanks, and implementing agencies (e.g., Renault, 2002; SIWI, 2005; IWMI, 2007; Sojamo et al., 2012), and may complement established notions of food security (FAO, 2009). The link between water and energy security is relatively less developed. The competition for water between crops for food and crops for biofuels ('water for energy') is directly related to the demand (and cost) of fossil fuels (Berndes, 2002; Lundqvist et al., 2007, p56). The water footprint of biomass is 70 to 400 times larger than that of conventional fuels (Gerbens-Leenes, Hoekstra, and van der Meer, 2008, p5), raising issues of allocation, particularly in India and China. Concerns about energy use for the treatment, production, and transmission of water ('energy for water') is also receiving attention (e.g., King, Holman, and Webber, 2008; Rothausen and Conway, 2011).

Water and human (or community) security is most frequently discussed in relation to water and sanitation concerns, or individual access to water[2] (e.g., Barlow, 2007). The 'bottom-up' approach has been explored in relation to armed conflict through water and climate issues (Smith and Vivekananda, 2007), while the emerging concept of climate security has developed in relation to national security (CNA, 2007; WBGU, 2008) and human security (Adger et al., 2006).

Cross-fertilisation between these concepts is not yet well-developed. Such examinations of the intersection of two water-related 'security' areas appear to be giving way to studies of water 'nexus'; that is, the intersection of water processes with multiple resources. Climate–water–national security links have been discussed in relation to the Middle East (Brown and Crawford, 2009), for example, and more tangentially by water and agriculture think tanks concerned about impacts of climate change on food production (e.g., IWMI, 2009). Houdret, Kramer, and Carius (2010) connect water–human–state security, while Magsig (2010, p62) refers to the 'security triad' of environment, energy, and food. The water–food–climate nexus has also been explored in the Middle East and North Africa (FAO, 2008). The water–energy–food nexus has generated very interesting research (such as Lundqvist et al., 2007; Hellegers et al., 2008; McCornick, Awulachew, and Abebe, 2008). More recently, these issues have attracted high-level policy meetings (for instance at the World Economic Forum [WEF, 2011] and the German government–hosted Water Energy and Food Security Resource Platform[3] [see www.water-energy-food.org]).

Tony Allan (Chapter 20) encourages us to think even beyond the three-dimensional nexus to the 'big nexus' (to include food trade) and the 'mega-nexus' (which includes financing). Further broad and encompassing water security work is investigating the links between water and human security, food security, economic security, and health security (e.g., FAO, 2000;

Hellegers et al., 2008; McCornick, Awulachew, and Abebe, 2008). Developments in the legal perspective on multidimensional water security may eventually form the legal platform from which a deeper understanding of the challenges (and eventually framework) may depart (e.g., Tarlock and Wouters, 2010; Leb and Wouters, Chapter 3).

On the whole, however, this broad body of work suffers many of the same shortcomings as the purely 'water resources security' research, that is, an unfounded sense of certainty in biophysical processes and a neglect of social processes. The 'web' of sustainable water security complements these research streams in developing the interdisciplinary political ecology and political economy lenses, and a critical interpretation of both politics and natural resource science.

The Web of Sustainable Water Security

The web[4] shown in Figure 2.1 is offered as a conceptual tool that can help guide water security research and policy formation through the shortcomings identified. The tool centres on the interdependencies reviewed to provide a combined reading of how social and physical processes combine to create or deny water security.

The web metaphor positions six 'security areas' related to water security. These include the intimately associated and interdependent natural 'security resources' discussed (water resources,[5] energy, climate, food), and the social groups concerned (individual/community, nation/state). The latter obliges asking (and answering) the question, 'Water security for who?', and thus addressing the pitfalls seen with the strictly biophysical interpretations of water security.

The range of groups caught up in the web (from the individual, through to river basin and global levels) furthermore obliges exploration of the connections between them, which are possibly best understood through political economy (the selectively sticky filaments of the web). The researcher pursuing the analysis in this way can demonstrate, for example, that an individual's water security may coexist with national water insecurity. This is the case of wealthy Yemeni farmer-sheikhs with the deepest wells (who may be temporarily water secure) in the dry highlands of the country (which itself is not, on the whole, water secure) (e.g., Lichtentaehler, 2002). The interdependencies that exist between both social groups and the 'security resources' are themselves intertwined. Consider the extent to which UK (i.e., national) food security is partly based on water insecurity of communities in Peru (Hepworth, Postigo, and Güemes Delgado, 2010) or the West Bank (Nazer et al., 2008).

The relationship between the various relevant security areas requires considerably more testing and theorisation, as a brief testing of the upper and right sides of the figure shows. The interdependency between social groups, created by transboundary waters flowing through or beneath state borders, poses a direct challenge to the traditional view of national security

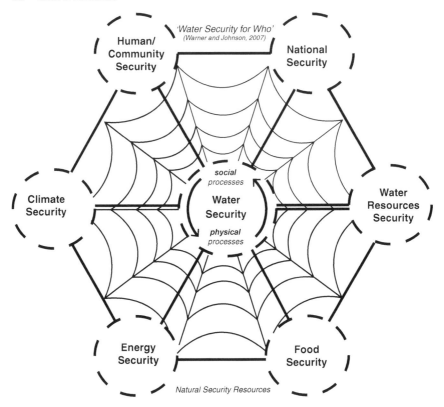

Figure 2.1 The 'web' of national water security. Sustainable water security is interpreted as a function of the degree of equitability and balance between the six related security areas, as this plays out within a web of socioeconomic and political forces at multiple spatial levels.

assured through sovereignty and independence. Indeed, attempts to exert full sovereign territorial control over what has been termed a 'fugitive resource' (Frederick, 1996) is antithetical with thinking on, and implementation of, natural resource management.[6] Adaptive natural resources and water management (see, e.g., Ostrom, 1990) developed from recognition that a sense of security is possible without stability and full control, as most recently reflected by the work on global limits (Rockström et al., 2009; Foley et al., 2012). Infrastructure built on the logic that variability in rainfall, river flow, or aquifer recharge is a source only of insecurity goes against the grain of the adaptive approach (Palmer et al., 2008; Lebel et al., 2010) and precludes alternative methods of dealing with the variability (e.g., Lankford, 2004). When the resource in question is transboundary, the options set is even less flexible (Zeitoun, Goulden, and Tickner, forthcoming) and the political economy of the web is stickier.

Use of the web obliges questioning of the structural causes of water insecurity. This is important, as the sociopolitical and economic context within which transboundary (or other) water dynamics occur is just as fluid as the resource itself, and a firm understanding of the interdependencies for any country is difficult to pin down for long. For example, a 'web' reading of Egyptian water security would factor in (where the NBI did not) the shifting political context between Nile Basin states (Granit et al., 2010) and the previously mentioned soil water that makes up so much of the Nile Basin water balance. Seen in this light, the decision by Cairo (along with Khartoum) not to sign the Cooperative Framework Agreement in May 2010 is interpreted as a choice to seek water security through independence over security through (more or less equitable) interdependence. The national water security achieved through the position is tenable only so long as the power asymmetry that sustains it is maintained (Nicol and Cascão, 2011). Given the way power moves to and away from states, any assurance on water issues achieved in this way is more appropriately labelled short-term and selective water security, and is not sustainable in the longer term.

Towards Sustainable Water Security

Given the gaps in knowledge and the very different approaches taken to water security, our task to develop a robust but flexible understanding of water security is considerable, and the accompanying research agenda is just as broad. The examples covered demonstrate that in the pursuit of water security, for instance, trade-offs occur as much between social groups as they do between uses of natural 'security resources'. Long-term water security, it is proposed, may be built on the principles of sustainability and of 'sustainable security'. The intergenerational concerns at the heart of any entity that seeks to endure—particularly in the face of uncertainty and unpredictability—would be the real test against temporary water security. In applying the concept of sustainable security to national and water security, Ben Zala (Chapter 17) recognises that elimination of uncertainty and control over all the consequences of insecurity (that is, the trade-offs) are impossible, and suggests instead the resolution of their root causes.[7]

Investigating the Interdependencies

Sustainable water security builds on these observations, and is further assisted by the web shown in Figure 2.1. Further consideration suggests that research supporting sustainable water security policy should follow two streams, and that the policy should be guided by two principles. The first stream of research required is the investigation of the nature and volume of interdependencies between natural 'security resources'. This stream would serve to deepen the nexus research already discussed through natural resource science and quantitative research informed by the social sciences. For example, to what degree does (or would) the production of biofuels generate national energy security at the expense of increased vulnerability of national water resources security[8] (and, hence, national water security)?

Pursuit of 'balance' in the use and effect of the 'security resources' is the suggested guiding principle for national water security here, if any semblance of sustainability is sought.

Similarly, effective study of the water–energy–food–climate nexus requires considerable quantification of the interdependencies, and consideration of the social consequences of large infrastructure (e.g., Skinner, Niasse, and Haas, 2009). Furthermore, more study of the co-evolving effect of the hydrological cycle on the carbon cycle could serve to determine the influence of intensive irrigated agriculture on rainfall patterns of neighbouring countries. Technological developments in climate modelling and earth observation[9] are helping us rise to this challenge in ways that were once unthinkable. The previously noted continent-level work of van der Ent et al. (2010) on the national provenance of atmospheric moisture is a step forward in this process. Further steps may be best assisted by social scientists working alongside the climate modellers, to ensure consequential socioeconomic political issues are not left behind. For transboundary water conflicts, for instance, the implications of comprehensively considering water–food–carbon means questioning beyond what may currently be seen as a curious debate about 'who owns the rain' (Falkenmark, in Allan, 2001, Ch. 3) to one of 'who owns the clouds'. As McCaffrey (2007, p170) points out, determining the 'equitable share of hydrologically disadvantaged states' will be a thorny issue indeed.

Investigating Inequity

The second stream of research required to help achieve sustainable water security would investigate how and to what extent the water security of different social groups in the web is affected by the interdependencies. Broad investigation of consequences of actions taken to ensure water security for one state (or nation, or community) may reveal several that were unintended. Security studies and political science theory would be assisted here by an indication of the magnitude of the concerns (see, e.g., Stetter et al., 2011), with an understanding that instability and uncertainty are affected by greater codependence (in ecosystems as much as in the European Union) in different ways than they are by independence (for discussion, see Buzan, Waever, and de Wilde, 1998). 'Equitability' is suggested as the guiding principle in the case of sustainable water security of interdependent social groups (just as it is for 'sustainable security') (see Zala, Chapter 17).

The science required to investigate the merit of equitability in the distribution of water security is much less cutting edge, and for water is linked with the mid-20th century move towards human rights. Dinar, Rosengrant, and Meinzen-Dick (1997) suggest placing it at the heart of their economic water allocation mechanisms, for example, and it is of course a founding pillar of IWRM and WDM (Zeitoun and McLaughlin, 2013). Widespread resistance to these more adaptive management paradigms has been well-documented, however, and is found to be due in part to the tensions stemming from prioritisation of the principle of water efficiency over concerns for justice (see, e.g., Brooks, 2005; Syme and Nancarrow, 2006).

Research into the development of international norms may serve the cause of justice and sustainable water security. The groundbreaking work of environmental and social activists has been impressive, and has been strengthened by developments in international law that may eventually find their way into policy. The UN General Assembly's adoption of the Legal Framework of the Human Right to water, for instance, supports the idea of access to safe water for all and is, very much in theory, a step towards greater individual water security. The 1997 UN Watercourses Convention builds on the Agenda 21 process to reinforce notions of justice and sustainability, and is developing the useful concept of limited territorial sovereignty (McCaffrey, 2007; McIntyre, 2010), which may serve eventually to reach the 'community of interests' approach required to reconcile the dissonance between river basin boundaries and political borders.

Understanding the resistance to the development of international norms will help interpret steps taken towards selective short-term water security, and away from sustainable water security. The UK government has stated its reluctance to accede to the 1997 UN Watercourses Convention, partly based on the fact that it would be ineffectual, as Brazil, China, Egypt, India, and Israel do not support it (DFID, 2008).[10] These same states have, however, supported the 2008 Draft Aquifer Articles, which have been noted for the 'retreat to sovereignty' (McCaffrey, 2009) and are a step away from the collective responsibility required to achieve sustainable water security. In order to advance in a particular direction, then, an awareness of the way international politics work should inform legal and natural science research. Even if the physical models of the interdependencies of 'security resources' were perfectly accurate (and they cannot be), they will not assist the task at hand without adequate information (and agreement) on guiding principles.

With the web as guide, 'sustainable water security' may also be assisted by the research and action on human security (e.g., Pachova, Nakayama, and Jansky, 2008). The power asymmetries that enable short-term and 'selective' water security suggest that greater examination of the mediating potential of international water law (WWF-DFID, 2010; Rieu-Clarke, Moynihan, and Magsig, 2012) or injustices meted out through international food trade (e.g., Via Campesina, 2006) will remain important research to pursue.

There also appears to be considerable merit in linking water security with the FAO's (2009) four-level definition of food security. Though the approach to food security is just as prone to downplay social processes and distribution (see, e.g., Schmidhuber and Tubiello, 2007), the insecurity deriving from a lack of access to food is at least explicitly acknowledged. As we have seen, the insecurity deriving from a lack of access to water is embedded in the concept of social scarcity, which the more influential biophysical studies (as ill-conceived thresholds develop into environmental orthodoxies) have managed to ignore.

Conclusion

The 'web' of water security has been proposed here as a response to the formulation of research agendas and policy currently based on unacknowledged weaknesses in biophysical science and a lack of engagement with social science. Viewing social and physical processes as inseparable, the political ecology underlying the web also sees the political economy as influential in determining who achieves water security, and who receives water insecurity. By drawing our attention to the importance of asking 'water security for who', and the interdependencies with other 'security resources' such as food, energy, and climate, the 'web' can serve to pinpoint analysis towards the structural causes of water insecurity.

The concept of 'sustainable water security' has been proposed to draw policy attention away from efforts that lead to selective and short-term water security. Reconciling the web to Zala's application of 'sustainable security' to water, this approach shifts the focus of future water security understanding away from temporary and selective water security, towards resolution of the root causes of insecurity and water security for the long term.

The water security research and policy agenda remains large, with several remaining gaps related to epistemological views on 'risk', or the philosophical foundations of uncertainty and 'security' itself. In the absence of this exploration, our collective ability to develop and implement long-term water security will remain constricted.

Notes

1. The term is analogous to Allan's (2003) use of 'sanctioned discourse' and Hajer's (1997) 'coalition of discourses', as well as to Gramsci's ideas about cultural power and hegemony (see, e.g., Scott, 2001, p90).
2. Vörösmarty et al. (2010) also use the term 'human water security' with the biophysical aspects of river flows, although it is not clearly defined.
3. These latter initiatives have been critically investigated for their tendency towards exclusive processes and infrastructure (e.g., the dams that can use water to create energy and food) and away from equity and other social considerations (see SOAS, 2012).
4. The 'web' metaphor comes from a draft World Economic Forum report by the Global Agenda Council on Water Security: "Water security is the gossamer that links together the web of food, energy, climate, economic growth and human security challenges that the world economy faces over the next two decades" (WEF, 2009, p5).
5. Water resources security may be understood to comprise the same principles and ideas about environmental 'quality' as does 'environmental security' (e.g., Dalby, 2006), and may be informed by the lessons drawn from the considerable effort spent globally on Integrated Water Resources Management (IWRM) (Molle, 2008).
6. This has evolved from seeing all of 'nature' as static (which leads to attempts to 'conserve' it) to an appreciation of global biophysical processes as both resilient and in 'non-equilibrium' (Milly et al., 2008).

7. In Zala's words, "the 'sustainable security' approach prioritizes the resolution of the interconnected and underlying drivers of insecurity and conflict, with an emphasis on preventive rather than reactive strategies".
8. Similarly, the 'planetary boundaries' suggested for water ignore its political economy and asymmetric distribution (Rockström et al., 2009; Steffen, 2009), thereby risking taking us down the same unhelpful path as Falkenmark's original indicators, even as they serve to raise awareness of issues.
9. For instance, the measurement of gravitational anomalies, which can give an indication of volumes of surface water and soil water storage.
10. In a response to a call for the UK government to accede to the Convention, the Department for International Development stated that such a move will not 'translate into action' since 'None of the large (geographically and/or economically) countries that share water with their neighbours (Brazil, China, Egypt, India, and Israel) in low- and middle-income regions, except South Africa, have ratified or acceded to the Convention.'

References

Adger, W. Neil, Paavola, J, Huq, S., and Mace, M.J. (Eds). (2006) *Fairness in Adaptation to Climate Change*, MIT Press, Cambridge, MA.

Allan, J.A. (1998) 'Watersheds and problemsheds: explaining the absence of armed conflict over water in the Middle East', *Middle East Review of International Affairs*, vol 2, no 1.

Allan, J.A. (2001) *The Middle East Water Question: Hydropolitics and the Global Economy*, I.B. Tauris, London, UK.

Allan, J.A. (2003) 'IWRM/IWRAM: Integrated water resources allocation and management: a new sanctioned discourse?', in P.P. Mollinga, A. Dixit, and K. Sthukorala (eds) *Integrated Water Resource Management in South Asia: Global Theory, Emerging Practice and Local Needs*, SaciWATERs, Hyderabad.

Amit, R.K. and Ramachandran, P. (2009) 'A fair contract for managing water scarcity', *Water Resources Management*.

Bakker, K. (2000) 'Privatizing water, producing scarcity: the Yorkshire drought of 1995', *Economic Geography*, vol 76, no 1.

Barlow, M. (2007). *Blue Covenant: The Global Water Crisis and the Coming Battle for the Right to Water*. Mclelland & Stewart Ltd, Toronto.

Bates, B.C., Kundzewicz, Z.W., Wu, S. and Palutikof, J.P. (Eds). (2008) *Climate Change and Water, Technical Paper of the Intergovernmental Panel on Climate Change*. IPCC Secretariat, Geneva.

Berndes, G. (2002) 'Bioenergy and water—the implications of large-scale bioenergy production for water use and supply', *Global Environmental Change*, vol 12, pp253–271.

Briscoe, J. (2009) 'Water security: why it matters and what to do about it', *Innovations*, vol 4, no 3, pp3–28.

Brooks, D.B. (2005) 'Beyond greater efficiency: the concept of water soft paths', *Canadian Water Resources Journal*, vol 30, no 1, pp83–92.

Brown, O. and Crawford, A. (2009) *Rising Temperatures, Rising Tensions: Climate Change and the Risk of Violent Conflict in the Middle East*, International Institute for Sustainable Development, Winnipeg.

Buzan, B., Waever, O. and de Wilde, J. (1998) *Security—A New Framework for Analysis*, Lynne Rienner Publishers, Inc., London.

CNA (2007). *National Security and the Threat of Climate Change*, CNA Corporation, Alexandria, VA.

Dalby, S. (2006) 'Introduction to Part Four', in G. Ó Tuathail, S. Dalby, and P. Routledge (eds) *The Geopolitics Reader* (Second edition), Routledge, London.

DFID (2008) Open letter on Accession to the UN Convention on the Law of Non-Navigational Uses of International Watercourses. Letter by UK Department for International Development of 28 April 2008, in response to a letter signed by over 20 UK-based academics.

Dinar, A., Rosengrant, M. and Meinzen-Dick, R. (1997) 'Water Allocation Mechanisms: Principles and Examples', Policy Research Working Paper 1779, The World Bank, Agriculture and Natural Resources Department, and International Food Policy Research Institute, Washington, DC.

Falkenmark, M., Berntell, A., Jägerskog, A., Lundqvist, J., Matz, M. and Tropp, H. (2007) *On the Verge of a New Water Scarcity: A Call for Good Governance and Human Ingenuity.* Brief, Stockholm International Water Institute, Stockholm.

FAO (2000) *New Dimension in Water Security: Water, Society and Ecosystem Services in the 21st Century,* AGL/MISC/25/2000, Food and Agriculture Organization of the United Nations, Land and Water Division, Rome.

FAO (2008) *Climate Change, Water and Food Security,* Technical Background Document from the Expert Consultation held on 26 to 28 February 2008, Food and Agriculture Organization of the United Nations, Rome.

FAO (2009) *Food Security and Agricultural Mitigation in Developing Countries: Options for Capturing Synergies,* Food and Agriculture Organization of the United Nations, Rome.

FAO (2011) *The State of the World's Land and Water Resources for Food and Agriculture: Managing Systems at Risk,* Food and Agriculture Organization, Rome and Routledge, Abingdon.

Foley, J.A., Ramankutty, N., Brauman, K.A., Cassidy, E.S., Gerber, J.S., Johnston, M., Mueller, N.D., O'Connell, C., Ray, D.K., West, P.C., Balzer, C., Bennett, E.M., Carpenter, S.R., Hill, J., Monfreda, C., Polasky, S., Rockstom, J., Sheehan, J., Siebert, S., Tilman, D. and Zaks, D.P.M. (2012) 'Solutions for a cultivated planet', *Nature,* vol 478, pp337–342.

Forsyth, T. (2003) *Critical Political Ecology—The Politics of Environmental Science,* Routledge, London.

Frederick, K.D. (1996) 'Water as a source of international conflict'. *Resources* 123: 9–12.

Gerbens-Leenes, P.W., Hoekstra, A.Y. and van der Meer, Th. H. (2008) 'Water Footprint of Bio-Energy and Other Primary Energy Carriers', Value of Water Research Report Series No. 29, UNESCO-IHE, Delft.

Granit, J., Cascao, A., Jacobs, I., Leb, C., Lindstrom, A. and Tignino, M. (2010) 'The Nile Basin and the Southern Sudan Referendum', Regional Water Intelligence Report, Paper 16, Stockholm International Water Institute, Stockholm.

Hajer, M.A. (1997) *The Politics of Environmental Discourse: Ecological Modernization and the Policy Process,* Oxford University Press, Oxford Scholarship Online, Oxford.

Hellegers, P., Zilberman, D., Steduto, P. and McCornick, P. (2008) 'Interactions between water, energy, food and environment: evolving perspectives and policy issues', *Water Policy,* vol 10, Supplement 1, pp1–10.

Hepworth, N., Postigo, J.C. and Güemes Delgado, B. (2010) *Drop by Drop: A Case Study of Peruvian Asparagus and the Impacts of the UK's Water Footprint,* Progressio, in association with Centro Peruano De Estudios Sociales, and Water Witness International, London.

Houdret, A., Kramer, A. and Carius, A. (2010) 'The Water Security Nexus: Challenges and Opportunities for Development Cooperation', GTZ International Water Policy and Infrastructure Concept Paper, GTZ/Adelphi, Eschborn, Germany.

IWMI (2007) *Water for Food, Water for Life: A Comprehensive Assessment of Water Management in Agriculture,* International Water Management Institute, Colombo/ Earthscan, London.

IWMI (2009) *Flexible Water Storage Options and Adaptation to Climate Change,* Water Policy Brief Issue 31, International Water Management Institute, Colombo.

Johnson, V., Fitzpatrik, I., Floyd, R. and Simms, A. (2011) *What Is the Evidence That Scarcity and Shocks in Freshwater Resources Cause Conflict Instead of Promoting Collaboration,* CEE Review 10-010, Collaboration for Environmental Evidence.

King, C.W., Holman, A.S. and Webber, M.E. (2008) 'Thirst for energy', *Nature Geoscience,* vol 1, no 55, pp283–286.

Lankford, Bruce A. (2004). Resource-centred thinking in river basins: should we revoke the crop water requirement approach to irrigation planning? *Agricultural Water Management* 68: 33–46.

Lebel, L., Xu, J., Bastakoti, R.C. and Lamba, A. (2010) 'Pursuits of adaptiveness in the shared rivers of Monsoon Asia', *International Environmental Agreements,* vol 10, pp355–375.

Lichtentaehler, G. (2002) *Political Ecology and the Role of Water: Environment, Society and Economy in Northern Yemen,* Ashgate Publishing Ltd, Fareham.

Lundqvist, J., Barron, J., Berndes, G., Berntell, A., Falkenmark, M., Karlberg, L. and Rockstrom, J. (2007) *Water Pressure and Increases in Food & Bioenergy Demand: Implications of Economic Growth and Options for Decoupling,* Swedish Environmental Advisory Council Memorandum 1, Chapter 3, Stockholm, Sweden.

Magsig, B.-O. (2010) 'Introducing an analytical framework for water security: a platform for the refinement of international water law', *The Journal of Water Law,* vol 20, pp61–69.

McCaffrey, S. (2007) *The Law of International Watercourses,* Oxford University Press, Oxford.

McCaffrey, S.C. (2009) 'The International Law Commission adopts draft articles on transboundary aquifers', *The American Journal of International Law,* vol 103, pp272–293.

McCornick, P.G., Awulachew, S.B. and Abebe, M. (2008) 'Water-food-energy-environment synergies and tradeoffs: major issues and case studies', *Water Policy,* vol 10, Supplement 1, pp23–36.

McIntyre, O. (2010) 'International water law: concepts, evolution and development', in A. Earle, A. Jägerskog and J. Ojendal (eds) *Transboundary Water Management: Principles and Practice,* Earthscan, London.

Mehta, L. (2011a) 'Introduction', in Mehta, L. (ed) *The Limits to Scarcity: Contesting the Politics of Allocation,* Earthscan, London.

Mehta, L. (ed) (2011b) *The Limits to Scarcity: Contesting the Politics of Allocation,* Earthscan, London.

Milly, P.C.D., Betancourt, J., Falkenmark, M., Hirsch, R.M., Kundzewicz, Z.W., Lettenmaier, D.P. and Stouffer, R.J. (2008) 'Stationarity is dead: whither water management?', *Science,* vol 319, no 5863, pp573–574.

Molle, F. (2008) 'Nirvana concepts, narratives and policy models: insights from the water sector', *Water Alternatives,* vol 1, no 1, pp131–156.

Mollinga, P.P. (2008) 'Water, politics and development: framing a political sociology of water resources management', *Water Alternatives,* vol 1, no 1, pp7–23.

Nazer, D.W., Siebel, M.A., Van der Zaag, P., Mimi, Z. and Gijzen, H. (2008) 'Water footprints of the Palestinians in the West Bank', *Journal of American Water Resources Association,* vol 44, no 2, pp449–458.

Nicol, A. and Cascão, A.E. (2011) 'Against the flow—new power dynamics and upstream mobilisation in the Nile Basin', *Review of African Political Economy,* vol 38, no 128, pp317–325.

Ohlsson, L. (1999) 'Water conflicts and social resource scarcity', *Physics and Chemistry of the Earth, Part B: Hydrology, Oceans and Atmosphere,* vol 25, no 3, pp213–220.

Ostrom, Elinor (1990). *Governing the Commons: The Evolution of Institutions for Collective Action.* Cambridge: Cambridge University Press.

Pachova, N.I., Nakayama, M. and Jansky, L. (2008) 'National sovereignty and human security: changing realities and concepts in international water management', in N.I. Pachova, M. Nakayama and L. Jansky (eds) *International Water Security: Domestic Threats and Opportunities*, United Nations Press, Tokyo.

Palmer, M.A., Reidy-Liermann, C.A., Nilsson, C., Florke, M., Alcamo, J., Lake, P.S. and Bond, N. (2008) 'Climate change and the world's river basins: anticipating management options', *Frontiers in Ecology and the Environment*, vol 6, no 2.

Perveen, S. and James, L. A. (2010) 'Scale invariance of water stress and scarcity indicators: Facilitating cross-scale comparisons of water resources vulnerability', *Applied Geography*, doi:10.1016/j.apgeog.2010.07.003.

Renault, D. (2002) 'Value of virtual water in food: principles and virtues', *UNESCO-IHE Workshop on Virtual Water Trade*, 12–13 December, Delft, Netherlands.

Rieu-Clarke, A., Moynihan, R. and Magsig, B.-O.(2012) *UN Watercourses Convention: User's Guide*, IHP-HELP Centre for Water Law, University of Dundee.

Rockström, J., Steffen, W., Noone, K., Persson, Å., Chapin, F.S., Lambin, E.F., Lenton, T.M., Scheffer, M., Folke, C., Schellnhuber, H.J., Nykvist, B., de Wit, C.A. , Hughes, T., van der Leeuw, S., Rodhe, H., Sörlin, S., Snyder, P.K., Costanza, R., Svedin, U., Falkenmark, M., Karlberg, L., Corell, R.W., Fabry, V.J., Hansen, J., Walker, B., Liverman, D., Richardson, K., Crutzen, P. and Foley, J.A. (2009) 'A safe operating space for humanity', *Nature*, vol 461, pp472–475.

Rothausen, S.G.S.A. and Conway, D. (2011) 'Greenhouse-gas emissions from energy use in the water sector', *Nature Climate Change,* vol 1, pp210–219.

Schmidhuber, J. and Tubiello, F. N. (2007) 'Global food security under climate change', *Proceedings of the National Academy of Sciences*, vol 104, no 50, pp19703–19708.

Scott, J. (2001) *Power*, Polity Press, London.

SIWI (2005) *Let it Reign: The New Water Paradigm for Global Food Security.* Final Report to CSD-13, Stockholm International Water Institute, with IFRPI, IUCN, and IWMI, Stockholm.

Skinner, J., Niasse, M. and Haas, L. (Eds) (2009) *Sharing the Benefits of Large Dams in West Africa*, International Institute for Environment and Development, London.

Smith, D. and Vivekananda, J. (2007). *A Climate of Conflict: The links between climate change, peace and war*, International Alert, London.

SOAS(2012)'NotAnotherNexus?CriticalThinkingonthe"NewSecurityConvergence" in Energy, Food, Climate and Water'. London, 26 October 2012, School of Oriental and African Studies, and the STEPS Centre of the Institute for Development Studies, http://steps-centre.org/2012/project-related/the-new-security-agenda-in-water-energy-and-food.

Sojamo, S., Keulertz, M., Warner, J. and Allan, T. (2012) 'Virtual water hegemony: the role of agribusiness in global water governance', *Water International,* vol 37, no 2, pp169–182.

Steffen, W. (2009) 'Tipping elements, planetary boundaries and water', *SIWI Water Front*, no 3–4, pp10–12.

Stetter, S., Herschinger, E., Teichler, T. and Albert, M. (2011) 'Conflicts about water: securitizations in a global context', *Cooperation and Conflict*, vol 46, pp441–459.

Syme, G. J. and Nancarrow, B. E. (2006) 'Achieving sustainability and fairness in water reform: a Western Australia case study', *Water International,* vol 31, no 1, pp23–30.

Tarlock, D. and Wouters, P. (2010) 'Reframing the water security dialogue', *The Journal of Water Law*, vol 20, pp53–60.

Taylor, R. (2009) 'Rethinking water scarcity: the role of storage', *Transactions of the American Geophysical Union*, vol 90, no 28, pp237–238.

UNDP (2006) *Beyond Scarcity: Power, Poverty and the Global Water Crisis*, Human Development Report, United Nations Development Programme, New York.

van der Ent, R.J., Savenije, H.H.G., Schaefli, B. and Steele-Dunne, S.C. (2010) 'Origin and fate of atmospheric moisture over continents', *Water Resources Research*, vol 46, no 9.

Via Campesina (2006) *Rice and Food Sovereignty in Asia Pacific*, La Via Campensina, Jakarta.

Vörösmarty, C.J., McIntyre, P.B., Gessmer, M.O., Dudgeon, D., Prusevich, A., Green, P., Glidden, S., Bunn, S.E., Sullivan, C.A., Reidy Liermann, C. and Davies, P.M. (2010) 'Global threats to human water security and river biodiversity', *Nature*, vol 467, pp555–561.

Walker, G. and King, D. (2008) *The Hot Topic: How to Tackle Global Warming and Still Keep the lights On*, Bloomsbury, London.

WBGU (2008). *Climate Change as a Security Risk*. German Advisory Council on Global Change. German Advisory Council on Global Change, Earthscan, London.

WEF (2009) 'The Bubble is Close to Bursting: A Forecast of the Main Economic and Geopolitical Water Issues Likely to Arise in the World during the Next Two Decades' (Draft for Discussion at the World Economic Forum Annual Meeting 2009). World Economic Forum Initiative: Managing Our Future Water Needs for Agriculture, Industry, Human Health and the Environment.

WEF (2011) *Water Security: The Water-Food-Energy-Climate Nexus*, Island Press Washington.

WRG (2010) 'Charting our water Future: economic frameworks to inform decision-making', Water Resources Group: The Barilla Group, The Coca-Cola Company, The International Finance Corporation, McKinsey & Company, Nestlé S.A., New Holland Agriculture, SABMiller plc, Standard Chartered Bank, and Syngenta AG (also known as 'the McKinsey Report').

WWF-DFID (2010) *International Architecture for Transboundary Water Resources Management: Policy Analysis and Recommendations*, Worldwide Fund for Nature and the UK Department for International Development, with Pegasys Strategy and Development, and the UNESCO Centre for Water Law, Policy and Science, London.

Zeitoun, M., Allan, J.A. and Mohieldeen, Y. (2010) 'Virtual water "flows" of the Nile Basin, 1998—2004: a first approximation and implications for water security', *Global Environmental Change*, vol 20, pp229–242.

Zeitoun, M., Goulden, M. and Tickner, D. (forthcoming). *Current and Future Challenges Facing Transboundary River Basin Management: Review and Analysis*, WIREs Climate Change.

Zeitoun, M. and McLaughlin, K. (2013) 'Basin justice: using social justice to address gaps in river basin management', in Sikor, T. (ed) *The Justices and Injustices of Ecosystem Services*, Routledge, Abingdon and New York.

3 The Water Security Paradox and International Law

Securitisation as an Obstacle to Achieving Water Security and the Role of Law in Desecuritising the World's Most Precious Resource

Christina Leb and Patricia Wouters

Introduction

Lack of access to adequate quantities and qualities of water resources is linked to most, if not all, development issues—poverty, poor health, diminished livelihoods, child mortality, and conflict. It is within this context that we explore the issue of water security. 'Water insecurity' derails pathways to economic efficiency, social equity, and environmental sustainability, all of which are necessary for national security (GWP). Growing water scarcity and competition for shared water resources driven by population growth, expansion of human economic activity, and climate change are at the heart of the debate over the securitisation of water and water security.

National security threats linked to water, or the securitisation of water, are one strand of the debate. In that domain, the lack of access to water is viewed primarily as a political issue, as the very object of securitisation. Distinct, but connected to this discourse, is the notion of water security, which is concerned primarily with resource management. The former approach places water at the centre of the politicisation process and favours isolationist resource planning and potential zero-sum games, while the latter approach, the achievement of water security, revolves around resource management objectives and evokes a range of environmental, economic, and social equity considerations best addressed jointly among the nation-states sharing the resource.

Securitisation and Water Security

The process of securitisation emphasises the urgency of the scarcity threat in order to legitimise or make the adoption of measures that go beyond the normal bounds of politics acceptable to society (Sinha, 2005; Buzan et al., 1998). Water, as the reference object, is politicised and pulled away from being an object of scientific/technocratic dialogue and is transformed into an issue of national security with attendant military considerations. In the Indus Basin, as one example, the tendency towards the securitisation of water found expression in the rallying cry of extremist groups in Pakistan responding to Indian dam construction on upstream tributaries with the slogan: 'Water flows or blood!' (Brulliard, 2010). While this process remains, at present, outside of official government rhetoric on the topic, some members of parliament have raised the issue in alarmist ways. The national governments of India and Pakistan have preferred to have recourse to international law as a tool to deescalate the matter. Disagreements over the Indus related to hydropower generation, as a key area of discord, have been addressed through the dispute settlement procedures[1] jointly agreed to under the 1960 Indus Waters Treaty (IWT). That international agreement, which has survived three wars between these two states, has proven to be a robust vehicle to manage international tensions over the transboundary waters shared by these two riparian watercourse states (Alam, 2002).

Indeed, the negotiations and the conclusion of the IWT facilitated and resulted in a sustained process of desecuritisation of the Indus waters matter and have served to increase water security for both states. The partition of 1947, resulting in the creation of one largely Muslim and one largely Hindu country, caused mass-migration and led to a volatile security situation across the region. Indian East Punjab gained control over the headwaters of important irrigation canals that fed Pakistan's bread basket, the West Punjab. The temporary stoppage of water supplies to these canals—and thus to the lifeline of a large number of people—increased the risk of regional water-related conflicts. This instance provided the initial impetus for bilateral negotiations over the use of the newly internationalised transboundary water resources. Following some 10 years, and with the help of the World Bank, India and Pakistan agreed to the Indus Water Treaty, which aimed at increasing the amount of water available to the two states (Salman and Uprety, 2002). The treaty partitioned the use and control over six key Indus tributaries (allocating three to each country), and provided a platform for financing infrastructure to ensure adequate water supply and storage for food and energy production in both countries.

This example of the Indus reveals some of the ways that international law is embedded within the securitization/desecuritisation process of transboundary water resources management and in the promotion of water security. The case study also highlights that national (political) securitisation approaches to water are not required in the quest for achieving enhanced water security. Militarisation and extreme measures, such as stopping or critically reducing the flow of transboundary waters to a downstream neighbour to secure national supplies will, in fact, be more likely to provoke greater regional insecurity as a result of the adverse impacts downstream. A sudden lack of vital water supplies could trigger civil strife, terrorist attacks, or—in the extreme, although unlikely, case—armed insurgency and conflict. Military security and water security considered within this paradigm are incongruent goals. One of the underlying factors particular to the water security discourse is the hydrological interdependence of watercourse states arising from shared transboundary water resources (international rivers, lakes, and aquifers) and from the atmospheric water streams of the global water cycle. This interconnectivity accentuates the need for co-management strategies, indeed for cooperation, as the bedrock for achieving true water security.

International Law and Water Security

Water security has been defined as the availability of an acceptable quantity and quality of water for people and the environment at an acceptable level of water-related risks.[2] The concept encompasses 'the availability of water for human development and ecological health along with the capacity to manage and adapt to multiple risks' (Garrick et al., 2012). The risks can have natural causes, such as changes in the hydrological cycle causing floods or droughts, or may be caused by humans, such as pollution, embankment failures, or water scarcity in one part of the system due to overconsumption someplace else. Climate change exacerbates these risks.

Recognizing the urgency of the matter, the ministers assembled at the 2nd World Water Forum in The Hague in 2000 declared water security as their common goal in the 21st century. In their Ministerial Declaration on Water Security, the first intergovernmental declaration on this issue, they listed seven core challenges to achieve this goal: (1) meeting basic human needs; (2) securing food supply and water for food production; (3) ensuring integrity of ecosystems; (4) sharing water resources and promoting peaceful cooperation; (5) managing risks; (6) valuing water to reflect its economic, social, environmental, and cultural values; and (7) ensuring good governance through stakeholder participation. While this approach might not go far enough, it provides an aspiration for the global community directly linked to water security objectives.[3] This topical focus has increased over the past decade, with a growing and diverse constituency of communities, from the public and private sectors, interested in water security. A recent meeting of the UN General Assembly in New York (25 September 2012) included a high-level side event on global water security, where the participants concluded that there is an urgent need for enhanced cooperation in this field, especially in a world comprised of sovereign nation-states who compete for the use of shared transboundary water resources.

International law plays an important role in addressing these challenges. Recent literature has developed the 3-A Analytical Framework on water security, which identifies the three legally relevant core elements of water security (Wouters et al., 2009):

1. Availability: an adequate quality and quantity of water at an adequate risk
2. Access: through enforceable rights to water for the broad range of stakeholders
3. Addressing conflicts-of-use: where there are conflicts-of-use arising from inadequate availability and/or access to transboundary water resources, these are to be resolved through dispute prevention and settlement mechanisms

Addressing these issues within transboundary water agreements goes a long way towards enhancing the potential for achieving water security. At the regional and global levels, international water law provides an identifiable body of treaty and customary norms and practices that regulate interactions among states, between states and international organizations, and, to some extent, between nation-states and people under their jurisdiction and control. A range of legal rules apply to various aspects of water security, intervening across the spectrum of water-related governance regimes, and with respect to diverse subject areas. Adopting a systemic approach, this chapter highlights how different areas of international law interact, inform, and mutually support each other across different governance levels in its promotion of the goal to achieve water security. The international legal system is analysed through the prism of international water law and traces its interaction with other areas of law at the three scales: (1) universal/regional, (2) basin, and (3) individual/user. The chapter assesses the status quo and identifies areas where the legal system might be improved in order to enhance its effectiveness in promoting water security.

Desecuritisation and Principles of Management Adopted at the Universal and Regional Levels

As the world is becoming increasingly interdependent—through multiplying trade links; growing labour migration and the economic importance of remittances; and ever-increasing networks of transport routes, canals, pipelines, and fibre-optic cables—its natural resource dependency has passed a critical tipping point, with attendant adverse impacts. Natural resources, such as water, are becoming increasingly scarce and their transboundary character heightens the potential for securitisation and regional tensions as a result. The imperative for states to cooperate across national borders to meet demand for diminishing resources is now emerging as one of the most pressing national and international policy agendas (McKinsey, 2009). Desecuritisation of transboundary resources, defined here as the process of extracting them from the realm of politics and focusing on their sustainable management, becomes an imperative. International water law feeds into this process supporting the management of shared resources and offering mechanisms for the prevention and settlement of potential disputes. Thus, where transboundary water resources are not available (in adequate quality and quantity) or accessible (to all stakeholders, including ecosystems) and conflicts of use may arise, international law provides a suite of rules and processes that guide states in addressing such water insecurities through desecuritisation.

The Law of Nations as a Platform for Regional Peace and Security

Concluded following two world wars, the UN Charter contains the fundamental tenets of the 'law of nations', promoting regional peace and security and advancing the fundamental freedoms of all.[4] A recent UN Resolution endorsing the outcomes of Rio+20 reiterated support for the law of nations and the importance of the rule in this context (UN Resolution 66/288, 2012). Anchored in the law of nations, the duty to cooperate in the peaceful management of freshwaters shared across national borders is linked directly to the core principles espoused in the UN Charter. It is timely that UN-Water (a platform for the close to thirty UN bodies that work on water) now takes forward its work on water security, but these efforts must be informed by international law. Often this aspect is missing from the work in the sector, and it is a missed opportunity for completing the water security puzzle. The next sections explore the contribution that international law makes in this regard.

Promoting Water Security through Universal and Regional Instruments: Rules of International Water Law

The rules that apply to international watercourses (broadly considered here to include shared rivers, lakes, and aquifers)[5] are elaborated in a body of international law generally referred to as *international water law*. Its principles, which have emerged from coherent state practice as customary norms and are codified largely in the 1997 Convention on the Law of the Non-Navigational Uses of International Watercourse (UNWC)[6] and supplemented in important ways under the 1992 UNECE Convention on the Protection and Use of Transboundary Watercourses and International Lakes (UNECE TWC), address the core elements of water security: water availability, access, and conflicts-of-use management.

The governing rules of customary and treaty law in this field include: the principle of equitable and reasonable utilization, the due diligence obligation not to cause significant harm, and the general duty to cooperate. The rule of equitable and reasonable utilization is the cornerstone norm that promotes adequate access to, and availability of, water for all watercourse states and stakeholders (Wouters et al., 2009). Under this guiding principle, all factors relevant to the equitable and reasonable use of the water resources of the shared watercourse are to be identified and considered together, with an evaluation reached 'on the basis of the whole'. Article 5 of the UNWC[7] provides that shared systems 'shall be used and developed . . . with a view to attaining optimal and sustainable utilization thereof and benefits there from, taking into account the interests of the [other] States [sharing the system], consistent with adequate protection of the [system]'. The factors to be considered include the circumstances of the respective system; geography, hydrology, climate, and other natural characteristics; social and economic needs; the water dependent population; existing and potential uses and available alternatives and the effects of the use on others; and the conservation, protection, and development of waters and their costs. Thus, in their assessment of equitable and reasonable use, watercourse states have to take into account issues related to water availability and access, including issues related to climate change and its interlinkage with the global water cycle and wider precipitation patterns. The list of factors under the UNWC is only indicative and is open to include the new range of challenges that arise under water insecurity scenarios.

The due diligence obligation not to cause significant harm[8] guards against instances where water security would be undermined in some states by, for example, instances of over-abstraction or significant pollution of shared waters by one state resulting in inadequate supplies of water quantities and qualities to other watercourse states on the transboundary system. This due diligence obligation not to cause substantial injury to other states is derived from the principle of good neighbourliness and applies equally to the transboundary protection of the environment more generally. Part of this due diligence obligation is the implementation of an environmental impact assessment (EIA) for new projects that may cause significant harm (International Court of Justice, 2010). EIAs facilitate the prevention of transboundary harm and help to maintain the health of ecosystems that provide important environmental services, such as the water purification and flood control function of wetlands, thus contributing to local water security. This approach is promoted under the UNECE TWC.

The general duty to cooperate intervenes in particular with respect to risk management and the peaceful development and use of shared water resources.[9] It facilitates compliance with other IWL principles and is the legal basis for the more specific customary law obligations, such as the obligations of regular information and data exchange and notification. Regular data exchange keeps riparian states informed about the condition of the shared system and thus not only facilitates assessment of equitable and reasonable use, but also enhances the preparedness for extreme hydrologic events. Human-induced and naturally caused water risks can be identified early on if states share their system-relevant data and notify each other of planned measures and emergencies.

The general principles of international water law are reiterated in a number of regional instruments that apply to the management of transboundary water resources, such as the 2000 Revised Protocol on Shared Watercourses of the Southern African Development Community (hereinafter, SADC Revised Protocol) and the 1992 UNECE TWC. In addition, these two instruments place a strong emphasis on the interlinkages between water and its surrounding ecosystems. The SADC Revised Protocol makes the achievement of a balance between resource development for people and the conservation and enhancement of the environment one of the guiding principles for the region, thus strongly supporting *inter alia* the element of environmental water needs of the water security concept. Similarly, the 1992 UNECE TWC has a particular focus on water quality aspects, which is also present in the 2000 Water Framework Directive of the EU whose member states are all parties to the 1992 UNECE TWC, except those of which are island states. The focus on pollution prevention and water quality management as a core aim of these instruments is further underlined by the inclusion of the precautionary and the polluter-pays principles, which also apply as customary principles in this region.

A factor contributing to the practical relevance and strength of the 1992 UNECE TWC is its strong institutional framework. An international secretariat is responsible for supporting the implementation of the convention and its protocols, and also supports the Meeting of the Parties (MOP) in its actions to achieve the objectives of the convention. Over time, the MOP established a number of working groups, ad-hoc expert groups, and a legal board that provide technical services to the parties. MOP decisions also led to an expansion of tasks and issue areas the convention bodies are engaged in beyond the original scope of the treaty. The Convention's Task Force on Water and Climate change, for example, prepared a unique guidance note on the topic of how states sharing transboundary water systems can improve resilience and water security through climate change adaptation (UNECE, 2009). This note, although written within the UNECE framework, also aims at supporting riparian states in other regions.

Universal Instruments That Contribute to Water Security

International water law is not the only area of the international legal system that promotes water security. As mentioned, the geographic and hydrological scope of international water law is restricted to transboundary water systems that form a unitary whole, and does not address interdependencies that exist beyond the geographic boundaries of basins. Hydrological interdependencies exist beyond the watershed and are determined by the health of the global environment; the atmospheric streams of the global water cycle connect states and determine the availability of water for people and the environment. A warming climate alters the hydrologic balance and precipitation patterns around the planet. This wider ambit of water security is addressed in particular by international environmental law instruments, which must be considered in this context.

A number of multilateral environmental agreements (MEAs), while not having water management as their primary objective, can affect water security. These instruments, *inter alia,* promote the maintenance and enhancement of ecosystem services that enhance water availability and quality for people and the environment. As one example, the 1971 Convention on Wetlands of International Importance Especially as Waterfowl Habitat (Ramsar Convention) contributes by protecting wetlands and thus promotes conservation of ecosystems that play an important role in regulating flow patterns, groundwater recharge, and water purification. Water is also an indirect object of the 1992 Convention on Biological Diversity (CBD); to protect biological diversity and to sustainably use its components and share its benefits, the lifeblood of biodiversity, namely freshwater, must be protected (Boisson de Chazournes et al., 2011). This is a two-way relationship: Up to the point where water quality can be provided through the conservation of biodiverse ecosystems, it will enhance water security through the protection of ecosystem capacity to absorb and recycle nutrients; when the tipping point is reached, interventions to enhance water quality are needed to ensure the necessary minimum level of biodiversity health to provide these ecosystem services. The 1994 Convention to Combat Desertification (UNCCD) is concerned with the conservation and sustainable management of land

and water resources in order to improve living conditions of people in the affected areas (Article 2, UNCCD). The United Nations Framework Convention on Climate Change (UNFCCC) contributes in a similar way to the promotion of water security, even though water is not much considered in the text of this convention. The UNFCCC forest initiative, the UN Collaborative Programme on Reducing Emissions from Deforestation and Forest Degradation in Developing Countries (UN-REDD), for example, contributes to surface runoff control and water retention through the protection and expansion of plant cover. The conservation of plant cover and wetlands reduces runoff velocity and increases the capacity of nature to soak up floodwaters and excessive runoff. In promoting the conservation and rehabilitation of diverse ecosystems, these legal instruments also contribute to water risk management and, thus, to water security.

Similar to the institutional approach adopted under the 1992 UNECE TWC, these MEAs are supported by comprehensive institutional mechanisms that, over time, have expanded the focus of their work on water in line with a general global trend to concentrate on issues of water scarcity and deterioration of ecosystems (Boisson de Chazournes et al., 2011). In addition, these conventions can turn to the same international financial mechanism, the Global Environment Facility (GEF), to mobilize funding for projects that implement their respective environmental objectives. Additional UNFCCC financing mechanisms facilitate funding of activities that respond to the water security challenges caused by climate change. More than half of the adaptation projects financed through these mechanisms at the national and transboundary levels to date concern the water sector. The contribution of MEAs to water security is significant, given their almost universal membership (Ramsar Convention: 162 Parties, Convention on Biological Diversity: 193 Parties, Convention to Combat Desertification: 193 Parties and UN Framework Convention on Climate Change: 195 Parties).

The universal and regional instruments discussed here are all framework treaties for the adoption of more specific agreements at the basin level and for national laws that implement their respective objectives. The interaction between universal, regional, and basin instruments can be regarded as an 'interaction continuum' across these different scales; the framework instruments inform the content of new basin-level agreements in the same way that the general principles included in regional and universal instruments have emerged from previous treaty practice at the basin level. The more general agreements then encourage new legal instruments that adapt the general principles and obligations to the specific basin and country conditions. This building-block approach to watercourse regulation provides increasingly innovative and comprehensive legal regimes, which provide a nested paradigm for water resources management. While this has been mostly positive, there are challenges related to treaty congestion and fragmentation, which require attention.

Enhancing Water Security at the Basin Level

The basin level is where agreements between watercourse states might be tailored to meet the specific needs and water development interests of the individual countries based on the circumstances particular to the hydrologic system. A large number of these basin treaties (more than 70%) are bilateral. This bilateral practice might be explained by the fact that the large majority of watercourses are shared by only two states or because the agreements are aimed at regulating specific issue areas, such as joint projects or localized flow regulation, as is the case for countries such as China and India (Chen, Rieu-Clarke, and Wouters, 2012). Where only two countries share the transboundary water resources, bilateral agreements can cover basin-wide management issues, which is the preferred approach from an IWRM perspective (GWP, 2012). The increase in the number of multilateral basin treaties over the past decades (Brown Weiss, 2007) bears evidence that water issues are becoming more pressing at the basin level. And scholarship on the issue demonstrates support for the view that the scope of agreements should be aligned with the basin as the unit of management, in line with the hydrology of the watercourse.

Adapting to Changing Circumstances in Managing Risks

Water security challenges have, in many cases, been the result of hydrologic variability that occurs due to natural cyclical flow variations. In this light, the general formulations of most rules of international water law provide space for adjustments to local conditions through more specific instruments. Depending on how treaties are crafted, they can build in the flexibility that is required to cope with hydrologic variability and extreme events, such as floods, droughts, and ice hazards.

Where drought risks threaten to undermine vital economic production processes, states refer to allocation agreements for low-flow periods, such as in the 1944 Treaty relating to Utilization of Water of the Colorado and Tijuana Rivers, and of the Rio Grande. Flow deficiencies that make it difficult for Mexico to provide agreed runoff volumes to the United States within one five-year cycle are made up in the subsequent five-year cycle (Article 4 of the 1944 Treaty). Such treaty formulae can address water availability risks as long as the fluctuations remain within the identified volume and flow change range. If the underlying hydrologic conditions change, the treaties may require revision or modification under the guidance of river basin organisations, such as has been the case between the United States and Mexico through the workings of the International Boundary Waters Commission.

Obligations to notify other watercourse states of emergencies help to prevent damage caused by sudden and extreme hydrologic events, such as the floods that occurred across Europe in recent years. Overall, resilience can be enhanced through agreements on regular monitoring and data exchange and the installation of early warning systems. The 1987 Agreement on Flood

Warning in the Catchment of the Moselle, a Rhine tributary, is a case in point. Based on the agreement, the riparian states Germany, France, and Luxembourg established a flood preparedness information system that provides remote data access digitally as well as via voice transmission in French and German at all times (Articles 2 and 3 of the 1987 Agreement). The Technical Committee managing the automatic information system is composed of representatives from the respective national navigation and water authorities. One of its limitations is that its mandate is restricted to the technical operations of the system. Thus, the scope of its interventions with regard to water risk management is also limited. More comprehensive risk management capacity is provided by treaties that establish permanent basin management mechanisms with multi-issue mandates and scientific bodies or working groups. These institutional structures can more effectively adapt to new challenges that may emerge only after treaty adoption and are, therefore, not explicitly considered in the respective constituent instruments. This is highlighted, for example, with respect to the International Commission for the Protection of the Rhine (ICPR) established on the basis of the 1999 Convention on the Protection of the Rhine, with its focus is on water quality, environmental protection, and flood control. Over time, the ICPR has taken on the task of adapting basin management to climate change impacts, a water challenge that was not explicitly considered at the time of its establishment. The commission now elaborates climate change response measures within the context of its Action Plan on Floods and is engaged in basin-wide hydrologic and meteorological data processing and renaturalisation of wetlands in order to mitigate for the expected increase in quantity and severity of flood events.

The management of hydrologic variability has long been one of the core concerns for treaty regulation at the basin level. The more than 2,000 water-related treaties adopted over the past 300 years (Brown Weiss, 2007), therefore, depict a large variety of solutions adopted by states to manage hydrologic risks on transboundary systems, from notification, information exchange, and water allocation formulae to the establishment of institutional mechanisms, which offer platforms for continuous cooperation, knowledge generation, and adaptive decision making for states. Each of these developments provides concrete platforms for addressing issues related to water security (access, availability, addressing conflicts-of-use).

Water for Human Development and Environmental Needs

In a manner similar to how states have dealt with the management and regulation of risk scenarios, there are multiple ways in which they have used basin-level treaties to address water security issues related to human development and environmental needs. Adapted to the specific circumstances and needs of individual basins, these agreements vary in their context; some, for example, are limited to regulating availability and quality issues linked to specific projects and hydraulic infrastructure, such as the 1959 agreement between Norway, Finland, and the Soviet Union concerning the regulation of flow in the Paatsjoki River Basin, or the 1996 treaty between India and Bangladesh on sharing of the Ganga/Ganges waters at Farakka. In the first case, the three states agree to releases from Lake Inari, which is located in Finland, to ensure the availability of flow for the operation of a series of downstream hydroelectric power stations in Russia and Norway and also to control spring floods. In the second example, the treaty establishes flow sharing formulae to allocate dry season flows of the Ganges between India and Bangladesh. This sharing agreement aims at satisfying, in particular, the water needs for agricultural production and maintenance of shipping and port operations downstream of the barrage. In both cases, decisions on discharge are based on jointly agreed high- and low-flow allocation formulae to ensure adequate water supply for economic activities.

Other treaty regimes are more comprehensive in scope; treaties such as the 1995 Agreement on the Cooperation for the Sustainable Development of the Mekong River Basin and the 2002 Senegal Water Charter, as just a couple of examples, establish jointly agreed principles of management, development, and protection of basin waters and regulate multiple uses across the entire or at least large parts of these two multistate basins. The 1995 Mekong Agreement, which reflects to a large part the principles included in the UNWC, promotes cooperation among the four lower riparian states Cambodia, Laos, Thailand, and Vietnam for the sustainable development and conservation of the river's water and its related resources. Covering a broad range of different water uses, the treaty aims at optimizing use and mutual benefits, while minimizing harmful effects that are due to natural or human causes. The parties are required to maintain minimum flows to protect the complex ecologic balance of the basin. As a joint management mechanism, the Mekong River Commission (MRC) was established to coordinate implementation. The MRC has implemented a number of procedures and guidelines, including some on data sharing and notification, which have served to deepen cooperation between states and engaged the two upstream countries, China and Myanmar, as dialogue partners (Wouters, 2003). This latter process resulted in a 2002 agreement with China on the provision of water level information during the flood season, enhancing the downstream riparian states' water management capacity. Cooperation with China appears to be increasing through processes enabled under the Mekong Agreement, where China attends as an observer.

The 2002 Senegal Water Charter adds to an already existing, elaborate legal framework comprising multiple treaties that serve as a basis for equitable utilization of the Senegal River's water resources and the sharing of benefits derived there from by means of joint infrastructure projects and navigation regulation. With the adoption of the charter by all four riparian states, this framework now covers the entire basin. The original purpose of the treaty regime, which dates back to the years immediately after decolonization, was to increase flood protection, hydropower, irrigation, and navigation benefits; it is, thus, an early example of riparian states addressing the water–energy–food security nexus. The Senegal Water Charter has placed additional emphasis on the parties' objective to develop the river's resources in a sustainable way, taking into consideration, in particular, environmental and ecosystem needs, as well as individuals' right to water. Similar to the Mekong, this treaty regime also benefits from the existence of a basin-wide management mechanism, the Organisation pour la Mise en Valeur du fleuve Sénégal (OMVS), composed of the Conference of Heads of States and Government, the Council of Ministers, the High Commissariat, and the Permanent Water Commission. The OMVS has a comprehensive river basin management mandate, including decision-making powers on key operating decisions of jointly owned infrastructure, such as the Manantali and Diama dams, which provide electricity, irrigation, and flow regulation benefits to the riparian states (Ba and Mbengue, 2006). Benefits derived from the use of the shared water resources are distributed among the basin states based on a complex benefit sharing formula (the 'key'), which is regularly reviewed to adjust to changing needs and circumstances (Yu, 2008).

Many basin treaties address water security without explicitly mentioning the concept. Only the Agreement on the Nile River Basin Cooperative Framework (CFA), not in force, refers expressly to 'water security'. The instrument defines the term as 'the right of all Nile Basin States to reliable access to and use of the Nile River system for health, agriculture, livelihoods, production and environment' (Article 2). However, Article 14, which aims at regulating the rights and obligations of the Nile Basin states with regards to respecting each other's water security, is not acceptable to all riparian states, resulting in the current impasse in the basin-wide adoption of the CFA. The negotiating parties could not reach agreement on this point and referred the issue to resolution once and if the agreement would enter into force (annex on Article 14(b)). This experience has provoked the argument that rather than introducing the concept into treaty texts, water security questions are more effectively addressed through the implementation of the general principles of international water law (Mekonnen, 2010). The CFA has been signed by six countries: Ethiopia, Tanzania, Rwanda, Uganda, Burundi, and Kenya. However, it will enter into force only once six countries have ratified the agreement, and none of the current signatories has ratified it to date (as of December 2012). A basin-wide agreement on the Nile would facilitate cooperation and contribute to water security objectives; as one regional expert notes with regard to the direct link between access to water on the Nile and poverty: 'The higher water access you have the less the poverty profile. . . . This is not only in comparison between Egypt and upstream countries: within Ethiopia itself, 22 per cent less poor were observed in those communities who have access to water' (*Africa Review*, 2012). And the reality of climate change exacerbates water insecurity across the basin, highlighted in the recent World Bank Report on climate change: 'River basins dominated by a monsoon regime, such as the Ganges and Nile, are particularly vulnerable to changes in the seasonality of runoff, which may have large and adverse effects on water availability' (PICIRCA, 2012).

Basin level treaties breathe life into the general principles of international water law, especially where effective institutional mechanisms are created and supported; these bodies provide operational meaning to implement the general principles contained in treaties as solutions to achieve water security in specific local contexts. The treaties give these principles a basin-specific character and reflect jointly agreed responses on matters related to water availability, access, and dispute prevention, thus addressing important legal issues linked to water security. Preidentified water-sharing formulae for fluctuating flow scenarios are one of the frequently used tools to introduce an ongoing in-built flexibility mechanism capable of responding as needed to hydrologic variability and other water risks; basin-wide institutions (or bilateral bodies), and in particular those with standing technical secretariats, enhance the basin states' capacity to react and adapt more effectively to changing conditions and extreme events. The legal framework of the Senegal Basin, as only one example, illustrates the merit of basin-level instruments and international water law with regards to promoting water security also at the individual and household levels.

Individual Water Security

Individual stakeholders at the local level are those that are most directly affected by water resources management decisions at the national and transboundary levels. Individual households, private enterprises, communities, and ecosystems are at the receiving end of the impacts of these decisions. International law sets forth a body of rights and obligations that also promotes water security objectives at this level.

The Helping Hand of Human Rights Law

Human rights law is one area of international law that deals with the legal relationship between states and its peoples; the rights of individuals are to be 'respected, protected and fulfilled' by the national governments under which jurisdiction and control they find themselves. The duties and obligations of states related to water were considered by the UN Committee on Economic, Social and Cultural Rights (CESCR), which clarified that states have to promote, facilitate, and provide individuals with access to adequate water in terms of availability, quality, and accessibility, and must refrain from interfering in, and prevent other parties from interfering in, ensuring this basic right (CESCR, 2002). Given its vital importance, some find it surprising that the right to water was not explicitly included in the early human rights documents adopted at the universal level; neither the 1966 International Covenant on Civil and Political Rights (ICCPR) nor the 1966 International Covenant on Economic, Social and Cultural Rights (ICESCR) refers to water. However, as confirmed later by the CESCR in 2002 and the UN General Assembly (GA) and Human Rights Council (HRC) in their respective resolutions recognizing the human right to water and sanitation in 2010,[10] these covenants comprise rights that are inherently linked to the availability of safe and sufficient supplies of water, such as the right to life (Article 6, ICCPR), the right to an adequate standard of living (Article 11, ICESCR), and the right to health (Article 12, ICESCR). Human rights obligations with regards to water were also included in later universal and regional human rights conventions, including in the 1979 Convention on the Elimination of All Forms of Discrimination against Women, the 1989 Convention on the Rights of the Child, the 2006 Convention on the Rights of Persons with Disabilities, the 1990 African Charter on the Rights and Welfare of the Child, and the 2003 Protocol to the African Charter on Human and People's Rights on the Rights of Women in Africa. Rights and obligations with regards to enhancing water security at the individual level therefore already exist across the body of human rights law. And even if the recognition of the human rights to water and sanitation by the GA and HRC does not come with absolute legal guarantees (apart from reporting) and is not accompanied by the establishment of institutions that would monitor compliance or ensure enforcement at the universal level, the international debate on these rights has by itself contributed to improvements of implementation at the national, basin, and regional levels. An increasing number of state constitutions and national legislation have come to recognise and support the operational of the right to water. These rights are also considered in international water law. At the regional and basin levels, the rights and obligations have become a subject of at least one regional water treaty and a small number of basin-level instruments.

The Supporting Role of International Water Law

Human water security has gained attention in international water law with the special consideration attributed to it in Article 10 of the UNWC, which provides that special regard shall be given to 'the requirements of vital human needs' in the event of conflict between different uses of international watercourses.[11] However, the scope of the concept of vital human needs is less comprehensive than that of human rights obligations with regards to water. It encompasses the provision of sufficient water to 'sustain human life, including both drinking water and water required for production of food in order to prevent starvation' (ILC, 1994), while human rights obligations include an additional progressive state obligation to continuously improve standards beyond the provision of a minimum, essential level of safe water and to respect the cultural and religious value of water (CESCR, 2002).

Other water law instruments have gone beyond this recognition of the need to protect and provide vital human water needs at the universal level. The 1999 Protocol on Water and Health to the 1992 UNECE TWC translates the core content of the human right to water into an international water treaty to promote human health and well-being. The protocol establishes international cooperation obligations to support parties that lack the necessary resources and capacity to achieve domestic implementation plans and targets. These international cooperation obligations are the value-added duties and entitlements that water law brings to the achievement of human rights objectives within the UNECE framework (Tanzi, 2010). In contrast to human rights law, where the existence of international cooperation obligations remains debated, international water instruments frequently include firm cooperation obligations in the spirit of the general duty to cooperate and through a range of operational processes at the basin level.

The recognition of the right to water has also found entry into basin-level treaties since the beginning of the new millennium. This is unusual because water treaties traditionally regulate state-to-state relations; only occasionally are state obligations towards individuals or nonstate actors considered, such as in the 2002 Framework Agreement on the Sava River Basin, in which states are asked to facilitate monitoring of treaty implementation by providing information to the general public. The 2002 Senegal Water Charter and the 2008 Niger Basin Water Charter are pioneers in linking human rights and water law at the basin level. The Niger Basin States consider the right to water as a principle guiding their cooperation in managing the shared resources, while for the Senegal Basin the enjoyment of the right to water is recognized as an explicit objective of any repartition of the river's water (Article 4 of the 2002 Senegal Water Charter). Moreover, the Senegal Water Charter provides even more robust guarantees stipulating that the satisfaction of vital human needs takes priority over all technical and economic considerations of water allocation (Articles 6 and 7). One advantage of having the right to water recognized at the regional level of the UNECE and in the basin-wide Senegal and Niger examples is that in all three instances there exist institutions that promote monitoring and support compliance of the agreed regimes. The UNECE TWC Secretariat provides funding and technical assistance to member states that do not have sufficient resources to achieve the objectives of the UNECE TWC and provides general support for compliance measures.

Legal issues related to the promotion of water security at the individual level highlight the dynamics of the complementarity of different areas of international law and their collective contribution to water security matters. Just as environmental law informs and supports water law at the universal level and vice versa, human rights law and international water law interact with respect to rights of access to water for individuals. International water law contributes with international cooperation obligations and institutional frameworks that ensure the realization of these rights, while human rights law informs both the content and scope of state obligations in this regard. This topic also illustrates the interconnectivity across various water-related international instruments and their cumulative promotion of water security across the various scales.

Observations and Summary Conclusions: The Grand Paradox of Security and Water

Water security is one of the world's greatest challenges. Enhancing water security within a world of sovereign states continues to be an aspiration, complicated by compounding problems related to economic, social, and environmental interests and demands and exacerbated by uncertainties affecting all of these.

This work has demonstrated the water security paradox—how cooperation and not securitization is at the heart of achieving effective water security. Addressing issues related to water availability, access, and conflicts of use, at a range of scales, is the bedrock of international water law. Based on the fundamental tenets of the law of nations, the norms and processes that govern the uses of transboundary water resources shared across national borders originates from, and revolves around, the duty to cooperate. Achieving water security in the transboundary water context—where water resources of appropriate quality and adequate quantities are available to all stakeholders—requires 'rules of the game' that are implemented jointly and not unilaterally imposed. The global community under the Hague 'water security' declaration set itself seven key challenges: (1) meeting basic human needs; (2) securing food supply and water for food production; (3) ensuring integrity of ecosystems; (4) sharing water resources and promoting peaceful cooperation; (5) managing risks; (6) valuing water to reflect its economic, social, environmental, and cultural values; and (7) ensuring good governance through stakeholder participation. Whilst these are interconnected and may even overlap, setting the scene for increased competition, the response must be a collective one. Water, the world's most precious resource, is a catalyst for peace, not an object of military conflict.

International law is an integral component of the water security puzzle, providing a comprehensive system of norms and processes, and rights and obligations that can be (and have been) employed to achieve water security at the universal, basin, and individual levels. A significant number of international agreements that promote water security, or certain aspects of it, are already in place. While clearly international law cannot be a panacea and the current horizon of legal regulation might not yet adequately address the interlinkage of surface, groundwater, and atmospheric water streams, the system as a whole does provide a solid foundation that can be built upon in order to promote water security at the universal level. An important aspect when using international law to pursue this goal is the responsibility to not only define rights and obligations, but also to establish the institutions and processes (such as prior notice and exchange of information) that are necessary to achieve effective compliance. Across the spectrum of international law, there is a cognate body of laws and state practices that contribute to achieving water security, from multilateral environmental treaties to human rights instruments at international and national levels. There is room for improvement—climate change instruments must be made more robust; the fragmentation across the sector continues to cause challenges; and a more inclusive and comprehensive reading of all relevant treaties would be beneficial.

> Water is fluid, soft, and yielding.
> But water will wear away rock, which is rigid and cannot yield.
> As a rule, whatever is fluid, soft, and yielding will overcome
> whatever is rigid and hard.
> This is another paradox: what is soft is strong.
>
> Lao-Tzu

Notes

1. The term *dispute settlement procedures* has to be understood in this context as technical term generally employed in international law, because the 1960 Indus Waters Treaty distinguishes more specifically between procedures to settle 'differences' and 'disputes,' and the use of the term dispute settlement procedures may therefore be considered by some as misleading.
2. This definition is an adapted version of the definition provided by Grey and Sadoff (2007). Literature provides a broad range of different definitions that capture similar elements or are focused on the human-centric aspects of water security. For an overview, see Wouters, Vinogradov, and Magsig (2009).
3. 'While pioneering on its own right, this list does not identify water as the key factor binding nature and society. The connections between nature and engineered water infrastructure, the water–energy–food security nexus, the high rates of freshwater biodiversity loss, and linkages between water and land use must all be addressed in the quest for sustainability' (Bogardi et al., 2011).
4. Article 1 of the UN Charter: 'The Purposes of the United Nations are:
   1. To maintain international peace and security, and to that end: to take effective collective measures for the prevention and removal of threats to the peace, and for the suppression of acts of aggression or other breaches of the peace, and to bring about by peaceful means, and in conformity with the principles of justice and international law, adjustment or settlement of international disputes or situations which might lead to a breach of the peace;
   2. To develop friendly relations among nations based on respect for the principle of equal rights and self-determination of peoples, and to take other appropriate measures to strengthen universal peace;
   3. To achieve international co-operation in solving international problems of an economic, social, cultural, or humanitarian character, and in promoting and encouraging respect for human rights and for fundamental freedoms for all without distinction as to race, sex, language, or religion; and
   4. To be a centre for harmonizing the actions of nations in the attainment of these common ends.'
5. Article 2, Convention on the Law of Non-Navigational Uses of International Watercourses, adopted 21 May 1997, UN Doc. A/RES/51/299; and Article 2, Draft Articles on the Law of Transboundary Aquifers, adopted 11 December 2008, UN Doc. A/RES/63/124.
6. The Convention on the Law of the Non-Navigational Uses of International Watercourses, known also as the UN Watercourses Convention, was adopted by the General Assembly in 1997 and has yet to enter into force. As of September 30, 2012, 28 states ratified the Convention, but 35 ratifications are needed for

its entry into force. As customary rules, the IWL principles apply between states, even in the absence of binding treaties.

7. Article 5, Equitable and reasonable utilization and participation:
   1. Watercourse States shall in their respective territories utilize an international watercourse in an equitable and reasonable manner. In particular, an international watercourse shall be used and developed by watercourse States with a view to attaining optimal and sustainable utilization thereof and benefits therefrom, taking into account the interests of the watercourse States concerned, consistent with adequate protection of the watercourse.
   2. Watercourse States shall participate in the use, development and protection of an international watercourse in an equitable and reasonable manner. Such participation includes both the right to utilize the watercourse and the duty to cooperate in the protection and development thereof, as provided in the present Convention.'

8. Article 7, Obligation not to cause significant harm:
   1. Watercourse States shall, in utilizing an international watercourse in their territories, take all appropriate measures to prevent the causing of significant harm to other watercourse States.
   2. Where significant harm nevertheless is caused to another watercourse State, the States whose use causes such harm shall, in the absence of agreement to such use, take all appropriate measures, having due regard for the provisions of articles 5 and 6, in consultation with the affected State, to eliminate or mitigate such harm and, where appropriate, to discuss the question of compensation.'

9. Article 8, General obligation to cooperate:
   1. Watercourse States shall cooperate on the basis of sovereign equality, territorial integrity, mutual benefit and good faith in order to attain optimal utilization and adequate protection of an international watercourse.
   2. In determining the manner of such cooperation, watercourse States may consider the establishment of joint mechanisms or commissions, as deemed necessary by them, to facilitate cooperation on relevant measures and procedures in the light of experience gained through cooperation in existing joint mechanisms and commissions in various regions.'

10. UN GA Resolution 64/292, The Human Right to Water and Sanitation: 1. *Recognizes* the right to safe and clean drinking water and sanitation as a human right that is essential for the full enjoyment of life and all human rights. . . .'

11. Article 10(2) of the UNWC provides, 'In the event of a conflict between uses of an international watercourse, it shall be resolved with reference to articles 5 to 7, with special regard being given to the requirements of *vital human needs*'[emphasis added].

References

2030 Water Resources Group (2009) *Charting our Water Future*, McKinsey & Company, www.2030waterresourcesgroup.com/water_full/Charting_Our_Water_Future_Final.pdf.

Africa Review (2012) 'Enough water in the Nile to share, little to waste', IRIN, 23 November 2012 www.africareview.com/Special-Reports/Enough-water-in-the-Nile-to-share/-/979182/1627692/-/26pp5g/-/index.html.

Alam, U. (2002) 'Questioning the water wars rationale: a case study of the Indus Waters Treaty', *The Geographic Journal*, vol 168, no 4, pp341–353.

Ba, A.S. and Mbengue, M. (2006) 'Le régime juridique du fleuve Sénégal: aspects du droit des cours d'eau dans un contexte régional', *African Yearbook of International Law*, vol 12, pp309–347.

Bogardi, J.J., Dudgeon, D., Lawford, R., Flinkerbusch, E., Meyn, A., Pahl-Wostl, C., Vielhauer, K. and Vörösmarty, C. (2011) 'Water security for a planet under pressure: interconnected challenges of a changing world call for sustainable solutions', *Current Opinion in Environmental Sustainability*, vol 4, pp1–9, www.gwsp.org/fileadmin/documents_news/Elsevier.pdf, accessed 8 October 2012.

Boisson de Chazournes, L., Leb, C. and Tignino, M. (2011) 'Environmental protection and access to water: the challenges ahead', in M. Van der Valk and P. Keenan (eds) *The Right to Water and Water Rights in a Changing World*, UNESCO-IHE, Delft.

Brown Weiss, E. (2007) 'The evolution of international water law', Recueil de cours, *Hague Academy of International Law*, vol 331, pp163–404.

Brulliard, K. (2010) 'Rhetoric grows heated in water dispute between India, Pakistan', *Washington Post*, 28 May, www.washingtonpost.com/wp-dyn/content/article/2010/05/27/AR2010052705393.html, accessed 30 June 2012.

Buzan, B., Waever, O. and de Wilde, J. (1998) *Security: A New Framework for Analysis*, Lynne Rienner Publishers, Boulder.

Committee on Economic, Social and Cultural Rights (CESCR) (2002) *General Comment No. 15, The right to water (arts. 11 and 12 of the International Covenant on Economic, Social and Cultural Rights)*, UN Doc. E/C.12/2002/11.

Garrick, D., Connell, D. and Pittock, J. (2012) 'Water security and federal rivers', *Brief No. 9*, Oxford University, www.eci.ox.ac.uk/watersecurity/downloads/briefs/9-garrick-2012.pdf, accessed 30 June 2012.

Global Water Partnership (GWP) (2012) 'Increasing water security—a development imperative', *Perspectives Paper*, www.gwp.org/Global/About%20GWP/Publications/Perspectives%20Paper_Water%20Security_final.pdf, accessed 20 September 2012.

Grey, D. and Sadoff, C. (2007) 'Sink or swim? Water security for growth and development', *Water Policy*, vol 9, pp545–557.

International Court of Justice (2010) *Case concerning Pulp Mills on the River Uruguay (Argentina v. Uruguay)*, Judgment. *I.C.J. Reports 2010*, p14.

International Law Commission (ILC) (1994) 'Draft articles on the law of the non-navigational uses of international watercourses and commentaries thereto and resolution on transboundary confined groundwater', *Yearbook of the International Law Commission, 1994*, vol. II (part two), pp89–135.

Mekonnen, D.Z. (2010) 'The Nile Basin Cooperative Framework Agreement negotiations and the adoption of a "water security" paradigm: flight into obscurity or a logical cul-de-sac?', *European Journal of International Law*, vol 21, no 2, pp421–440.

Potsdam Institute for Climate Impact Research and Climate Analytics (PICIRCA) (2012) *Turn Down the Heat: Why a 4°C Warmer World Must be Avoided*, The World Bank, Washington.

Salman, M.A. and Uprety, K. (2002) *Conflict and Cooperation on South Asia's International Rivers—A Legal Perspective*, Kluwer International Law, The Hague.

Sinha, U.K. (2005) 'Water security: a discursive analysis', *Strategic Analysis*, vol 29, pp317–331.

Tanzi, A. (2010) 'Reducing the gap between international water law and human rights law: the UNECE Protocol on Water and Health', *International Community Law Review*, vol 12, no 3, pp267–285.

UN Economic Commission for Europe (UNECE) (2009) *Guidance on Water and Adaptation to Climate Change*, United Nations, Geneva

UN Resolution 66/288. The future we want (11 Sept 2012), http://daccess-dds-ny.un.org/doc/UNDOC/GEN/N11/476/10/PDF/N1147610.pdf?OpenElement.

Wouters, P. (2003) 'Universal and regional approaches to resolving international disputes: what lessons learned from state practice?', in International Bureau of the Permanent Court of Arbitration (ed) *Resolution of International Water Disputes*, Kluwer Law International, Leiden, pp111–154.

Wouters, P., and Chen, H., China's 'soft-path' to transboundary water cooperation, *Journal of Water Law*, (2013) 22: 222–247.

Wouters, P. Vinogradov, S. and Magsig, B.-O. (2009) 'Water security, hydrosolidarity, and international law: a river runs through it. . . ', *Yearbook of International Environmental Law,* vol 19, pp97–134.

Yu, W.H. (2008) *Benefit Sharing in International Rivers: Findings from the Senegal River Basin, the Columbia River Basin, and the Lesotho Highlands Water Project,* Africa Region Water Resources Unit Working Paper 1, World Bank, Washington.

Part II

Perspectives and Principles

4 Debating the Concept of Water Security

Christina Cook and Karen Bakker

Introduction

The concept of water security has received increased attention over the past decade. A range of international organizations—notably the Global Water Partnership and the World Economic Forum—has promoted multiple definitions of the concept.[1] Other groups identifying the importance of water security include UNESCO's Institute for Water Education, which has made water security one of its research themes (UNESCO-IHE, 2009) and the Asia-Pacific Water Forum, which in 2007 held its first summit entitled 'Water Security: Leadership and Commitment' (Asia Pacific Water Forum, 2007). Water security has also come to the fore of some domestic water management agendas in the past decade, particularly associated with (bio)terrorism concerns, leading some to characterize it as 'a key objective of a range of governmental and nongovernmental agencies across the spectrum of governance levels' (Jansky et al., 2008, p289). However, while a growing number of scholars, policymakers, and international organizations have employed the concept of water security, the divergence between different framings of water security has become apparent, sparking debate over analytical approaches to, and definitions of, water security (Bakker, 2012).

This chapter presents a comprehensive review of the concept of water security in academic debates; specifically, the chapter analyses the differences and commonalities in approaches to water security across academic disciplines (natural, social, applied, and medical sciences). In this chapter, we review water security–related research; identify distinct differences in methods, scale, and framings of water security in the different disciplines surveyed; and discuss the implications of these differences for theory and practice. We also show that framings of water security have become more diverse, expanding from an initial focus on quantity and availability of water for human uses to include water quality, human health, and ecological concerns. We conclude by suggesting that a broad and integrative definition of water security has utility, despite challenges that arise in the context of implementation and management.

Framings of Water Security across the Physical and Social Sciences

Based on analysis of results from a comprehensive search of 'water security' in the English-language academic literature, we discerned that (1) use of the term is increasing rapidly; (2) uptake of the term has occurred across a wide range of disciplines; (3) framings vary considerably; and (4) studies vary in their methods and scale of application.[2] To analyse the trends, we used a subset of our findings: 198 articles in peer-reviewed journals on the topic of water security from January 1, 1990, to September 30, 2012, in the Web of Science database.

We note that, recently, the concept of water security has been more explicitly linked to food security and energy security in the articulation of a 'nexus' of related concerns (cf. Stucki and Sojamo, 2012; World Economic Forum, 2011). Recent commentary on the Rio+20 sustainable development goals explicitly calls for goal development to be based on 'cross-disciplinary themes such as food, water and energy security' rather than on the traditional three-pillar sustainability approach (Glaser, 2012, p35). However, the further exploration of the 'nexus' concept is beyond the scope of this chapter.

Increasing Use of 'Water Security' across the Disciplines

Concern with the link between water and security is not new; as a basic life resource, water has long been connected to security for humans. Searches of academic databases find the term *water security* first emerged in the early 1990s in the context of Middle Eastern geopolitics (Starr, 1991). The 1990s did not see sustained use of the term. However, in the past decade, there is a clear trend of increased use of across a wide range of disciplines. Figure 4.1

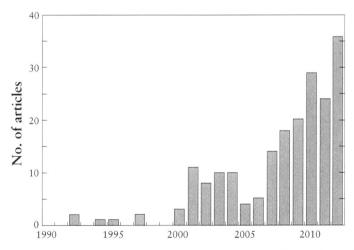

Figure 4.1 Articles containing the term *water security* in the academic literature (1990–2012)

Source: Web of Science database; analysis conducted on September 30, 2012

illustrates the increasing frequency with which the term *water security* is used in academic, peer-reviewed journals. (Our review of the policy literature indicated a similar trend.)

Our analysis shows that a diverse group of disciplines is engaged in academic research on water security (Figure 4.2).[3] Using the Web of Science database 'subject areas' analysis tool, we sorted the 198 articles into disciplinary groups, amalgamating complementary subject areas to produce combined categories of cognate disciplines.[4] The top five most-cited articles (as of September 2012) containing the term *water security* in the Web of Science database originate from various disciplines. In descending order of citation frequency, the articles are from: water resources, fisheries science, multidisciplinary sciences, public health, and environmental sciences (Döll et al., 2003; Hrudey et al., 2003; Schindler, 2001; Vorosmarty

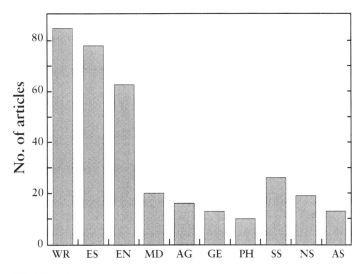

WR=Water Resources
ES=Environmental Studies, Sciences, and Ecology
EN=Engineering (Civil, Environmental, Chemical, Multidisciplinary)
MD=Geosciences, Multidisciplinary sciences
AG=Agriculture, Agronomy
GE=Geography
PH=Health (Public, Environmental, Occupational)
SS=Social Science (International Relations, Law, Planning and Development,
    Anthropology, Area Studies, Ethics, Economics, Operations Research and
    Management Science, Sociology)
NS=Natural/Physical Science (Biology, Computer Science, Fisheries,
    Food Science, Limnology, Biodiversity Conservation, Tropical Medicine,
    Plant Science, Parasitology)
AS=Meteorology and Atmospheric Sciences

Figure 4.2 Disciplinary grouping of articles containing the term *water security* (1990–2012)

Source: Web of Science database; analysis conducted on September 30, 2012

et al., 2010; Wagnerer et al., 2010). As explored below, different disciplinary perspectives offer distinct framings and methodologies for the analysis of water security.

We also found that scholars study water security using different methods and different scales (Table 4.1). Most of the 198 Web of Science water security articles studied water security empirically (88) or through models (76). The remaining studies were mostly conceptual (23), with a small number of lab-based studies (11). The scalar variability of hydrology means that choice of scale will affect assessment of water security (see Vorosmarty et al., 2010). A national scale is useful for intercountry comparisons, but it can hide significant variability in intracountry water security. Such coarse spatial resolution classifies Canada, for example, as 'water secure', yet decreasing water availability in the Prairie region is a growing concern and long-term water quality issues in Aboriginal communities have been well documented (Phare, 2009).

Unsurprisingly, different disciplines have a tendency to focus on different scales. Development studies tend to use nation-state scales, hydrologists often focus on watershed scales, and social scientists most frequently work at the local community scale. Across the literature, as identified in Table 4.1,

Table 4.1 Articles containing the term *water security* in the academic literature (1990–2012) sorted into type and scale of study

Type of Study	Scale of Study	Total No. of Articles
Empirical	No specific scale	3
	Community/municipal/hydraulic infrastructure	9
	Subnational watershed/drainage basin	13
	Regional (province/state/subnational)	25
	Nation-state	19
	Supranational (two or more countries)	19
Modelling	No specific scale	11
	Community/municipal/hydraulic infrastructure	24
	Subnational watershed/drainage basin watershed	16
	Regional (province/state/subnational)	2
	Nation-state	17
	Supranational (two or more countries)	6
Conceptual		23
Lab-based		11
	Total No. of Articles	198

Source: Web of Science database; analysis conducted on September 30, 2012

political boundaries are used more frequently than are hydrologic boundaries for empirical and modelling studies. This is of note, given that many of the studies are based in subject areas (such as water resources and environmental science and studies) that tend to privilege the watershed.

Drawing from our water security database, we selected examples of water security framings and sorted them into the subject areas used in Figure 4.2. Table 4.2 illustrates the variables of analysis and the degree of differentiation in scope used by different disciplines and organizations.

The content cloud (Figure 4.3) shows the key concepts associated with water security in the academic community (for a discussion of content

Table 4.2 Scope of approaches to water security, selected examples

Subject Area(s)	Water Security Focus or Definition
Agriculture	• Input to agricultural production and food security
Engineering	• Protection against water-related hazards (floods, droughts, contamination, terrorism)
	• Supply security (percentage of demand satisfied)
Environmental science, environmental studies	• Access to water functions and services for humans and the environment
	• Water availability in terms of quality and quantity
	• Minimizing impacts of hydrological variability
Fisheries, geology/ geosciences, hydrology	• Hydrologic (groundwater) variability
	• Security of the entire hydrological cycle
Public health	• Supply security and access to safe water
	• Prevention and assessment of contamination of water in distribution systems
Anthropology, economics, geography, history, law, management, political science	• Drinking water infrastructure security
	• Input to food production and human health/ wellbeing
	• Armed/violent conflict (motivator for occupation or barrier to cooperation and/or peace)
	• Minimizing (household) vulnerability to hydrological variability
Policy	• Interdisciplinary linkages (food, climate, energy, economy, and human security)
	• Protection of water systems and against floods and droughts; sustainable development of water resources to ensure access to water functions and services
Water resources	• Water scarcity
	• Supply security (demand management)
	• 'Green' (versus 'blue') water security—the return flow of vapour

Figure 4.3 Content cloud of *Water security keywords*
Source: Web of Science database using TagCrowd.com

clouds as a method for qualitative data analysis, see Cidell, 2010). The relative frequency of words used in the 198 academic articles in our subdatabase (including the title, abstracts, and keywords) suggests convergence around a set of core concepts: agriculture, areas, climate, change, countries, development, environmental, food, global, health, management, model, policy, region, resources, and system. We note that the core concepts have changed since we first produced a content cloud of our water security database in December 2010 (see Cook and Bakker, 2012).

This analysis shows that water security is indeed a debated concept. A diversity of definitions of and approaches to analysing water security are deployed across the natural and social sciences. Below, we identify key themes and compare and contrast the usage of the term in different disciplines.

Defining Water Security

The widespread uptake of the term *water security* is relatively recent. Contemporary framings of water security are highly diverse and vary with context and disciplinary perspectives on water use. For example, from a legal perspective, water security has generally been associated with allocation rules that seek to secure entitlements to desired quantities of water (Tarlock and Wouters, 2009). In contrast, from an agricultural perspective, protection from flood and drought risk is generally considered a key determinant of water security. Thus, framings of water security are dependent on perspective, as reflected in the diversity of framings put forth in the academic and policy literature.

Within this diversity, some common themes and trends can be identified. In general, the definitions of water security used in the 1990s were linked to specific human security issues, such as military security, food security, and (more rarely) environmental security. This changed at the Second World Water Forum in 2000; the Global Water Partnership introduced an integrative definition of water security that considered access and affordability of water, as well as human needs and ecological health. After the Second World Water Forum, scholars and policymakers began to use the term; some developed discipline-based definitions, and others advanced an integrative, interdisciplinary approach. Four interrelated themes dominate the published research on water security: water availability, human vulnerability to hazards, human needs (development related, with an emphasis on food security), and sustainability.

First, framings of water security that focus on quantity and availability of water are often linked to water security assessment tools. One of the best-known assessment tools to date combines two indices—for water stress and water shortage—in the measurement of water scarcity (Falkenmark et al., 2007; Falkenmark and Molden, 2008). The first index of water stress evaluates the ratio of water use to availability and estimates demand-driven scarcity by measuring how much water is withdrawn from rivers and aquifers—the blue water resources. The second index of water crowding or water shortage estimates population-driven real water shortages by measuring the number of people that have to share each unit of blue water resource (Falkenmark et al., 2007; Falkenmark and Molden, 2008). From this perspective, sufficiency of water supply for humans is the primary gauge of water security. For an individual, water security exists when he or she has access to sufficient safe and affordable water to satisfy individual needs for drinking, washing, and livelihood (Rijsberman, 2006).

A second theme of the academic literature on water security is the issue of water-related hazards and vulnerability. For example, the UNESCO-Institute for Water Education advocates an infrastructure and systems approach to water security that 'involves protection of vulnerable water systems, protection against water related hazards such as floods and droughts, sustainable development of water resources and safeguarding access to water functions and services' (UNESCO-IHE, 2009). The U.S. Environmental Protection Agency defines water security as prevention and protection against contamination and terrorism (Crisologo, 2008; Minamyer, 2008; Morley et al., 2007). Of course, this is directly linked to broader concerns over state or 'homeland' security; indeed, United States federal law has made 'drinking water infrastructure security . . . a cornerstone of homeland security' (Shermer, 2005, p359). In implementing this concept, water engineers have developed an understanding of water security as 'guns, gates, and guards' to ensure potable water and drinking water infrastructure security (see especially, the journal—*American Water Works Association*; Staudinger et al., 2006).

A third theme found in water security literature is *human needs,* a term that covers a broad range of issues, including access, food security, and human development–related concerns. For example, one framing of water security from the 1990s focuses on the human need for water: '[W]ater security is a condition where there is a sufficient quantity of water at a quality necessary, at an affordable price, to meet both the short-term and long-term needs to protect the health, safety, welfare and productive capacity of position (households, communities, neighborhoods, or nation)' (Witter and Whiteford, 1999, p2). The United Nations Development Program's approach to human security underpins many of these definitions (UNDP, 1994); for example, Janksy et al. (2008) defined water security as 'all aspects of human security pertaining to the use and management of water' (p289). Of course, the anthropocentrism of such a framing of water security risks neglecting the importance of the ecosystem as an integral component of both human and water security.

Within the human needs approach, there is a tendency to frame water security as a component or subset of food security (Biswas, 1999; FAO Land Division Water Development, 2000; White et al., 2007). The Food and Agricultural Organization (FAO) linked the concept of water security to food security, in which water security was the ability to provide adequate and reliable water supplies for populations living in the world's drier areas to meet agricultural production needs (Clarke, 1993). The FAO has maintained an agricultural focus of water security—'crop water security'—where water quantity is highly relevant (FAO Land Division Water Development, 2000). In many countries, reservoir storage for the purposes of irrigation is the salient feature of water security (El Saliby et al., 2009). This focus on water quantity also holds true for framings that widen concern from reservoir storage to consider the entire hydrological cycle (Johansson et al., 1999; Oki and Kanae, 2006; Tuinhof et al., 2005). From this perspective, water security is threatened by either water scarcity or risk of inundation that can be attributed to an inability to manage water.

A fourth theme in the water security literature is that of sustainability. According to the Global Water Partnership (GWP) (2000), for example, 'water security at any level from the household to the global means that every person has access to enough safe water at affordable cost to lead a clean, healthy and productive life, while ensuring that the natural environment is protected and enhanced' (p1). This broad framing includes seven variables: meeting basic needs, securing the food supply, protecting ecosystems, sharing water resources, managing risks, valuing water, and governing water wisely. This, the GWP argues, implies the need for baseline requirements for water resources management in a watershed on a continuous basis—for 'life'—and demands access to adequate quantities of acceptable quality of water for both humans and the environment.

Somewhat surprisingly (given the attention paid to 'water wars' in popular media), there is relatively little emphasis in the water security literature on military security or on the concept of environmental security ('green wars'), a concept that emerged in the 1990s to refer to the links between violent conflict and environmental degradation (Homer-Dixon, 1999; Kaplan, 1994; Stern, 1999). Of course, these issues have received significant scrutiny from academics (Giordano et al., 2002; Gleick, 1993; Wolf, 1999), but these scholars do not appear to have adopted the term *water security*, even where their nuanced approach to the integrative nature of environmental issues leads them to voice parallel issues, such as the links between multiple scales or the importance of good governance[5] (e.g., Dalby, 2002). The one exception is the Middle East and North Africa, where early uses of the term *water security* explicitly focused on geopolitical security concerns (Anderson, 1992; Savage, 1991; Shuval, 1992; Starr, 1991).

Discussion

This review indicates that approaches to water security are diverse and evolving. The potential compatibility (and incommensurability) between these approaches raises a series of questions.

First, how does water security overlap with Integrated Water Resources Management (IWRM)—arguably the dominant water management paradigm, particularly in international water policy (Conca, 2006)? Like IWRM, water security—at least in its broad, integrative framings—offers a paradigmatic approach to the analysis of water systems, which integrates across scales (from the local to the global) and incorporates both quality and quantity concerns (including hazards and water access). Many definitions of IWRM overlap with the four themes of water security identified in this chapter: water availability, human vulnerability, human needs, and sustainability; moreover, both water security and IWRM suggest the need to balance human use and development with ecosystem needs. To the extent that framings of IWRM and water security display similarity, the two concepts appear to be complementary; however, perhaps neither concept is broad enough to capture the complexity of water-related issues. And, a broad and integrative framing has a significant potential pitfall: Multiple variables tend to increase the technical complexity of water security assessment and raise the risk of conflating water status (e.g., ecosystem health) with stressors (e.g., the quality of good governance regimes).

Second, how might the concept of water security be coherently implemented in both scientific research and water management? As suggested by the previously discussed literature review, one of the most significant challenges in implementing the concept of water security is the diversity of potential variables and methods. For this reason, we view broad and integrative framings of water as conceptual and paradigmatic, and, as such, as a useful complement to narrow framings, which are more likely to be focused on management issues and linked to policy, modelling, empirical research, and/or lab-based studies.

To give a concrete example, consider the approach adopted in one of the most influential papers on water security to date. In this study, which originates in the discipline of development studies, water security is framed broadly as 'the availability of an acceptable quantity and quality of water for health, livelihoods, ecosystems and production, coupled with an acceptable level of water-related risks to people, environments, and economies' (Grey and Sadoff, 2007, p545). However, when applied, the framing is narrowed to the ability of a country to harness the productive potential of water and to limit its destructive impact. Grey and Sadoff (2007) separate countries into three categories: those that have harnessed hydrology, those that are hampered by hydrology, and those that are hostage to hydrology. This allows a conclusion that countries, such as Canada, that have successfully harnessed hydrology have achieved water security—a somewhat spurious conclusion, as we have already noted.[6] A true picture of country water security requires assessment at multiple scales—from the local to the national—for both human and ecosystem needs.

We suggest that the advantage inherent in narrower framings of water security is that they enable precise identification and assessment of specific issues of concern. When managers try to implement a concept, they necessarily must narrow it, focussing on the primary concerns in the management area. An obvious critique of these narrow framings is the failure to recognize or integrate the multiple stressors that affect water security. We suggest that narrow framings would be usefully allied with broader, integrative framings of water security, such that overarching issues are also taken into account.

Our analysis indicates that implemented definitions of water security are also likely to vary geographically. Specific definitions of water security have emerged in regions where particular water security concerns are acute. For example, in Australia, well known as the world's most arid continent, water security has been defined predominantly as a concern of water availability (quantity) to be addressed by the national and state governments through a variety of mechanisms, as detailed in A National Plan for Water Security (Government of Australia, 2007, 2010). Framings of water security are varied to suit particular geographies.

In other words, 'narrow' and 'broad' framings of water security are complementary rather than mutually exclusive. Implementing water security at the management level will likely require specific, and sometimes narrow, framings of the concept. In this context, both will be useful and may be necessary; however, integrative framing of water security still needs to happen at the policy level and in governance processes in which priorities are established and decisions made between competing uses and users.

Conclusions

The critical review of the concept of water security provided in this chapter indicates that the concept of water security emerged in the 1990s, and has evolved significantly since then. Until two decades ago, the term was rarely used. When introduced in the 1990s, it was linked to other types of security, especially food and military security. Since the Second World Water Forum in 2000, where the Global Water Partnership introduced an integrative definition of water security, a variety of a scholars and policymakers have taken up the term and given it various meanings. Some have developed discipline-based definitions, and others have advanced an integrative, inter-disciplinary approach.

We have reviewed these approaches and suggest that broad and integrative conceptual framings of water security are useful for establishing priorities and facilitating analysis of tradeoffs between competing uses and users, whereas narrower framings of water security will be essential for implementation. Finally, we note that scepticism regarding the uptake of water security as a concept is merited. In particular, those interested in implementing the water security concept would do well to reflect on the reasons why this term has become popular and on the agendas it might serve. For example, it might be appropriate to examine the links between an increased uptake of the term *water security* and recent reforms in water governance, notably decentralization, devolution, and/or greater participation of communities in water governance (Blomquist and Schlager, 2005; Conca, 2006; Irvin and Stansbury, 2004; Leach and Pelkey, 2001; Reed and Bruyneel, 2010; Sabatier et al., 2005; Singleton, 2002). These changes in water governance have occurred for several reasons: shifting views over the role and mandate of governments; new legal requirements; increased expectation for public participation; a desire to draw on expertise available outside of government; concern regarding low efficiency of water use, associated with ineffective management of resources and supply systems; and increased emphasis on integrated management of environmental issues (Brick et al., 2001; Gleick, 2000; Sabatier et al., 2005; UNWWAP, 2006, 2009).[7]

Notes

1. At its Summit on the Global Agenda in Dubai (November 2008), the Network of Global Agenda Councils of the World Economic Forum discussed water security extensively (World Economic Forum Network of Global Agenda Councils, 2008).

2. We searched water security and cognate concepts ('water vulnerability', 'water stress', 'water index/ices', 'water frameworks', 'water sustainability', and 'secure water') in the following databases: Geobase/GeoRef, PAIS, EconLit, Worldwide Political Science Abstracts, International Political Science Abstracts, JSTOR, Web of Science, and LegalTrac. In addition, policy references were gathered via Internet (Google) searches for 'water security'.

3. It is important to note that Web of Science offers patchy coverage of the social sciences and humanities, which may, thus, be underrepresented.

4. The articles were sorted into 10 subject areas, based on the Web of Science primary categorization: WR = Water Resources; ES = Environmental Studies, Sciences, and Ecology; EN = Engineering (Civil, Environmental, Chemical, Multidisciplinary); MD = Geosciences, Multidisciplinary Sciences; AG = Agriculture, Agronomy; GE = Geography; PH = Health (Public, Environmental, Occupational); SS = Social Science (Agricultural Economics Policy, Anthropology, Area Studies, Economics, International Relations, Law, Management, Operations Research and Management Science, Planning and Development); NS = Natural/Physical Science (Biology, Computer Science Interdisciplinary Applications, Food Science Technology, Limnology, Marine Freshwater Biology, Nutrition Dietetics); AS = Meteorology and Atmospheric Sciences.

5. According to UNESCAP (2011), good governance 'has 8 major characteristics. It is participatory, consensus oriented, accountable, transparent, responsive, effective and efficient, equitable and inclusive and follows the rule of law. It assures that corruption is minimized, the views of minorities are taken into account and that the voices of the most vulnerable in society are heard in decision-making. It is also responsive to the present and future needs of society'.

6. For an exploration of Grey and Sadoff's (2007) thesis that water security is necessary for rapid economic development, see Merrey (2009).

7. These developments are articulated with, and to some extent illustrative of, a more general shift from 'government' to 'governance' in which nongovernmental actors play a more significant role than in the past (Jessop, 2004; Pierre, 2000; Pierre and Peters, 2000; Rhodes, 1996; Strange, 1996).

References

Anderson, E.W. (1992) 'The political and strategic significance of water', *Outlook on Agriculture,* vol 21, no 4, pp247–253.

Asia Pacific Water Forum (2007) 'Asia-Pacific Water Summit', www.apwf.org/project/result.html, accessed 3 September 2011.

Bakker, K. (2012) 'Water security: research challenges and opportunities', *Science,* vol 337, pp914.

Biswas, M.R. (1999) 'Nutrition, food, and water security', *Food and Nutrition Bulletin,* vol 20, no 4, pp454–457.

Blomquist, W. and Schlager, E. (2005) 'Political pitfalls of integrated watershed management', *Society and Natural Resources,* vol 18, pp101–117.

Brick, P., Snow, D. and Van de Wetering, S. (eds) (2001) *Across the Great Divide: Explorations in Collaborative Conservation and the American West,* Island Press, Washington, DC.

Cidell, J. (2010) 'Content clouds as exploratory qualitative data analysis', *Area,* vol 42, no 4, pp514–523.

Clarke, R. (1993) *Water: The International Crisis,* MIT Press, Cambridge, MA.

Conca, K. (2006) 'Expert networks: the elusive quest for integrated water resources management', in K. Conca (ed) *Governing Water: Contentious Transnational Politics and Global Institution Building,* MIT Press, Cambridge, MA.

Cook, C. and Bakker, K. (2012) 'Water security: debating an emerging paradigm', *Global Environmental Change*, vol 22, no 1, pp94–102.

Crisologo, J. (2008) 'California implements water security and emergency preparedness, response, and recovery initiatives', *Journal—American Water Works Association*, vol 100, no 7.

Dalby, S. (2002) *Environmental Security*, University of Minnesota.

Döll, P., Kaspar, F. and Lehner, B. (2003) 'A global hydrological model for deriving water availability indicators; model tuning and validation', *Journal of Hydrology*, vol 270, no 1–2, pp105–134.

El Saliby, I., Okour, Y., Shon, H.K., Kandasamy, J. and Kim, I.S. (2009) 'Desalination plants in Australia, review and facts', *Desalination*, vol 247, no 1–3, pp1–14.

Falkenmark, M. and Molden, D. (2008) 'Wake up to realities of river basin closure', *International Journal of Water Resources Development*, vol 24, no 2, pp201–215.

Falkenmark, M., Berntell, A., Jägerskog, A., Lundqvist, J., Matz, M. and Tropp, H. (2007) 'On the verge of a new water scarcity: a call for good governance and human ingenuity', *SIWI Policy Brief*, Stockholm International Water Institute (SIWI), Stockholm.

FAO Land Division Water Development (2000) *New Dimensions in Water Security: Water, Society and Ecosystem Services in the 21st Century*, Food and Agricultural Organization of the United Nations, Rome.

Giordano, M., Giordano, M. and Wolf, A. (2002) 'The geography of water conflict and cooperation: internal pressures and international manifestations', *The Geographical Journal*, vol 168, no 4, pp293–312.

Glaser, G. (2012) 'Policy: base sustainable development goals on science', *Nature*, vol 491, pp35.

Gleick, P.H. (1993) 'Water and conflict, fresh water resources and international security', *International Security*, vol 18, pp79–112.

Gleick, P.H. (2000) 'The changing water paradigm—a look at twenty-first century water resources development', *Water International*, vol 25, no 1, pp127–138.

Global Water Partnership (2000) *Towards Water Security: A Framework for Action*, Global Water Partnership, Stockholm.

Government of Australia (2007) *A National Plan for Water Security*.

Government of Australia (2010) *Water for the Future*, www.environment.gov.au/water/publications/action/water-for-the-future.html, accessed 29 November 2011.

Grey, D. and Sadoff, C.W. (2007) 'Sink or swim? Water security for growth and development', *Water Policy*, vol 9, no 6, pp545–571.

Homer-Dixon, T.F. (1999) *Environment, Scarcity, and Violence*, Princeton University Press, Princeton, NJ.

Hrudey, S.E., Payment, P., Huck, P.M., Gillham, R.W. and Hrudey, E.J. (2003) 'A fatal waterborne disease epidemic in Walkerton, Ontario: comparison with other waterborne outbreaks in the developed world', *Water Science and Technology*, vol 47, no 3, pp7–14.

Irvin, R.A. and Stansbury, J. (2004) 'Citizen participation in decision making: is it worth the effort?', *Public Administration Review*, vol 64, pp55–65.

Jansky, L., Nakayama, M. and Pachova, N.I. (2008) *International Water Security: Domestic Threats and Opportunities*, United Nations University Press, New York, NY.

Jessop, B. (2004) 'Hollowing out the "nation-state" and multilevel governance', in P. Kennett (ed) *A Handbook of Comparative Social Policy*, Edward Elgar Publishing, Cheltenham, UK.

Johansson, P.O., Scharp, C., Alveteg, T. and Choza, A. (1999) 'Framework for ground water protection—The Managua Ground Water System as an example', *Ground Water*, vol 37, no 2, pp204–213.

Kaplan, R.D. (1994) 'The coming anarchy (cover story)', *Atlantic Monthly*, vol 273, no 2, pp44–77.

Leach, W.D. and Pelkey, N. (2001) 'Making watershed partnerships work: a review of the empirical literature', *Journal of Water Resources Planning and Management*, vol 127, no 6, pp378–385.

Merrey, D.J. (2009) 'Will future water professionals sink under received wisdom, or swim to a new paradigm?', *Irrigation and Drainage*, vol 58, no S2, ppS168-S176.

Minamyer, S. (2008) 'Effective crisis communication during—water security emergencies', *Journal American Water Works Association*, vol 100, no 9, pp180–184.

Morley, K., Janke, R., Murray, R. and Fox, K. (2007) 'Drinking water contamination—warning systems: water utilities driving water security research', *Journal American Water Works Association*, vol 99, no 6, pp40–46.

Oki, T. and Kanae, S. (2006) 'Global hydrological cycles and world water resources', *Science*, vol 313, no 5790, pp1068–1072.

Phare, M.-A.S. (2009) *Denying the Source: the Crisis of First Nations Water Rights*, Rocky Mountain Books, Surrey, BC.

Pierre, J. (2000) 'Introduction: Understanding governance', in J. Pierre (ed) *Debating Governance: Authenticity, Steering, and Democracy*, Oxford University Press, Oxford, UK.

Pierre, J. and Peters, B.G. (2000) *Governance, Politics, and the State*, St. Martin's Press, New York, NY.

Reed, M.G. and Bruyneel, S. (2010) 'Rescaling environmental governance, rethinking the state: a three-dimensional review', *Progress in Human Geography*, vol 34, no 5, pp646–653.

Rhodes, R.A.W. (1996) 'The new governance: governing without government', *Political Studies*, vol 44, no 4, pp652–667.

Rijsberman, F.R. (2006) 'Water scarcity: fact or fiction?', *Agricultural Water Management*, vol 80, no 1–3, pp5–22.

Sabatier, P., Focht, W., Lubell, M., Trachtenberg, Z., Vedlitz, A. and Matlock, M. (2005) *Swimming Upstream: Collaborative Approaches to Watershed Management*, MIT Press, Cambridge, MA.

Savage, C. (1991) 'Middle East water', *Asian Affairs*, vol 22, no 1, pp3–10.

Schindler, D.W. (2001) 'The cumulative effects of climate warming and other human stresses on Canadian freshwaters in the new millennium', *Canadian Journal of Fisheries and Aquatic Sciences*, vol 58, pp18–29.

Shermer, S.D. (2005) 'The Drinking Water Security and Safety Amendments of 2002: is America's drinking water infrastructure safer four years later?', *UCLA Journal of Environmental Law & Policy*, vol 24, no 2, pp355–457.

Shuval, H.I. (1992) 'Approaches to resolving the water conflicts between Israel and her neighbors—a regional water-for-peace plan', *Water International*, vol 17, pp133–143.

Singleton, S. (2002) 'Collaborative environmental planning in the American West: the good, the bad and the ugly', *Environmental Politics*, vol 11, no 3, pp54–75.

Starr, J.R. (1991) 'Water security: the missing link in our Mideast strategy', *Current World Leaders*, vol 34, no 4, pp571–588.

Staudinger, T.J., England, E.C. and Bleckmann, C. (2006) 'Comparative analysis of water vulnerability assessment methodologies', *Journal of Infrastructure Systems*, vol 12, no 2, pp96–106.

Stern, E.K. (1999) 'Case for comprehensive security', in D. Deudney and R.A. Matthew (eds) *Contested Grounds: Security and Conflict in the New Environmental Politics*, State University of New York Press, Albany, NY.

Strange, S. (1996) *Retreat of the State: The Diffusion of Power in the World Economy*, Cambridge University Press, Cambridge, UK.

Stucki, V. and Sojamo, S. (2012) 'Nouns and numbers of the water-energy-security nexus in Central Asia', *International Journal of Water Resources Development*, vol 28, no 3, pp399–418.

Tarlock, A.D. and Wouters, P. (2009) 'Reframing the water security dialogue', *Journal of Water Law*, vol 20, no 2, pp53–60.

Tuinhof, A., Olsthoorn, T., Heederik, J.P. and de Vries, J. (2005) 'Groundwater storage and water security: making better use of our largest reservoir', *Water Science and Technology*, vol 51, no 5, pp141–148.

UNESCO-IHE (2009) 'Research themes. Water security', www.unesco-ihe.org/Research/Research-Themes/Water-security, accessed 3 September 2011.

United Nations Development Programme (UNDP) (1994) *Human Development Report 1994: New Dimensions of Human Security*, UNDP, New York, NY.

United Nations Economic and Social Commission for Asia and the Pacific (UNESCAP) (2011) 'What is good governance?', www.unescap.org/pdd/prs/ProjectActivities/Ongoing/gg/governance.asp, accessed 7 November 2012.

United Nations World Water Assessment Program (UNWWAP) (2006) 'Water a shared responsibility', *The United Nations World Water Development Report* (UNESCO), vol 2.

United Nations World Water Assessment Program (UNWWAP) (2009) 'Water in a changing world', *The United Nations World Water Development Report* (UNESCO), vol 3.

Vorosmarty, C.J., McIntyre, P.B., Gessner, M.O., Dudgeon, D., Prusevich, A., Green, P., Glidden, S., Bunn, S.E., Sullivan, C.A., Reidy Liermann, C. and Davies, P.M. (2010) 'Global threats to human water security and river biodiversity', *Nature*, vol 467, no 7315, pp555–561.

Wagener, T., Sivapalan, M., Troch, P.A., McGlynn, B.L., Harman, C.J., Gupta, H.V., Kumar, P., Rao, P.S.R., Basu, N.B. and Wilson, J.S. (2010) 'The future of hydrology: an evolving science for a changing world', *Water Resources Research*, vol 46, W05301.

White, D.M., Gerlach, S.C., Loring, P., Tidwell, A.C. and Chambers, M.C. (2007) 'Food and water security in a changing Arctic climate', *Environmental Research Letters*, vol 2, no 4.

Witter, S.G. and Whiteford, S. (1999) 'Water security: the issues and policy challenges', *International Review of Comparative Public Policy*, vol 11, pp1–25.

Wolf, A.T. (1999) '"Water wars" and water reality: conflict and cooperation along international waterways', in S. Lonergan (ed) *Environmental Change, Adaptation, and Human Security*, Academic, Dordrecht, The Netherlands.

World Economic Forum (2011) *Water Security: The Water-Food-Energy-Climate Nexus*, Island Press, Washington, DC.

World Economic Forum Network of Global Agenda Councils (2008) *Discussion Highlights on Water Security at the Summit on the Global Agenda*, Dubai, United Arab Emirates.

5   The Multiform Water Scarcity Dimension

Malin Falkenmark

> *Water security is tolerable water-related risk to society.*
>
> Professor David Grey
> University of Oxford

Introduction

Water is an essential component of the planet's life support system. Our aspirations for 'water security' depend on a set of features: a functioning hydrological cycle, reliable engineering schemes, awareness of threats and preparedness to face them as legal interventions, sensible policies, and effective governance (Global Water System Project, 2011). Water is vital both for basic ecological functions and processes, sustaining the life support system on the planet, and for the functioning of the societal system. Its crucial role as 'bloodstream' of the biosphere is based on balanced pairs of physical, chemical, and biological processes (Ripl, 2003). Water security is, therefore, essential for a society's survival, health, and prosperity. Scarcity of water, which will be discussed in this chapter, or difficulty to safeguard access, is consequently an obstacle and functions as a bottleneck in socioeconomic development.

Freshwater is provided to the continents by precipitation where it is partitioned between two types of water: infiltrated rain forming the base for plant production and, therefore, food (green water); and liquid water in rivers and aquifers (blue water), which is the base for water supply of human settlements, industry, irrigation, and energy production. Human water security depends both on biophysical phenomena and on successful human activities and policies, and can be seen as part of a natural resource security web (Zeitoun, 2011). Safe water supply both for people themselves and for crop production are essential steps in socioeconomic development. A reliable societal system, therefore, depends on some level of water security. Grey and Sadoff (2007), in fact, hypothesised that basic water security is a crucial element of socioeconomic development. They observed clear regional disparities between countries that had 'harnessed their hydrology' (basically industrial countries), those that are 'hampered by their hydrology' (basically emerging economies), and those that remain 'hostages of their hydrology' (basically low-income countries). Vörösmarty et al. (2010) high-lighted investments in the struggle towards water supply stabilization and basic water services, and showed a clear correlation between the benefits achieved in terms of reduction of threat to human water security, on the one hand, and the GDP, on the other.

Both human water stress and agricultural water stress constitute risks to human water security. Without water security, there will, in fact, be no food security. When compared with the 2 to 5 litres of water required for drinking, the water required for food supply amounts to 70 times as much (Rockström, Falkenmark, et al., 2009) is a critical issue as populations and wealth grow in a country. Water security is, in other words, an essential component of pov-erty- and hunger-alleviation strategies, and the difficulties that water scarcity involves contributes to an insecurity that must be overcome in the process of socioeconomic development.

Kummu and Varis (2011) have highlighted the large latitudinal differences in precipitation amounts north of 20 degrees N, where annual precipitation remains surprisingly stable, around 500 mm/yr. This is the latitudinal belt with the largest population, a situation that has a large impact on human living conditions, especially as tradeoffs are needed between water for dif-ferent purposes: between societal and economic production (water supply of population, industry and energy) and, therefore, for societal water security; plant production and, therefore, food and timber supply; and protection of natural ecosystems—terrestrial as well as aquatic—and, therefore, provision of ecological goods and services.

In this chapter, we will look more closely at key water-security risks tied to how water scarcity constrains both human water security and agricul-tural (and, therefore, food) water security. The chapter summarises 25 years of studies by the author and her colleagues (Falkenmark, 1986, 1997; Falkenmark and Lundqvist, 1997; Falkenmark and Rockström, 2004, 2008; Rockström et al., 2007; Falkenmark and Molden, 2008; Falkenmark and Lannerstad, 2010).

Two Categories of Water That Can Be Scarce

At the ground, the precipitation over the continents is being partitioned between two water flows: one by infiltrating into the ground and return-ing to the atmosphere as evaporation, either from moist surfaces or after uptake by plants, transpiring from their foliage (so-called *green water flow*); the other through runoff generation or groundwater recharge (so-called *blue water flow*). Both types of water have the same origin, the precipitation falling within a water divide in the landscape. This means that this water input is being shared between green water–dependent and blue water–dependent processes and activities. In dry climates such as in Kenya, the high evaporative demand consumes most of the rainfall, leaving only very limited amounts to form the blue water runoff (see Figure 5.1).

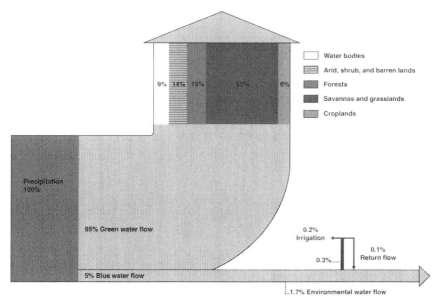

Figure 5.1 Rain water partitioning over Kenya between green water flow through different terrestrial ecosystems, blue water flow in rivers and aquifers, and blue water withdrawals resulting in respectively consumptive water use and return flow back to the river system.

Source: Stockholm Environment Institute

Blue Water Scarcity Complicates Water Management

Blue water scarcity has been studied from two complementary aspects: population-driven scarcity, implying that more and more people together depend on each unit of water (water crowding), and demand-driven scarcity, when mobilization of even more of the resource gets increasingly difficult and costly. The conceptual development is still causing problems and misunderstandings, especially regarding the vague concept of water stress, as will be discussed later.

Water Crowding

Starting in the 1970s, attention has been drawn to the implications of the continuing population growth in the naturally water-poor latitudes, and the resulting differences between water competing sectors in different countries (Falkenmark and Lindh, 1976; Falkenmark, 1986). Blue water may be scarce in relation to societal water requirements (i.e., water supply of household, industry, and irrigation) and in relation to its role as habitat for aquatic ecosystems. In the mid-1980s, the author analysed population-driven blue water scarcity, in terms of water competition level—in today's vocabulary, *water crowding* (people per flow unit of 1 million cubic meters per year [Mm^3/yr]

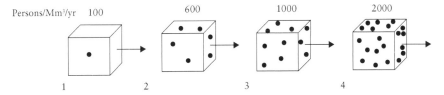

Figure 5.2 Water crowding: Growing population pressure on a finite blue water availability. Each cube represents 1 Mm³/yr, each dot 100 p.

Source: M. Falkenmark, Stockholm International Water Institute.

of blue water availability) (Falkenmark, 1986). The author empirically noted that, on a country/state level, water management tended to get stressful around 600 p/Mm³/yr and difficult around 1,000 p/Mm³/yr (see Figure 5.2). In a recent study, Kummu et al. (2010) showed that water crowding is in fact a rather recent, post–World War II phenomenon, now expanding in response to the continued population growth.

Water Stress, Use-to-Availability

Demand-driven scarcity is generally measured as use-to-availability, also called *criticality,* and often referred to as *water stress* (Kummu and Varis, 2011). This scarcity dimension was studied in the 1970s by the European scholar Balcerski (Falkenmark and Lindh, 1976), who explained water resources management differences at different levels of use-to-availability: already at the 20% level, infrastructure investments are costly from a national economic perspective, and have to be incorporated in national economic planning. In the 1997 Comprehensive Assessment of the Freshwater Resources of the World (UN, 1997), the UN set the withdrawal of 40% as the line distinguishing the situation of high water stress. Later, International Water Management Institute introduced the concept *economic water scarcity* for the low water-stress levels typical for many developing countries (Molden et al., 2007).

Increasingly, a certain part of the river flow is today being reserved to protect aquatic habitats in South Africa, of the order of 20 to 30% (Falkenmark, 2003). Beyond the 70% level, the basin can be seen as closed, in the sense that additional water commitments cannot be made because the basin would be overcommitted. Basin closure is already globally widespread with more than 1.4 billion living in basins with this dilemma (Falkenmark and Molden, 2008).

Combined Water Scarcity Predicament Studies

Figure 5.3 links the two blue water scarcity coordinates to give an idea of the degree of blue water scarcity in a country or basin. Even if data quality and degree of representativity may be imprecise (SIWI, 2007), the diagram indicates principal differences in exposure between different world regions and allows preliminary comparisons between regions, countries, and river

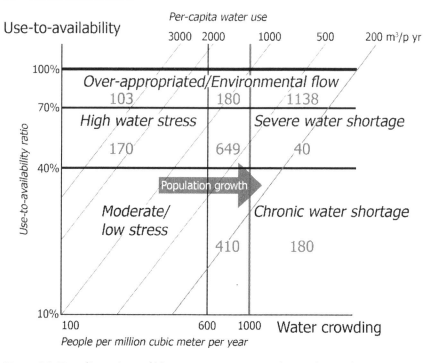

Figure 5.3 Two dimensions of blue water scarcity: population-driven shortage/water crowding (horizontal axis), and demand-driven scarcity/water stress (vertical axis), assuming an environmental flow reserve of 30%. Diagonal lines show per capita water use. The vulnerable triangle is the area between the two characteristic lines 1,000 p/flow unit and 200 m³/p/yr. Numbers indicate millions of people living under different water scarcity conditions.

Source: SIWI (2008)

basins based on publicly accessible water data (Falkenmark, 1997; Falkenmark and Rockström, 2004; Falkenmark and Molden, 2008).

The Falkenmark Indicator—A Blue Water Scarcity Coordinate

The earlier-mentioned tentative intervals for water crowding are known under the connotation 'Falkenmark indicator' and have attracted considerable international interest, as well as misunderstandings. What the 1,000 p/Mm³/yr (p/flow unit) signifies is a degree of complexity in water resources management. Through human ingenuity and socioeconomic resources, the difficulties can fortunately be overcome and development secured, even in countries with large-scale irrigation. Israel and Saudi Arabia are evident examples. Yang et al. (2007) observed that food trade tended to increase with water crowding beyond some 700 p/ flow unit (equivalent to below some 1,500 cubic meters per person per year).

Water scarcity literature has revealed quite exaggerated expectations of this indicator, taking it for a predictor rather than a water scarcity coordinate. Chenoweth (2008), for instance, interprets the water scarcity indicator as 'measuring the ability of individual countries to satisfy their basic water requirements'. Rather than this ability, it indicates difficulties that have to be overcome to achieve tolerable risks. It specifies just how much water is passing through a basin/country/state as part of the water cycle, and how many people depend on that amount. In a similar way as a child's age is an indicator of the number of years it has been growing, the water indicator specifies no more than how many people have to share each water unit. What can be achieved under those basic circumstances evidently depends on topography, seasonality, technical ability, peoples' expectations, economy, as well as governance, and to what degree it allows sound management.

When combining the water crowding indicator with how much of the resource has been mobilized for societal use (i.e., so-called *water stress*) and entering the per capita level of water use corresponding to different combinations of coordinates, some very preliminary conclusions may be drawn in terms of possible constraints, as indicated by the analytical diagram in Figure 5.3. Since the ability of individual countries to satisfy their basic water requirements evidently depends on management and governance capacity, which is not measured by these indicators, water scarcity can only be expected to make the task of satisfying water requirements more demanding.

In Figure 5.3, it is possible to point out a vulnerable triangle, defined by two characteristic lines: the 1,000p/flow unit and the 200m^3/p/yr lines. Beyond the former there is a situation of chronic water shortage. The 200m^3/p/yr is a tipping point; it was identified in the mid-1990s as what is required for a non-wasteful water supply of municipalities and industry (Lundqvist and Gleick, 1997). Today, more than 1 billion people are living in the river basins in east and south Asia situated in that zone, pushed by population growth towards the 200 line. Beyond that line, even urban/industry water supply will become a complex task, and a shift in thinking will be called for in water policy.

Green Water Scarcity

Turning now to agricultural water security, it is susceptible to green water scarcity. Crop production critically depends on plant roots taking up soil moisture, transporting it up to the leaves to balance transpiration losses during the photosynthesis intake of airborne carbon dioxide. Plant growth is thus a function of green water accessibility in the root zone. Under conditions of deficiency, blue water may be added as a complement through irrigation to achieve green water security.

Dryland Agriculture

Green water can be scarce for several reasons (SIWI, 2007). Some reasons are climate and soil related, while others are related to people's activities. There may be too little rain; most of the rain may evaporate leaving the soil dry; there may be problems with infiltration, such as soil crusting, and rain quickly runs off; and the soils may have poor water holding capacity so that water

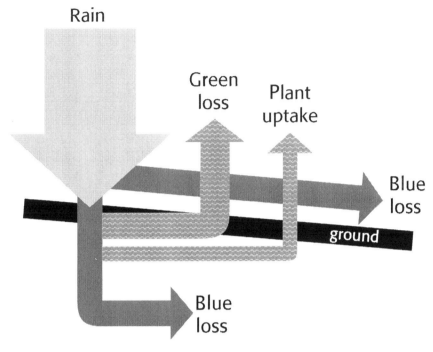

Figure 5.4 Partitioning of rainfall on a sub-Saharan farmer's field, showing large blue water losses (surface runoff, percolation) and large green water loss (nonproductive evaporation).

Source: M. Falkenmark

percolates to recharge groundwater layers. Through field studies in the Sahel, Rockström (2003) showed that the root uptake capacity may be disturbed, resulting in very low crop yields. In a typical situation, the roots are disturbed by dry spells so that the root systems cannot take up the water in fact available in the root zone. This may result in huge water losses. Figure 5.4 shows that the combination of green evaporation losses and blue losses in terms of both flash floods and percolation losses typically resulted in crop yields on the 1 ton/hectare level only, far below the potential yield level for that particular hydroclimate. In a systematic series of case studies, International Center for Agricultural Research in the Dry Areas (ICARDA) has recently shown that with improved land and water management practices, yield gaps in the Mediterranean region can be reduced by half (Pale et al., 2011).

Drought May Cause Famine

Rain water variability in terms of droughts adds largely to water security problems. Different categories of droughts cause different types of disturbances (Falkenmark and Rockström, 2008): *intraseasonal dry spells*, which can be compensated by water harvesting and supplementary irrigation; *interannual droughts,* resulting in different degrees of crop failure to be compensated by irrigation, and climate change–related situations of slow *aridification,* necessitating altered water policies (Lundqvist and Falkenmark, 2011).

The African famine in the mid-1980s offers an illustration. Falkenmark and Rockström (2004) showed that famine was linked to a set of simultaneous disturbances to water security: interannual drought and climatic green water scarcity, exacerbated by soil degradation, to which were added blue water limitations in the sense that small water courses go empty except during rainstorms.

This set of water-related challenges clearly exposes the vulnerability of the people living on rain-fed agriculture on the sub-Saharan savannahs. Dominating farming systems are rain fed and nonmechanised. Yields tend to be low and linked to a number of biophysical and socioeconomic factors; during critical crop development stages, water storage in the root zone is a fundamental issue in this region.

Combined Green/Blue Water Constraints

Water security analysis and potential water constraints are now increasing in relevance as the world population continues to expand and socioeconomic development proceeds. Many issues are water scarcity related, including river basin closure, hydrological effects of land use change, river flow depletion, global food production prospects, and even potential global vulnerability to continental scale alterations in water partitioning (Rockström et al., 2012).

Water Partitioning Disturbances

Green–blue water partitioning is easily disturbed by land use change. Consumptive water use may decrease when land use changes, reflected in altered blue water generation. L'vovich and White (1990) showed that the global scale deforestation for development of agriculture during the period between 1680 and 1980 resulted in a reduced vapour flow and an increased runoff generation of the order of some 2,500 km³/yr. Gordon et al. (2005) later found this to be, on the global scale, more or less compensated by increased consumptive use linked to irrigation, although the two effects tend to dominate in different regions.

A similar reflection of altered water balance is the effects of consumptive water use during irrigation as demonstrated by the widespread river flow depletion on 15% of the continental land area—a phenomenon to which IWMI drew international attention at the World Water Forum in Kyoto in 2003. Sharpening blue water scarcity has generated a call for higher water use efficiency in irrigation. However, since the result will evidently be reduced return flow, river depletion might in fact be further exacerbated.

A third aspect highlighted by ecologists is the call for protection of aquatic ecosystems by securing a minimum river flow, a so-called *environmental flow,* which, besides a certain average minimum flow, also involves a certain annual variability/flow regime (Falkenmark, 2003).

Limited Food Production

In a recent global study, Rockström et al. (2010) analysed a country's food self-reliance potential on current cropland as a function of green and blue water preconditions. The green–blue water framework in Figure 5.5 was used to characterise country situations. Total food water requirements were based on a generic food supply of 3,000 kcal/p/d (kilocalorie per person per day), out of which 20% was animal protein (current supply levels in Mexico, Brazil, and China) and water productivity of 1,300 m³/p/yr (Rockström, Lannerstad, and Falkenmark 2007; Rockström, Falkenmark, et al. 2009). The diagram shows a country-by-country diagnosis of green and blue water availability for food production by 2050, if restricted to current cropland area. The horizontal axis shows available blue water with the vertical line indicating the chronic water shortage indicator 1,000 p/flow unit; the vertical axis shows the water available on current cropland areas with the upper slanting line indicating the total water requirement for food production on the current water productivity level. Different country preconditions are indicated by the letters a, b, c, d. Class a countries lack the water required for food self-reliance, in that they lack enough green water

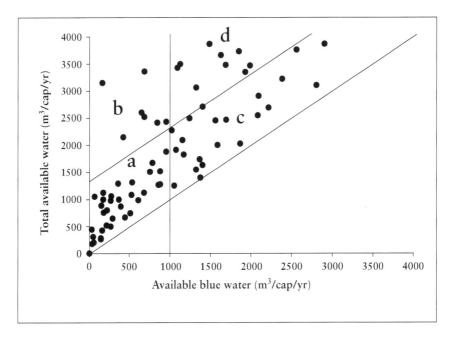

Figure 5.5 Country-by-country diagnosis of food water potentials on current croplands in water scarce countries by 2050, categorized by water availability (horizontally blue, vertically blue plus green availability). Assumed water requirements based on 3,000 Kcal/p / d, 20% animal food. Climate change according to the SRES A2 scenario; UN medium population projection.

Source: Redrawn from Rockström et al. (2010)

to depend on rain-fed agriculture and suffer from chronic blue water shortage, complicating irrigation. Class c countries lack enough green water but are assumed to be able to compensate by irrigation. Class b and d countries have enough green water.

The diagram gives an idea of the fact that the world community can be foreseen to be meeting very severe water constraints in the next 40 years in its efforts to eradicate hunger, in line with the first of the Millennium Development Goals. The primary options will look different for the four different water shortage situations, as indicated in Table 5.1. This table also shows the percentage of the population that falls into each of the different regions. The analysis suggests that almost half the world population would by 2050 live in countries suffering from both green and blue water shortage for food production. This will call for early awareness and great care in the world's coming efforts towards global food security.

In regions where water shortages do not allow the production of enough food for the population (water requirements beyond 100% of the combined green–blue water resource), there are a set of options for achieving food security: in zone a, food import; in zone c, vapour shift or irrigation. It is worth noting that many African countries belong to categories b and d and should be able to make much better use of their green water resources.

The ultimate level of carrying capacity overshoot will evidently depend on the earlier mentioned in-field water losses (cf. Figure 5.4). With plausible improvements in water productivity, the gross water deficit in water short countries by 2050 might be limited to around 2,100 km^3/yr (Rockström, Karlberg, and Falkenmark, 2010). This remaining water deficit, in water-scarce regions where most of the future population expansion is predicted to take place, could theoretically be met by increased food imports from water-rich regions (Falkenmark and Lannerstad, 2010). In many water-scarce countries where the economical situation constrains import opportunities,

Table 5.1 Some policy implications. Global food supply based on water productivity 1,300 m^3/p/yr

GREENBLUE	Green shortage<1,300m^3/p/yr	Green abundance>1,300m^3/p/yr
Blue shortage <1,000m^3/p/yr	a 46 % of world population • horizontal expansion • food import • radical water productivity increase	b 14% of world population • upgrading rain-fed agricul ure/rainwater harvesting
Blue abundance >1,000m^3/p/yr	c 21% of world population • irrigation expansion	d 19% of world population • upgrading rain-fed agriculture • irrigation expansion

Source: Falkenmark and Rockström (2010)

the additional water requirement will have to be met by national solutions, such as horizontal expansion of agriculture into grasslands, lowered dietary expectations, food aid, and so forth.

Risks Related to Future Water Scarcity

Water scarcity and water security are closely linked from many perspectives.

River Basin Vulnerability

Varis et al. (2011) recently analysed river basin vulnerability in 10 major monsoon basins in the Asia-Pacific region by combining average water stress with five other vulnerability indicators (governance, economy, social, environmental, hazards). They showed that water stress was high in the Yellow River Basin and very high in the Indus Basin. They combined these indicators into river basin vulnerability profiles, and concluded that Ganges and Indus have the highest river basin vulnerability. It is worth noting that all three basins—Yellow, Ganges, and Indus—are located in the vulnerable triangle identified in Figure 5.3, combining high use-to-availability value with a high level of water crowding.

Where water crowding is high, economic development is particularly challenging. One extreme example is the Limpopo basin in the Southern African Development Community region, which is expected to approach almost 5,000 p/Mm^3/yr in water crowding in 2025 (Turton and Botha, 2013), at the same time expecting water demands to more than double. To approach water security under such extreme water shortage, water governance will critically depend on strong leadership. Otherwise, social instability might be triggered. Turton (2011) has even hypothesized that severe water shortage might raise frustration levels and even result in xenophobic violence; the organization Genocide Watch International recently placed South Africa on stage 6 ('preparation').

Exposure to Climate Change

Also in terms of exposure to climate change, the analytical diagram in Figure 5.3 may give some guidance by distinguishing four contrasting regions with rather different water scarcity dilemmas (Falkenmark, 2012):

- The vulnerable triangle between 1,000 p/Mm^3/yr in water crowding (chronic water shortage) and the diagonal water use line 200 m^3/p/yr, is a region sensitive to both climate aridification and droughts, where competing water demands will be particularly challenging.
- The upper left-hand corner of the graph hosts well-developed regions, like Australia or California, with advanced irrigation-dependent economies, high water demand (more than 2,000–3,000 m^3/p/yr), and large-scale export of agricultural products that is vital for the national

economy. The dominating challenge is long periods of drought years when irrigation water needs cannot be met, and policymakers' efforts to alter water allocations in response cannot meet social acceptance.

- Regions that are low or moderate in both directions, shown in the low left-hand part of the diagram, include poor developing countries with low stress and economic water scarcity, especially in Africa. These countries are dependent on dry-land agriculture and remain highly sensitive to rainfall variability and green water problems. Only a few such countries have been able to develop their rivers to provide agricultural water security through irrigation.
- The lower right-hand area hosts some poor and highly crowded basins. In this position, we find exceptional water shortages and overpopulated basins.

Carbon Sequestration

Potential methods for climate change mitigation include carbon sequestration through water-consumptive biomass production to reduce the increasing overload of CO_2 in the atmosphere, which continuously exacerbates global warming. Rockström et al. (2012) recently assessed the sequestration feasibility from a planetary water boundary perspective, applying a green–blue water approach, highlighting two main modes for sequestration: increased carbon storage in respectively soil and forest plantations. In both cases, the increase in consumptive water use would result in a reduced runoff generation and, therefore, deplete the river.

Sequestration in soil refers to the potential for carbon uptake on croplands, irrigated lands, restored degraded soils, and rangelands in tropical countries. Lal (2004) estimated that globally it would be possible, after 40 years, to reach a total accumulation of organic soil carbon content of around 25 to 50 gigaton (Gt). This would by 2050 involve an annual consumptive water use beyond current croplands in the order of 250 km^3/yr, reducing the annual runoff generation by a similar amount. Carbon can also be sequestered in living forest biomass by converting grasslands to forests, and could potential accumulating about 50 Gt C by 2050 (Hansen et al., 2008). The additional consumptive water use involved has been estimated at around 1,300 km^3/yr. These calculations indicate a need to allocate large amounts of additional green water for these bioresources needs and accept the reductions in runoff generation. The expected increase would, in fact, surpass the recently suggested planetary freshwater boundary (Rockström et al., 2012; Rockström, Steffen, et al., 2009a, 2009b).

In other words, what these assessments indicate is that the expected increases in water scarcity will be calling for large care when addressing pathways to the future of humanity. The water cycle itself is sending warning signals and many human activities involve altered runoff generation, threatening not only the habitats for aquatic ecosystems but also human downstream societies.

Discussion and Conclusions

This chapter has focused on key water scarcity–related risks to water security, highlighting constraints to both human water security and agricultural (and, therefore, food) security. It looked at water scarcity on the one hand for humans (blue water scarcity), and on the other for agriculture/food production (green water scarcity). From a blue water perspective, it analysed the implications of both increasing water crowding and sharpening water stress. It identified hot spots of severe water competition. From the green water perspective, it discussed low crop yields as related to water deficiency in different time scales: intraannual dryspells as opposed to interannual droughts and climate aridification. It also looked at drought-driven famine threats interacting with manmade green water scarcity as the effect of land degradation, often spoken about as *desertification*. It highlighted the implications of population growth and climate change in terms of cropland water deficiency, carrying capacity overshoot, increasing trade dependence, and a particular predicament of 1.5 billion people in poor countries lacking purchasing power.

We may conclude that the achievement of water security may be disturbed by the increasing pressure to be foreseen, as climate changes and population increases in regions with severe water shortage (i.e., both high water crowding and high use-to-availability, cf Figure 5.3). In such regions it will be increasingly difficult to find additional raw water to meet expanding water supply in rapidly growing cities, since almost all water is already in use.

Water scarcity has both an ecosystem and a social system dimension on the local and regional scales, which are relevant for the complications of achieving water security in a region. In the ecosystem dimension, aquatic ecosystems are vulnerable to river flow depletion generated by consumptive (evaporative) water use, altering the ecosystems habitats in terms of both water quantity since the water flow decreases, and water quality since there is less water to dilute introduced pollutants. This means that a basin may move towards closure with serious challenges in terms of protection of aquatic ecosystem habitats. In the social system dimension, rising water shortage will increasingly limit the remaining manoeuvring space. Serious strains can be foreseen, especially in the 'vulnerable triangle' where river basins are approaching the 200 m^3/p/yr line. In the mid-1990s, this was seen as what was needed for a nonwasteful municipal/industrial water demand. Continued population growth and climate change–related alterations of rainfall patterns and variability will greatly increase the challenges of water governance and management. Solutions will have to focus on waste water reuse and recirculation. Alternative pathways towards water security may also include rainwater harvesting, storage of water from urban rainstorms, and so forth. At high levels of water crowding, threatening social conflicts make good leadership a fundamental aspect.

While current food trade is basically driven by comparative advantages (Seekell, D'Odorico, and Pace, 2011), analyses of the situation by 2050 suggest that global food security will be strongly coloured by carrying capacity overshoot in water-short countries with continuing population growth, and, therefore, there will be a massive dependence on food import (Falkenmark and Lannerstad, 2010). Later comparative analyses based on alternative levels of food supply (2,200–3,000 kcal/p/d, 5% animal food) indicate that 3.3 to 3.5 billion people will be living in high- or medium-income countries that will depend on food import to compensate for water deficiency. In addition there may be some 1.5 billion people living in water-short low-income countries that lack purchasing power. These trade dependencies of water-short countries may involve large-scale connectivities not yet identified. One sign is the ongoing land acquisition in foreign countries ('land grabbing').

Final Remarks

This chapter has illuminated the central role of water scarcity risks when analysing the global water security challenge. It is interesting, if not depressing, to note that even more than 20 years into the discussion of water scarcity, there remains—as recently stressed by Zeitoun (2011)—a lot of confusion with flawed understandings of water scarcity and simplistic classifications. This long period of floating understanding of the phenomenon—and lack of distinction between whether the scarcity is driven by demand increase and related to difficulties of mobilising even more, or a true water shortage driven by population growth and, therefore, very real (SIWI, 2007)—has in the meantime allowed a water shortage exposure of humanity to expand to almost alarming levels.

References

Chenoweth, J. (2008) 'A re-assessment of indicators of national water scarcity', *Water International*, vol 33, pp5–18.

Falkenmark, M. (1986) 'Fresh water—time for a modified approach', *Ambio*, vol 15, pp192–200.

Falkenmark, M. (1997) 'Meeting water requirements of an expanding world population', *Philosophical Transactions Biology, The Royal Society, London*, vol 352, pp929–936.

Falkenmark, M. (2003) *Water Management and Ecosystems. Living with Change*, TAC, Report No. 9, Global Water Partnership, Stockholm.

Falkenmark, M. (2012) 'Adapting to climate change. Towards societal water security in dry climate countries', International Journal of Water Resources Development, Volume 29, Issue 2, June 2013, pages 123–136.

Falkenmark, M. and Lannerstad, M. (2010) 'Food security in water short countries—coping with carrying capacity overshoot', in L. Martinez-Corina et al. (eds) *Rethinking Water and Food Security*, Fourth Botín Foundation Water Workshop, CRC Press, Taylor and Francis Group.

Falkenmark, M. and Lindh, G. (1976) *Water for a Starving World*, Westview Press.

Falkenmark, M. and Lundqvist J. (1997) *World Freshwater Problems—Call for a New Realism*, Comprehensive Assessment of the Freshwater Resources of the World, Background report 1.

Falkenmark, M. and Molden, D. (2008) 'Wake up to realities of river basin closure', *Water Resources Development*, vol 24, pp201–215.

Falkenmark, M. and Rockström, J. (2004) *Balancing Water for Humans and Nature: The New Approach in Ecohydrology*, Earthscan, London.

Falkenmark, M. and Rockström, J. (2008) 'Building resilience to drought in desertification-prone savannas in Sub-Saharan Africa: the water perspective', *Natural Resources Forum*, vol 32, pp93–102.

Falkenmark, M. and Rockström, J. (2010) 'Back to basics on water as constraint for global food production: opportunities and limitations', in A. Garrido and H. Ingram (eds) *Water for Food in a Changing World* (2nd volume), Contributions from the Rosenberg International Forum on Water Policy, Routledge: London.

Global Water System Project, 2011. Accessed 11 August 2013. http://www.gwsp.org/

Gordon, L.J., Steffen, W., Jönsson, B.F., Folke, C., Falkenmark, M. and Johannessen, Å. (2005) 'Human modification of global water vapor flows from the land surface', *Proceedings of the National Academy of Sciences of the United States of America*, vol 102, pp7612–7617.

Grey, D. and Sadoff, C. (2007) 'Sink or swim? Water security for growth and development', *Water Policy*, vol 9, pp545–571.

Hansen, J., Satol, M., Kharechal, P., Beerling, D., Berner, R., Masson-Delmotte, V., Pagani, M., Raymo, M., Royer, D.L. and Zachos, J. C. (2008) 'Target atmospheric CO_2: where should humanity aim?', *The Open Atmospheric Science Journal*, vol 2, pp217–231.

Kummu, M., Ward, P. J., de Moel, H. and Varis, O. (2010) 'Is physical water scarcity a new phenomenon? Global assessment of water shortage over the last two millennia', *Environmental Research Letters*, vol 5, 034006 doi:10.1088/1748-9326/5/3/034006.

Kummu, M. and Varis, O. (2011) 'A world by latitudes: a global analysis of human population, development level and environment across the north-south axis over the past half century', *Applied Geography*, vol 31, pp 495–507.

Lal, R. (2004) 'Soil carbon sequestration impacts on global climate change and food security', *Science*, vol 304, pp1623–1627.

Lundqvist, J. and Falkenmark, M. (2011) 'Adaptation to rainfall variability and unpredictability', *Water Resources Development*, vol 26, pp595–612.

Lundqvist, J. and Gleick, P. (1997) *Sustaining Our Waters into the 21st Century*, Stockholm Environment Institute, Stockholm.

L'vovich, M.I. and White, G.F. (1990) 'Use and transformations of terrestrial water systems' in B.L. Turner II (ed) *The Earth as Transformed by Human Action: Global and Regional Changes in the Biosphere over the Past 300 Years*, Cambridge University Press, Cambridge, UK.

Molden, D., ed. (2007) *Water for Food, Water for Life: A Comprehensive Assessment of Water Management in Agriculture*, Earthscan, London and International Water Management Institute, Columbo.

Pala, M., Oweis, T.; Benli, B.; Jamal, M. and Zencirci, N. (2011) 'Assessment of wheat yield gap in the Mediterranean: Case studies from Morocco, Syria, and Turkey', ICARDA, Aleppo, Syria.

Ripl, W. (2003) 'Water: the bloodstream of the biosphere', *Philosophical Transactions of the Royal Society of London Series B—Biological Sciences*, vol 358, pp1921–1934.

Rockström, J. (2003) 'Water for food and nature in drought prone tropics: vapour shift in rainfed agriculture', in *Philosophical Transactions Biology*, The Royal Society, London, pp1997–2010.

Rockström, J., Falkenmark, M., Karlberg, L., Hoff, H., Rost, S. and Gerten, D. (2009) 'Future water availability for global food production: the potential of green water for increasing resilience to global change', *Water Resources Research*, vol 45.

Rockström, J., Falkenmark, M., Lannerstad, M. and Karlberg, L. (2012) 'The planetary water drama: Dual task of feeding humanity and curbing climate change', in Geophysical Research Letters. Vol. 39, No. 15, doi:10.1029/2012GL051688.

Rockström, J., Karlberg, L. and Falkenmark, M. (2010) *Global food production in a water-constrained world: Exploring green and blue challenges and solutions*, Cambridge University Press.

Rockström, J., Lannerstad, M. and Falkenmark, M. (2007) 'Assessing the water challenge of a new green revolution in developing countries', *Proceedings of the National Academy of Sciences*, vol 104, pp6253–6260.

Rockström, J., Steffen, W., Noone, K., Persson, A., Chapin, F.S., Lambin, E.F., Lenton, T.M., Scheffer, M., Folke, C., Schellnhuber, H.J., Nykvist, B., de Wit, C.A., Hughes, T., van der Leeuw, S., Rodhe, H., Sorlin, S., Snyder, P.K., Costanza, R., Svedin, U., Falkenmark, M., Karlberg, L., Corell, R.W., Fabry, V.J., Hansen, J., Walker, B., Liverman, D., Richardson, K., Crutzen, P. and Foley, J.A. (2009a) 'A safe operating space for humanity', *Nature*, vol 461, pp472–475.

Rockström, J., Steffen, W., Noone, K., Persson, A., Chapin, F.S, Lambin, E., Lenton, T.M., Cheffer, M., Folke, C., Schellnhuber, H:J:, Nykvist, B., de Wit, C.A., Hughes, T., van der Leeuw, S., Rodhe, H., Sörlin, S., Snyder, P.K., Costanza, R., Svedin, U., Falkenmark, M., Karlberg, L., Corell, R.W., Fabry, V.J., Hansen, J., Walker, B., Liverman, D., Richardson. K., Crutzen, P., Foley, J. (2009b) 'Planetary boundaries: exploring the safe operating space for humanity', *Ecol. Soc.*, vol 14.

Seekell, D.A., D' Odorico, P. and Pace, M.L. (2011) 'Virtual water transfers unlikely to redress inequality in global water use', *Environmental Research Letters*, vol 6, no 2.

SIWI (2007) *On the verge of a new water scarcity: A call for good governance and human ingenuity*, SIWI Policy Briefs, SIWI, Stockholm.

Turton, A. (2011) 'Resource allocation and xenophobic violence in S Africa', in H.E. Purkitt (ed) *African Environmental and Human Security in the 21st Century*, Amherst, New York: Cambria Press.

Turton, A.R. and Botha, F.S. (2013) New thinking on an anthropocenic aquifer in South Africa, in S. Eslamian, (ed.) *Handbook of Engineering Hydrology*, Vol. 3: Environmental Hydrology and Water Management, CRC Group, Francis and Taylor, USA.

UN (CSD) (1997) *Comprehensive Assessment of the Freshwater Resources of the World*. Economic and Social Council, 5th session, 5–25 April, E/CN.171997/9/.

Varis, O., Kummu, M. and Salmivaara, A. (2011) 'Ten major rivers in monsoon Asia-Pacific: an assessment of vulnerability', *Applied Geography*, vol 32, pp441–454.

Vörösmarty, C.J., Mcintyre, P.B., Gessner, M.O., Dudgeon, D., Prusevich, A. and Green, P. (2010) 'Global threats to human water security and river biodiversity', *Nature*, vol 467, pp555–561.

Yang, H., Wang, L. and Zehnder, A.J.B. (2007) 'Water scarcity and food trade in the southern and eastern Mediterranean countries', *Food Policy*, vol 32, pp585–605.

Zeitoun, M. (2011) 'The global web of national water security', *Global Policy*, vol 2, no 3, pp286–296.

6 Water Security in a Changing Climate

Declan Conway

The Challenge: Securing Water in a Changing Climate

The hydrological cycle is a fundamental component of the global climate system. Over millennial and century timescales, natural variations in the climate system have caused major changes in the state (i.e., snow or rain) and spatial and temporal patterns of precipitation leading to periods of significant aridification or humidity. Many of the most important environmental and societal effects of anthropogenic climate change are likely to be caused by changes in the hydrological cycle. Changes in the timing, amount, intensity, and state of precipitation and changes in atmospheric evaporative demand resulting from the combined effects of changes in temperature, radiation, humidity, and windspeed will modify surface and subsurface water availability. Land surface characteristics mediate the hydrological response and will further complicate hydrological outcomes due to the confounding effects of land use and land cover changes in response to, among other things, higher temperatures, soil moisture availability, groundwater connectivity, and human activity.

Anthropogenic climate change will occur as a complex mix of slow onset trends (unidirectional, such as warming), changes in variability (from daily through seasonal to decadal timescales), fluctuations in mean conditions (places may become wetter or drier), changes in the frequency and intensity of extreme events, and the possibility of rapid shifts in large-scale hydrometeorological systems associated with thresholds or tipping points (e.g., Lenton and Schellnhuber, 2007). Whilst there is very high confidence that anthropogenic climate change is ongoing and set to continue, there is much lower confidence in climate model projections of how key elements of the hydrological cycle will evolve at the spatial and temporal detail often desired for adaptation. Uncertainty is particularly high in relation to future precipitation patterns due largely to differences between climate model results, exacerbated by the fact that changes may not be unidirectional; it is possible for precipitation to fluctuate over decadal timescales, inducing wet and dry periods in the same location. This uncertainty has and continues to have important bearing on research directions and practical approaches to adaptation in the water sector. Sea level rise is another slow onset process with major implications for coastal freshwater systems, through extreme events, coastal erosion, inundation, and saltwater incursion; however, limited chapter length precludes detailed analysis of these processes here.

Climate change undermines the basic assumptions and principles of water resources management—the practice of relying on observations from the past to develop statistical distributions to represent current and future conditions (i.e., assuming that catchment behavior is invariant over time, or 'stationary'). This is at the heart of the problem for management, and was profiled in one of the earliest studies to focus attention on climate change and water resources, which contained many ideas, approaches, and results that remain highly relevant (Nemec and Schaake, 1982). Nemec and Schaake highlighted the issue of nonstationarity as, 'The importance of the sensitivity of the systems to such variations is demonstrated with respect to the stationarity of the mean and the extremes' (p327). Milly et al. (2008) characterized this challenge, highlighting the need to develop nonstationary probabilistic models to optimize water systems, incorporating temporal evolution of statistical distributions and uncertainty, coupled with enhanced information flow from the science community to water managers and linkages between modeling and observations. Yet, variability and uncertainty, expressed as nonstationary behavior, are ubiquitous in water resources management; for example, the consequences of persistence in Nile river flows (successive years with above- or below-normal streamflow) feature in accounts of ancient Egyptian society (Said, 1993). In modern times, this led to the famous hydrologist Hurst's influential work on persistence (the 'Hurst' phenomenon) (Hurst et al., 1959) that required design of overyear storage capacity in the Aswan High Dam to account for its effects on reliable yield. Other examples include the effects of the decline in rainfall across the Sahel in the early 1970s causing reductions in 30-year mean annual stream flow of up to 51% between 1931 and 1960 and 1961 and 1990 (Conway et al., 2009), and the effects of Australia's 'big dry' during the late 1990s and early 2000s on stream flow in the Murray-Darling Basin (Vernon-Kidd and Kiem, 2009).

In addition to nonstationary supply, changing patterns in the demand for water, manifest as both long-term trends and short-term fluctuations, mean the recent past can be an unreliable predictor of future conditions. The corollary of this is that many features of water resources management have evolved measures to deal with uncertainty and variability. This has generated considerable debate on the question of whether existing procedures and principles are sufficient or whether climate change adaptation requires planning criteria and management action beyond current measures. Some scholars have proposed that current systems will respond well and costs will be minimal, whilst others feel that adaptation will be challenging and, in some cases, extremely costly (Frederick, 1997).

However water security is defined, it is clear·that through potentially rapid and large changes in socially critical aspects of the hydrological cycle, climate change represents a major cross-cutting challenge, in terms of availability, exposure to hazard, management capacity (supply and demand), and individual well-being. From large-scale shifts in transboundary river basin stream flows to disruptions in local unimproved water sources, security of water availability is threatened. Changing patterns of extremes will affect flood risk of individuals and communities with consequences for well-being in terms of both physical and psychological effects of risk exposure and recovery from trauma. Moreover, the wide ranging secondary effects on food and energy production (e.g., see Allan, Chapter 20, and Froggatt, Chapter 8), land use, and so on are likely to cause even further disruption to water management systems.

This chapter first summarises the literature on water resources and climate change with respect to evidence of observed changes and impacts of future change. The next section traces developments in research on water resources and climate change during the last three decades using selective examples of benchmark papers identified through the author's own experience and consideration of successive Intergovernmental Panel on Climate Change (IPCC) chapters on hydrology and water resources. This profiles a shift from issue identification, through quantification of climate impacts and exploration of management implications, to attempts to operationalise adaptation within the water sector. The penultimate section proposes a broad typology of studies to represent the explosion of research that occurred since the early 2000s, and the chapter ends with some brief conclusions.

Evidence of Change in Hydrological Systems

Links between the observed global warming of the last few decades and changes in the large-scale hydrological cycle have been identified, particularly in temperature-sensitive systems such as snow-melt and glacier-fed river systems (Bates et al., 2008). Analysis of global precipitation variability has identified a climate change signal (Zhang et al., 2007), and increasing flood risk in autumn in England and Wales has been associated with greenhouse gas emissions (Pall et al., 2011), as has observed intensification of daily and five daily precipitation amounts during the second half of the last century (Min et al., 2011). The IPCC Special Report on Extremes (SREX) notes existence of statistically significant trends in the number of heavy precipitation events in some regions and, with medium confidence, increases and decreases (depending on region) in drought frequency/intensity (IPCC, 2012).

Changes in hydrological regimes are harder to detect and attribute for a variety of reasons, and whilst changes in the hydrological cycle of the western United States have been observed for the latter half of the 20th century (Bates et al., 2008), wider evidence of changes in flood frequency and intensity is more equivocal. The IPCC SREX notes limited to medium availability of evidence (often constrained by gauge record quality, length, and distribution) and low agreement in that evidence, concluding that there is low overall confidence regarding 'even the sign of these changes' (IPCC, 2012, p6). High natural variability in rainfall in many regions also means that a climate change signal may not be discernable until the mid-21st century (e.g.,

Hawkins and Sutton, 2011). The confounding effects of human activities in river basins also hamper detection. Higher temperatures will likely lead to more evaporation; however, data and observational constraints are even more pronounced for evapotranspiration than for streamflow. An estimate of evaporation for the global land surface using meteorological and remotely sensed data with a biosphere model showed increase of 7.1mm per year per decade from 1982 to 1997 (Jung et al., 2010). The trend ended after 1997 until 2008, associated with soil moisture limitations in the southern hemisphere, but whether these patterns reflect natural variability or anthropogenic influences is as yet unanswered (Jung et al., 2010).

The socioeconomic impacts of floods clearly demonstrate the significance of future climate change. Flood damage constitutes about one-third of the economic losses inflicted by natural hazards worldwide (Munich Re, 2005), and the economic losses associated with floods globally have increased by a factor of five between the periods 1950 to 1980 and 1996 to 2005 (Kron and Berz, 2007). However, normalizing hazard and flood damages over time to avoid conflating the effects of changes in streamflow and socioeconomic exposure and sensitivity is difficult and sometimes contentious. Case study examples of damage and longer-term recovery costs can provide useful information to assess costs of changes in frequency and intensity of future extremes. For example, the cost to Kenya of the floods associated with the major 1997/98 El Niño event was roughly 11% of its GDP (World Bank, 2006). Such estimates are hard to produce and given their importance are surprisingly rare because few countries have established mechanisms to systematically document costs. These knowledge gaps are acknowledged in the United Nation's Framework Convention on Climate Change programme on 'Loss and Damage' which recognises 'the need to strengthen international cooperation and expertise in order to understand and reduce loss and damage associated with the adverse effects of climate change' (UNFCCC, 2012, p1).

Detection and attribution of change in flood risk are likely to remain difficult; further research is required to integrate climate and land use interactions and communication of uncertainty (Wilby et al., 2008). Analysis of management and policy changes associated with major flood crises can provide powerful surrogates for how adaptation might be approached in the future (Penning-Rowsell et al., 2006). Extreme events provide a window of opportunity to modify policy and practice whilst individuals and decision makers are temporarily more sensitized. A case in point is the European Flood Alert System, which was created extremely rapidly in response to the 2002 Elbe/Danube floods (Demeritt et al., 2012). A recent national survey in the UK found respondents that had experience with flooding were more concerned about climate change and more willing to reduce energy use to mitigate it (Spence et al., 2011). National differences in perceptions of individual responsibility for flood protection, between Ireland and England, highlight important questions about the role of the state, market, and individuals in providing physical protection and insurance (Adger et al., 2013). Beyond the physical system, there are now many instances in which adaptation and risk management actions are being developed in the water sector. A review of observed adaptation in the UK found greater effort in identifying impacts and adaptations in sectors associated with large-scale infrastructure, with higher levels of activity in the flood defence and water supply sectors than others (Tompkins et al., 2010).

Impacts of Climate Change on Water Resources

In the near term, the hydrological effects of climate change are likely to be similar to the effects of recent climate and extreme weather conditions. Atmospheric physics, observations, and modelling tell us that some underlying changes are going to occur; for example, warmer temperatures allow the atmosphere to hold more moisture, and this will contribute to higher-intensity but less-frequent precipitation events (see Trenberth, 2011, for a comprehensive review). The secondary impacts of climate change on stream flows have been widely documented and most follow approaches developed in the early 1980s (see next section), using results from global climate models (scenarios or projections) to drive hydrological models and simulate changes in streamflow characteristics. Such climate impact studies demonstrate differences in sensitivity to assorted climate parameters, according to climatic regime (semi-arid and/or low runoff coefficients show higher sensitivities) and catchment characteristics such as geology and baseflow index. Contrasting outcomes can also occur for the same catchment within and between studies due to:

- Differences in precipitation projections between climate models
- Hydrological model type (although this is seldom treated explicitly)

The IPCC 'Technical Report on Water' captures the global patterns as follows:

> By the middle of the 21st century, annual average river runoff and water availability are projected to increase as a result of climate change at high latitudes and in some wet tropical areas, and decrease over some dry regions at mid-latitudes and in the dry tropics. Many semi-arid and arid areas (e.g. the Mediterranean Basin, western USA, southern Africa and northeastern Brazil) are particularly exposed to the impacts of climate change and are projected to suffer a decrease of water resources due to climate change. (Bates et al., 2008, p3)

Areas of decreasing precipitation and surface water availability are likely to experience worse problems associated with water scarcity. Increases in surface water may have beneficial effects on water supply in northwest South America, east Africa, south and east Asia, and Indonesia (possibly offset by exacerbated problems from flooding) (Bates et al., 2008). Many of these are regions where population and economic growth are already increasing water scarcity, yet climate change could 'reduce overall water stress at the global level because increases in runoff are concentrated heavily in the most populous parts of the world' (Bates et al., 2008, p45).

Until very recently, most assessments, including those reviewed in Bates et al. (2008) and Milly et al.'s (2005) widely cited analysis of global runoff response to climate change, have used climate model results compiled for the IPCC Fourth Assessment Report (AR4) during the mid-2000s (the Coupled Model Intercomparison Project Phase 3, CMIP3) to simulate impacts. Recently published results using the new CMIP5 multimodel data set compiled in time for the IPCC 5th Assessment Report (due in late 2013) suggest 'severe and widespread droughts in the next 30–90 years over many land areas resulting from either decreased precipitation and/or increased evaporation' (Dai, 2013, p52).

The literature covering specific case studies is too large to review here and, fortunately, this is done periodically by IPCC Working Group II, which covers impacts, adaptation, and vulnerability. Reviewing the chapters on hydrology and water resources from the First Assessment Report in 1990 to 2008's 'Technical Report on Water' highlights areas of both continuity and change (Table 6.1 lists the chapter titles, lead authorship, and demonstrates their expanding length). Some remarkably consistent findings over time include:

- Uncertainties (related to differences between climate model projections of precipitation, climate model resolution, and process simulation of hydrological response) limit ability to determine impacts with confidence.

Table 6.1 Chapters on hydrology and freshwater in successive IPCC Working Group II reports

Authors	Chapter titles	Pages	References
Shiklomanov et al. (1990)	First Assessment Report, Chapter 4: Hydrology and Water Resources	42	100
Lins et al. (1992)	1992 Supplement, Hydrology and Water Resources	13	60
Arnell et al. (1996)	Second Assessment Report, Chapter 10: Hydrology and Freshwater Ecology	38	385
Kaczmarek (1996)	Second Assessment Report, Chapter 14: Water Resources Management	18	93
Arnell et al. (2001)	Third Assessment Report, Chapter 4: Hydrology and Water Resources	43	287
Kundzewicz et al. (2007)	Fourth Assessment Report, Chapter 3: Freshwater Resources and Their Management	38	377
Bates et al. (2008)	Climate Change and Water, IPCC Technical Paper VI	200	905

- Snowmelt-dominated basins are potentially highly sensitive and the main sites to exhibit effects of observed warming.
- Semi-arid and arid basins show high sensitivities, particularly to precipitation changes.

All the assessment reports call for more reliable and detailed estimates of future climatic conditions and increased understanding of relations between climatic variability and hydrologic response. The First and Third Assessment Reports note water quality will be affected by changes in water temperature and CO_2 concentration, runoff pathways, and timing/quantity of streamflow. The Second and Third Assessment Reports mention that changes in mean and variability of water supply (added uncertainty) will require systematic reexamination of water resource management approaches. Points that appear in later reports include accumulating evidence of change in many temperature-sensitive hydrological systems, but noting that background variability still hampers detection and attribution; inclusion of impacts on groundwater and water quality; greater focus on management implications and adaptation; and, by the time of publication of the Fourth Assessment Report in 2007, documented cases of adaptation.

Subsumed within global-level analyses, many potential hotspots exist; for example, the impacts of climate change on Himalayan glaciers have received attention because of their status as the 'third pole' and source area for major river basins of international significance and primary water resource for vast populations and irrigated food production. Melt rates were the subject of controversy after the IPCC Fourth Assessment Working Group II Report cited erroneous estimates for their rate of change, suggesting that Himalayan glaciers could disappear by 2035 (Berkhout, 2010). The contribution of meltwater varies considerably in the region; for example, it is extremely important in the Indus Basin and important for the Brahmaputra Basin, but much less so for the Ganges, Yangtze, and Yellow rivers (Immerzeel et al., 2010). Variance also exists between basins in the extent to which climate change is projected to affect water availability and food security. The Brahmaputra and Indus basins appear most likely to experience reductions in stream flow, and projections for the 2050s show that changes will also have substantial consequences for food security (Immerzeel et al., 2010). The crucial determinant will be how the Asian Monsoon system evolves in a changing climate, something which remains unclear (Turner and Anamalai, 2012).

Successive IPCC reports observe that the impacts on groundwater and water quality are two important but under-researched areas with potentially serious societal risks. A recent review of groundwater found substantial doubt remains about projected impacts on recharge due to climate model uncertainty and the land surface response to changing land cover and climate factors. Understanding is also greatly constrained by lack of observations; however, interesting possibilities exist for using groundwater in climate adaptation strategies (Taylor et al., 2012).

Framing climate change in terms of water security requires consideration of other socioeconomic and technological trends, recognizing that neither water nor climate change can be considered in isolation. The effects of population growth, globalization, and associated drivers often outweigh the effects of climate change; enhanced water stress is generally associated with socioeconomically driven increases in demand, especially per capita, and irrigation requirements for food and biofuels (e.g., Vörösmarty et al., 2000; Alcamo et al., 2007). Nevertheless, the numbers of people moving across categories of water stress because of climate change can be very large (Arnell, 1999) and as global population and economic growth stabilize around the 2050s, climate change begins to dominate other factors. This is demonstrated in a study of impacts of four degrees warming on global per capita water availability (Fung et al., 2011).

Research Foci on Climate Change and Water Resources: From Impacts to Adaptation

Figure 6.1 shows the number of publications per year since 1990 on the themes of 'climate change water impacts' and 'climate change water adaptation' in the Scopus database. Publications on impacts far outweigh those on adaptation, which only appear in quantity after the mid-1990s, as is generally the case beyond the water sector (e.g., Wilby et al., 2009). Several key papers in the 1980s defined methodological approaches to climate impact assessment and

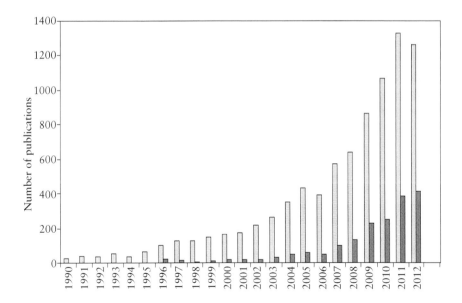

Figure 6.1 Number of publications per year since 1990 with word strings 'climate change water impacts' (grey bars) and 'climate change water adaptation' (black bars) in the Scopus database (February 2013)

profiled the potential significance of its impacts, in particular work by Peter Gleick (e.g., Gleick, 1986). The first of these was Nemec and Schaake's (1982) 'Sensitivity of Water Resource Systems to Climate Variation', which, though not the first to focus attention on the issue, was influential, containing many ideas, modeling approaches and results that captured the research agenda at the time and remain highly relevant today. The paper followed establishment of the World Climate Programme in 1979, which included a focus on climate and water resources variability and publication of a U.S. Academy of Sciences report on climate, climatic change, and water supply (U.S. National Academy of Sciences, 1977). Nemec and Schaake (1982) noted evidence of stationarity in long hydrological records for the United States, but they saw Manabe and Wetherald's (1980) groundbreaking paper on the regional effects of doubling atmospheric CO_2 concentrations as being one of several reasons for concern about nonstationarity. They used hypothetical changes in temperature ranging from −1°C to +3°C (converted using Budyko's relationship of 1°C corresponding to a 4% change in evapotranspiration) and precipitation variations of ±10 and 25% to drive the Sacramento Soil Moisture Accounting Model in two contrasting catchments, one very dry and one very humid. Their results showed much greater sensitivity of runoff to precipitation than to evapotranspiration, with increases of 250% (dry catchment) and 70% (humid catchment) in response to a 25% increase in precipitation and 1°C temperature. The effects of climate variations on storage-yield relations were explored showing nonlinear response, and they concluded by stressing 'the need for consideration of climate change impact on the design and operation of water resource systems' (Nemec and Schaake, 1982, p341).

The approach of linking climate model scenarios with hydrological models was consolidated in Gleick's (1986) paper, 'Methods for Evaluating the Regional Hydrologic Impacts of Global Climatic Changes', which had a much clearer focus on human-induced climate change than had Nemec and Schaake's study just four years before. Gleick noted even then there was evidence that changes could already be detected. The rationale for his recommended approach was limitations in knowledge of regional scale consequences of climate change due to coarse resolution and relatively crude parameterizations of land surface hydrology and the need to understand direct impacts on society. Gleick reviewed the advantages of using regional hydrologic models due to the diversity of model types, ability to fit characteristics of available data, ease of use relative to global climate models, and ability to simulate sensitivity of specific river basins with climate model or hypothetical scenarios.

Gleick's work set the ground rules for many subsequent impact studies (see reviews in IPCC Chapters) that during the 1980s and 1990s were primarily profiling the issue, or problematizing it, based on a 'predict then act' approach. These studies often fail to address management or practitioner needs as they consider time scales far into the future, often lack rigor in quantifying uncertainty, and exclude ongoing adaptation and wider socioeconomic changes to provide a context for interpreting impacts.

It was not until the mid-1990s that research began to integrate impacts assessments with their management implications and to frame these activities as adaptation to climate change. The lack of attention was noted in a set of papers edited by Frederick et al. (1997): 'relatively little has been done to review the adequacy of existing water planning principles and evaluation criteria and related impact procedures in the light of these potential changes' (p1). They observed the absence of standard understanding and assessment principles hampering the synthesis of numerous impact studies. The book was aimed as a first step in assessing whether and how water resources management should be revised in light of potential impacts of climate change by considering if existing procedures and principles developed to deal with climate variability were relevant under climate change uncertainty. They used the U.S. water planning guidelines of the time and IPCC Technical Guidelines (Carter et al., 1994) as reference points for their studies. The studies addressed issues in the United States such as geographic scale, risk and uncertainty, institutional context, engineering design, applications using nonstationary processes, decision making, and practical management case studies. Table 6.2 summarises the main issues and attempts to capture emergent key points. Most of the water resource systems investigated possessed enough resilience to withstand the impacts associated with climate model scenarios and enough institutional flexibility to adapt to changes in growth, demand, and also climate. Different types of systems exhibited different levels of sensitivity; highly integrated systems were less vulnerable than were isolated single-reservoir systems.

Frederick (1997), in the same volume, looked at adaptation through infrastructure, water supply management (integrated and flexible approaches, noting considerable institutional barriers), demand management, water planning, and institutional dimensions. The possibility was raised that although new infrastructure may be required as climate changes it would be difficult to justify projects solely on the basis of climate change, given uncertainties; 15 years on this is still the case in the UK (Charlton and Arnell, 2011).

Table 6.2 Subsection headings for Sections 1 and 2 in Frederick et al. (1997) and paraphrased key points

Section 1: Challenges posed by the prospect of climate change	Section 2: Adaptation to climate impacts: conceptual issues
Current views on climate change and water resources—Climate model projections.	**Non-climate sources of uncertainty**—Places climate change in context of other existing drivers; population, income levels, technology, etc., may outweigh effects of climate change; all are very difficult to predict.
Geographic scale factors—Climate model resolution mismatch, downscaling type issues.	**Discounting, intergenerational equity, and irreversibilities**—Maintaining options and building in flexibility, delaying major irreversible investments may be desirable.

(Continued)

Table 6.2 (Continued)

Section 1: Challenges posed by the prospect of climate change	Section 2: Adaptation to climate impacts: conceptual issues
Hydrologic non-stationarity—Background 'noise' in hydrologic systems makes detection and causal attribution difficult, better system understanding required.	Adaptation through infrastructure investments—Shift in approaches away from major projects, however, new projects could become appropriate. Use 'what-if' scenarios to assess robustness of design, effects of trends accounted for by using higher levels of variability in flow series. Concept of 'robustness' represents substantial change from existing methods. Where adding flexibility involves significant costs, confidence in impacts is required.
Environmental and ecological impacts—CO_2/stomatal response and transpiration complexities, how to model future social and economic baselines and value future ecological changes.	Adaptation through institutional change—Develop management and allocation institutions 'more flexible and responsive', this approach represents shift from earlier approaches to find the 'right' decision, demand-side management. Case studies cited of reservoir storage allocation in Washington State Green River basin between flood control and conservation of municipal water supply, and alternative allocation rules in the Colorado River. Both imply that current operating/allocating rules are efficient under example changed climate.
Impacts of climate on water demands and use—Flags irrigation water use as key sensitivity, but response uncertain.	Introducing expectations about climate change into project evaluation—Highlights case study approach of Great Lakes region using decision analysis to assess alternative investments which shows 'ignoring possibility of climate change can lead to significant losses' (p303). Key consideration is whether decision has long-term benefits and costs that could be affected by climate change.
Cascading uncertainties in climate impacts analysis—Translating from global climate model to catchment impacts, demand forecast also source of uncertainty.	Planning for shifts in snowmelt patterns and sea level rise—Short example of economic analysis of planning for shore protection. [NB—Given the high confidence in surface warming leading to changes in snowmelt and its importance in the U.S./Canada, it is surprising how little attention is given to this issue.]
Benefit and cost estimation methods—Use of contingent valuation for nonmarket dimensions, highlights as 'area in which new criteria for water planning and project evaluation could be developed' (p299).	Climate uncertainty and project design—Climate change has low relevance for decisions about small incremental capacity expansion as over short time scales impacts are likely to be quite small. This is not the case for large investments with long operating lifespans.

Research since the Early 2000s; Divergence and Specialization

Since the 2000s, research output grew substantially, making it difficult to easily capture the literature; however, Table 6.3 presents a cautious attempt to identify the main themes, of which there are at least five, recognizing some overlap between them. Many studies incorporate more than one of the themes. Almost 30 years ago, Peter Gleick was optimistic on the rate of progress in climate model resolution and scenario confidence: 'improvements in hydrologic parameterizations and grid resolution of these models over the next several years will improve the quality of regional climate evaluations' (Gleick, 1986, p110). Whilst climate models have followed a trajectory of increasing resolution and complexity of land surface hydrology, the outcome has not been marked convergence in precipitation changes or simulation of extremes. This has influenced research directions, which now give much greater attention to the quantification of uncertainty and its implications for management, as reflected in Table 6.3.

Several other areas of work are noteworthy. Research on complementary aspects of water and climate, beyond a focus on anthropogenic climate change, has made significant progress in several key areas: greater understanding of observed global and regional streamflow patterns and behavior (Peel et al., 2004; Conway et al., 2009; Dai et al., 2009), quantification of large-scale changes in groundwater storage using Groundwater Recovery and Climate Experiment (GRACE) satellite data (e.g., northwest India) (Rodell et al., 2009), and increase in the number, sophistication, and evaluation of global/regional hydrological models (Haddeland et al., 2011). Moreover, concern about climate change has contributed to much greater recognition and understanding of coupled hydro-meteorological systems, particularly regarding the hydrological effects of large-scale climate phenomena, such as the El Niño-Southern Oscillation, and their implications for water resources management (e.g., water availability for energy production and reservoir inflow forecasting). However, evidence of operational use of seasonal forecasts in the water sector is limited, for example in South Africa (Ziervogel et al., 2010) and other regions of the world (Kirchhoff et al., 2012). The significance of land use change has greater recognition, and high-profile studies of the CO_2 fertilization effect (increasing ambient CO_2 reduces stomatal conductance and tends to reduce transpiration) show contrasting results with some studies that suggest a discernable effect that can be simulated (Gedney et al., 2006) and others that a land use change signal is much stronger (Piao et al., 2007). This is an important issue that requires further integration of observations and modeling to clarify attribution of cause and effect and to support greater confidence in simulations of future hydrological conditions.

Coupled hydro-economic modeling is another critical area of development, but it is not included as a separate category in Table 6.3 because of the scarcity of studies specifically addressing climate change. One study,

Table 6.3 Five broad themes identified from research on climate change and water since the early 2000s

Research theme and indicative references	Problem addressed and methodological approach	Key issues
1. Downscaling and generation of technical information (e.g., Wilby et al., 2009)	Climate models have been used extensively to provide descriptive and quantitative information for impacts assessments, however, they have coarse spatial and temporal resolution. Downscaling includes a wide range of approaches to generate more detailed information on changes in variables such as temperature and precipitation. Application context is important for choice of approach.	Review papers highlight strengths and weaknesses of different approaches. All methods are subject to the underlying uncertainties of climate model outputs. Results can vary between approaches; time, cost, and technical requirements can be considerable. Limited evidence of use in real-world situations, with barriers related to uncertainty, complexity, communication issues, and problem specification.
2. Characterizing uncertainty in:		
2a. Climate scenarios (e.g., Prudhomme et al., 2003; New et al., 2007)	Climate scenarios vary between different climate models for reasons that are well understood. Differences in precipitation can be large, even in the direction of change. To capture the intermodel uncertainty (and other causes such as rates of greenhouse gas emissions), multiple scenarios need to be used. Where samples are large enough (i.e., large ensembles of GCMs) they can be presented in a probabilistic manner.	Rigorous approach to uncertainty, can incorporate uncertainty associated with hydrological impacts models and other sources. 'End to end' approach used by New et al. (2007) goes from scenarios to water resource planning end use and allows potential risks of impacts to be quantified but does not fully account for all sources of uncertainty (in demand). Scenario and impact distributions will change as new model results appear or experimental design is modified (i.e., we don't 'know' the true distribution, nor its sensitivity to new data and changed assumptions).
2b. Modelling impacts (Boorman and Sefton, 1997; New et al., 2007; Haddeland et al., 2011)	Hydrological model structure and parameterisation may affect the model response to any given change in climate. Differences in input datasets (precipitation, functions, and variables used to estimate potential evapotranspiration)	Whilst this has received less attention, these differences can be a major source of uncertainty. Trade-offs need to be considered between the extra demands to undertake this work (can be very high) and its added value to decision making (see next

(*Continued*)

Table 6.3 (Continued)

Research theme and indicative references	Problem addressed and methodological approach	Key issues
	can also influence results. Studies explore the causes of such differences and their significance for the interpretation of model results.	theme). Boorman and Sefton (1997) used two daily precipitation runoff models. Results sensitive to hydrological models and catchment characteristics.
3. Decision making under uncertainty, robust decision frameworks (Dessai and Hulme, 2007; Weaver et al., 2013)	Highlights the tension between 'predict then act' and 'seek robust solutions' paradigms. Recognises underutilization of climate models and failure to incorporate insights from decision sciences in climate impacts and adaptation processes; scenarios just one aspect of real-world decision-making context. Identifies 'core ideas . . . defining a proposed policy or policies, identifying vulnerabilities of these policies under multiple views of the future, seeking in particular those futures under which policies fail to meet their goals, identifying potential responses to these vulnerabilities, and organizing scenarios to help decision-makers determine the circumstances under which they would adopt these responses.' (Weaver et al., 2013 p44)	Dessai and Hulme (2007) develop a framework to test whether specific adaptation strategies are robust (or insensitive) to uncertainties in climate change projections. Uses a comprehensive approach to characterize uncertainty in projections tracking through emissions, climate sensitivity, and differences between GCMs. Fundamental to the analysis is to set out the decision-making context in the case study, in this case within UK policy, which requires that every five years water companies present water resource plans to ensure reliability of supply for the next 25 years. Sensitivity analysis on the elements of the modelling framework to determine whether a decision to adapt to climate change is sensitive to uncertainty in those elements. Arnell and Delany (2006) explain the 'headroom methodology provides a "risk envelope" against which companies should derive a balanced portfolio of supply and demand management options' (Arnell and Delany, 2006, p61).
4. Policy, institutional, and governance contexts for adaptation (e.g., Arnell et al., 1994; Frederick,	The context of the problem is critical to defining how to approach it. Describing climate impacts in the long-term future may add evidence to the need to mitigate (although current lack of progress suggests this really	Fredericks captures many aspects of these approaches. An early example of institutional climate screening assessment was done for the UK National Rivers Authority, linking its core water functions (resources, quality, floods, fisheries,

(*Continued*)

Table 6.3 (Continued)

Research theme and indicative references	Problem addressed and methodological approach	Key issues
1997; Subak, 2000; Pahl-Wostl, 2009; Charlton and Arnell, 2011; Huntjens et al., 2012)	is not working), however, such exercises are insufficient to decision making on water resources planning and management in the near term future. It is essential to embed climate concerns within the broader management and decision-making context for water to understand their significance. Studies use range of approaches drawing from quantitative and particularly qualitative analysis. Document review and key informant interviews, application of system models (for sensitivity analysis, identifying key vulnerabilities, and guiding decision support), interactive process between researchers and practitioners that may incorporate learning. Questions include: what are the theoretical (e.g., Pahl-Wostl, 2009) and empirical (e.g., Huntjens et al., 2012) principles and design considerations for governance of adaptation to climate change?	recreation, and conservation) with relevant dimensions of climate change (Arnell et al., 1994). Subak (2000) used interviews with managers in the main English and Welsh water companies to explore their perceptions of and response to extreme weather and future climate change. Regulatory and policy frameworks for water resources in England have developed sophisticated approaches to estimate climate change impacts and support adaptation through water management plans to balance supply and demand over the coming 25 years (Charlton and Arnell, 2011). By reviewing 21 draft plans, experiences were found to vary between companies; in some it is given low priority, and responses may be dominated by supply-side measures, creating tensions between companies and the regulator, The Water Services Regulation Authority (OFWAT). It is rare for management options to be planned solely in response to concerns about climate change (noteworthy that national guidelines require that climate change cannot be used to justify investment decisions). Identifies a need to improve methods, guidelines, and policy for consistent approaches between companies.
5. Multiple drivers(e.g., Vörösmarty et al., 2000; Alcamo et al., 2007)	Climate is one of many pressures facing water management which operates in a rapidly changing context of demand, policies and regulation. A major limitation of many impact studies is that they do not provide	For many parts of the world, population growth, through both absolute and per capita increases in demand for water, greatly outweighs climate change in defining numbers of people moving across thresholds of water stress (e.g., by 2025 in Vörösmarty et al., 2000).

(Continued)

Table 6.3 (Continued)

Research theme and indicative references	Problem addressed and methodological approach	Key issues
	a wider framing of water management in the future to provide context to interpret the consequences of climate change. Neither do they represent adaptation as an ongoing process, as impacts are generally compared between the present and some point in the future. Integrated studies link climate projections and hydrological models at regional and global scales with socio-economic scenarios incorporating population growth and sectoral demand for water. Many studies use quite crude indicators of water availability as proxies for water stress and water security.	Global analysis out to the 2050s, including effects of income, electricity production, and water use efficiency, produces increase in water stress in 62 to 76% and decreases in 20 to 29% of basin areas (Alcamo et al., 2007). Higher precipitation is generally the main cause of decreasing water stress (although some studies highlight the role that land cover change and CO_2 fertilization effect can have on this). The main cause of increasing stress is growing demand for water (Alcamo et al., 2007). Fung et al. (2011) ran a global hydrological model with climate scenarios representing 2°C and 4°C warming with population growth and found that population outweighed climate as a driver of water stress at 2°C, but with 4°C of warming climate change dominates.

which included climate driven variations in water availability in contrast to the usual steady-state functions used in general equilibrium models, found economy-wide impacts in Ethiopia reduced projected rates of economic growth by 38% per year over a 12-year period (World Bank, 2006). Such nonequilibrium approaches to the economic effects of climate show great potential and it is highly likely this area will see significant progress in the near future as it bridges critical knowledge gaps. Complementary research is required to develop economic and social data on costs to underpin the models.

Finally, the interconnections between climate and water through energy use and production are deservedly receiving more attention—water for energy (Hoff, 2011) and energy for water (Rothausen and Conway, 2011; Kenway et al., 2011; see also, for example, Allan, Chapter 20, and Froggatt, Chapter 8).

Concluding Comments

Water resources management and societal efforts to achieve water security have always been dynamic, attempting to balance multiple, sometimes conflicting, objectives within variable supply-and-demand situations. Climate change cuts across all aspects of water security and, whilst not intrinsically different to existing management challenges, its rate and extent of change may be unprecedented in the historical period bringing new combinations of weather and extreme events that may take us by surprise. The period since the early 1980s, when research on climate change and water began in earnest, has witnessed profound changes in the understanding and international profile of climate change, alongside extensive population, economic, and social change bringing with it massive environmental impact, not least on water resources. In the intervening time the IPCC was established, four major assessments were published, and among other things, climate change moved from a specialist area of research to a leading international focal point of global research and policy. Water as a global issue has seen the launch of a plethora of agencies and international conferences framed around contested notions of a global water crisis, with many players now advocating water's centrality in our response to climate change.

There is no doubt that the effects of climate change, whilst highly uncertain, will have far-reaching consequences for water resources across all dimensions of water security, particularly through changing frequency and intensity of extreme weather events such as floods and droughts. Key developments in research on climate change and water resources during the last three decades trace a shift of focus from climate impacts to broader management and policy concerns addressing the need for adaptation. Reflecting on these themes suggests there has been a tendency to problematize the issue of climate change for water resources management, which has highlighted challenges rather than identifying responses. During the last decade, research effort has grown substantially and developed specific issues in more detail, often involving greater technical and institutional complexity, and also addressing implications for policy and management. Key themes include: downscaling and the provision of more-detailed climate scenarios; attempts to characterize uncertainty more comprehensively; addressing the implications of uncertainty for decision making; case studies considering the broader context of management, encompassing policy, institutions, and governance; and assessments of multiple drivers of change. Uncertainty—how to characterize it and how to deal with it—remains a defining feature of research on adaptation and water. Our response will require a blend of these approaches, based on their strengths and weaknesses, in relation to the technical challenge and societal values about wider framings of water security, involving ownership and access and their importance for harnessing water's contribution to sustainable well-being.

References

Adger, W.N., Quinn, T., Lorenzoni, I., Murphy, C., and Sweeney, J. (2013) Changing social contracts in climate change adaptation. *Nature Climate Change* 3, 330–333.

Alcamo, J., Floerke, M., and Maerker, M. (2007) Future long-term changes in global water resources driven by socio-economic and climatic changes. *Hydrological Sciences Journal* 52, 247–275.

Arnell, N. (1999) Climate change and global water resources. *Global Environmental Change* 9, 31–49.

Arnell, N.W., Bates, B.C., Lang, H., Magnuson, J.J., and Mulholland, P. (1996) Hydrology and freshwater ecology. In: *Climate Change 1995: Impacts, Adaptations, and Mitigation of Climate Change: Scientific-Technical Analyses. Contribution of Working Group II to the Second Assessment Report of the Intergovernmental Panel on Climate Change* [R.T. Watson, M.C. Zinyowera, and R.H. Moss (eds.)]. Cambridge University Press, Cambridge and New York, pp. 325–363.

Arnell, N.W., Jenkins, A., and George, D.G. (1994) *The implications of climate change for the National Rivers Authority.* Final Report (R&D Report 12) to the National Rivers Authority, HMSO, London.

Bates, B., Kundzewicz, Z., Wu, S., and Palutikof, J. (2008) *Climate Change and Water.* Technical Paper of the Intergovernmental Panel on Climate Change, IPCC Secretariat, Geneva, p. 210.

Berkhout, F. (2010) Contested boundaries and reason in the climate debate. *Global Environmental Change* 20, 565–569.

Boorman, D. B., and Sefton, C.E.M. (1997) Recognising the uncertainty in the quantification of the effects of climate change on hydrological response. *Climatic Change* 35, 415–434.

Carter, T.R., Parry, M.L., Harasawa, H., and Nishioka, S. (1994) *IPCC Technical Guidelines for Assessing Climate Change Impacts and Adaptations.* University College, London and Centre for Global Environmental Research, Tsukuba, Japan.

Charlton, M.B., and Arnell, N.W. (2011) Adapting to climate change impacts on water resources in England—an assessment of draft water resources management plans. *Global Environmental Change* 21, 238–248.

Conway D., Persechino A., Ardoin-Bardin S., Hamandawana H., Dieulin C., and Mahe G. (2009) Rainfall and water resources variability in sub-Saharan Africa during the 20th century. *Journal of Hydrometeorology* 10, 41–59.

Dai, A., 2013: Increasing drought under global warming in observations and models. *Nature Climate Change* 3, 52–58. doi:10.1038/nclimate1633

Dai A., Qian T., Trenberth K.E., and Milliman J.D. (2009) Changes in continental freshwater discharge from 1949–2004. *Journal of Climate* 22, 2773–2791.

Demeritt D., Nobert S., Cloke H., and Pappenberger F. (2012) The European Flood Alert System (EFAS) and the communication, perception and use of ensemble predictions for operational flood risk management. *Hydrological Processes.* doi:10.1002/hyp.9253

Dessai S., and Hulme M. (2007) Assessing the robustness of adaptation decisions to climate change uncertainties: A case study on water resources management in the East of England. *Global Environmental Change* 17, 59–72.

Frederick, K.D. (1997) Adapting to climate impacts on the supply and demand for water. *Climatic Change* 37, 141–156.

Frederick, K.D., Major, D.C., and Stakhiv, E.Z. (1997) *Climate change and water resources planning criteria.* London, Kluwer.

Fung, F., Lopez, A., and New, M. (2011) Water availability in +2°C and +4°C worlds. *Philosophical Transactions of the Royal Society* A 369, 99–116.

Gedney, N., Cox, P.M., Betts, R.A., Boucher, O., Huntingford, C., and Stott, P. (2006) Detection of a direct carbon dioxide effect in continental river runoff records. *Nature* 439, 835–838.

Gleick, P.H. (1986) Methods for evaluating the regional hydrologic impacts of global climatic changes. *Journal of Hydrology* 88, 97–116.

Haddeland, I., and 23 others (2011) Multimodel estimate of the global terrestrial water balance: setup and first results. *Journal of Hydrometeorology* 12, 869–884.

Hawkins, E., and Sutton, R.T. (2011) The potential to narrow uncertainty in projections of regional precipitation change. *Climate Dynamics* 37, 407–418. doi:10.1007/s00382-010-0810-6.

Hoff, H. (2011) *Understanding the Nexus.* Background Paper for the Bonn 2011 Conference: The Water, Energy and Food Security Nexus, Stockholm Environment Institute, Stockholm.

Huntjens, P., Lebel, L., Pahl-Wostl, C., Schulze, R., Camkin, J., and Kranz, N. (2011) Institutional design propositions for the governance of adaptation to climate change in the water sector. *Global Environmental Change* 22, 67–81.

Hurst, H.E., Black, R.P. and Simaika, Y.M. (1959) *The Nile Basin, Volume IX. The hydrology of the Blue Nile and Atbara and the Main Nile to Aswan, with reference to some Projects*. Ministry of Public Works, Physical Department, Cairo.

Immerzeel, W.W., van Beek, L.P.H., and Bierkens, M.F.P. (2010) Climate change will affect the Asian water towers. *Science* 328, 1382–1385.

IPCC (2012) *Managing the Risks of Extreme Events and Disasters to Advance Climate Change Adaptation*. A Special Report of Working Groups I and II of the Intergovernmental Panel on Climate Change [C.B. Field, V. Barros, T.F. Stocker, D. Qin, D.J. Dokken, K.L. Ebi, M.D. Mastrandrea, K.J. Mach, G.-K. Plattner, S.K. Allen, M. Tignor, and P.M. Midgley (eds.)]. Cambridge University Press, Cambridge and New York.

Jung, M., and 33 others (2010) Recent decline in the global land evapotranspiration trend due to limited moisture supply. *Nature* 467, 951–954.

Kaczmarek, Z. (1996) Water resources management. In: *Climate Change 1995: Impacts, Adaptations, and Mitigation of Climate Change: Scientific-Technical Analyses. Contribution of Working Group II to the Second Assessment Report of the Intergovernmental Panel on Climate Change* [R.T. Watson, R.H. Moss, and M.C. Zinyowera (eds.)]. Cambridge University Press, Cambridge and New York, pp. 469–486.

Kenway, S.J., Lant, P., and Priestley, A. (2011) Quantifying the links between water and energy in cities. *Journal of Water and Climate Change* 2, 247–259.

Kirchhoff, C.J., Lemos, M.C., and Engle, N.L. (2012) What influences climate information use in water management? The role of boundary organizations and governance regimes in Brazil and the U.S. *Environmental Science & Policy*. doi: 10.1016/j.envsci.2012.07.001

Kron, W., and Berz, G. (2007) Flood disasters and climate change: trends and options—a (re)insurer's view. In: *Global Change: Enough Water for All?* [L. Lozan, H. Grasl, P. Hupfer, L. Menzel, and C.-D. Schonwiese (eds.)]. Wissenschaftliche Auswertungen/GEO, Hamburg, Germany.

Kundzewicz, Z.W., Mata, L.J., Arnell, N.W., Döll, P., Kabat, P., Jiménez, B., Miller, K.A., Oki, T., Sen, Z., and Shiklomanov, I.A. (2007) Freshwater resources and their management. In: *Climate Change 2007: Impacts, Adaptation and Vulnerability. Contribution of Working Group II to the Fourth Assessment Report of the Intergovernmental Panel on Climate Change* [M.L. Parry, O.F. Canziani, J.P. Palutikof, P.J. van der Linden, and C.E. Hanson (eds.)]. Cambridge University Press, Cambridge, pp. 173–210.

Lenton T.M., and Schellnhuber J. (2007) Tipping the scales. *Nature Reports Climate Change* 2, 97–98.

Lins, H., et al. (1990) Hydrology and water resources. In: *Climate Change: The IPCC Impacts Assessment* [W.J. McG. Tegart, W.J., G.W. Sheldon, and D.C. Griffiths (eds.)]. Australian Government Publishing Service, Canberra, Chapter 4.

Manabe, S., and Wetherald, R.T. (1980) On the horizontal distribution of climate change resulting from an increase of CO_2 content of the atmosphere. *Journal of Atmospheric Science* 37(1).

Milly, P., Dunne, K., and Vecchia, A. (2005) Global pattern of trends in streamflow and water availability in a changing climate. *Nature* 438, 347–350.

Milly, P.C.D., Betancourt, J., Falkenmark, M., Hirsch, R.M., Kundzewicz, Z., Lettenmaier, D.P., Stouffer, R.J. (2008) Stationarity is dead: Whither water management? *Science* 319, 573-574.

Min, S.-K., Zhang, X., Zwiers, F.W., and Hegerl, G.C. (2011) Human contribution to more intense precipitation extremes. *Nature* 470, 378–381.

Munich Re (2006) *Topics—Geo Annual review: Natural catastrophes 2005*. Munich Reinsurance Company, Munich, Germany.

Nemec, J., and Schaake, J. (1982) Sensitivity of water resource systems to climate variation. *Hydrological Sciences Journal* 27, 327–343.

New, M., Lopez, A., Dessai, S., and Wilby, R. (2007) Challenges in using probabilistic climate change information for impact assessments: an example from the water sector. *Philosophical Transactions of the Royal Society A: Mathematical, Physical and Engineering Sciences* 365/1857, 2117.

Pahl-Wostl, C. (2009) A conceptual framework for analysing adaptive capacity and multi-level learning processes in resource governance regimes. *Global Environmental Change* 19, 354–365.

Pall, P., Aina, T., Stone, D.A., Stott, P.A., Nozawa, T., Hilberts, A.G., Lohmann, D., and Allen, M.R. (2011) Anthropogenic greenhouse gas contribution to flood risk in England and Wales in autumn 2000. *Nature* 470, 382–386.

Peel, M.C., McMahon, T.A., and Finlayson, B.L. (2004) Continental differences in the variability of annual runoff—update and reassessment. *Journal of Hydrology* 295, 185–197.

Penning-Rowsell, E., Johnson, C., and Tunstall, S. (2006) 'Signals' from pre-crisis discourse: lessons from UK flooding for global environmental policy change? *Global Environmental Change* 16, 323–339.

Piao, S., et al. (2007) Changes in climate and land use have a larger direct impact than rising CO_2 on global river runoff trends. *Proceedings of the National Academy of Sciences.* 104, 15242–15247.

Prudhomme C., Jakob D., and Svensson C. (2003) Uncertainty and climate change impact on the flood regime of small UK catchments. *Journal of Hydrolology* 277, 1–23.

Rodell, M., Velicoga, I., and Famiglietti, J.S. (2009) Satellite-based estimates of groundwater depletion in India. *Nature* 460, 999–1002.

Rothausen, S.G.S.A., and Conway, D. (2011) Greenhouse-gas emissions from energy use in the water sector. *Nature Climate Change* 1, 210–219.

Said, R. (1993) *The River Nile. Geology, Hydrology and Utilization.* Elsevier Science, Oxford.

Shiklomanov, I., Lins, H., and Stakhiv, E. (1990) Hydrology and water resources. In: *The IPCC Impacts Assessment* [W.J. McG. Tegart, G.W. Sheldon, and D.C. Griffiths (eds.)]. Australian Government Publishing Service, Canberra, pp. 4.1–4.42.

Spence, A., Poortinga, W., Butler, C., and Pidgeon, N. F. (2011) Perceptions of climate change and willingness to save energy related to flood experience. *Nature Climate Change* 1, 46–49.

Stakhiv, E., Lins, H., and Shiklomanov, I. (1992) Hydrology and water resources. In: *The Supplementary Report to the IPCC Impacts Assessments* [W.J. McG. Tegart and G.W. Sheldon (eds.)]. Australian Government Publishing Service, Canberra, pp. 71–83.

Subak, S. (2000) Climate change adaptation in the UK water industry: managers' perceptions of past variability and future scenarios. *Water Resources Management* 14, 137–156.

Taylor, R., and 25 others (2012) Groundwater and climate change. *Nature Climate Change* 3, 322–329.

Tompkins, E.L., Boyd, E., Nicholson-Cole, S., Adger, W.N., Weatherhead, K., and Arnell, N. (2010) Observed adaptation to climate change: UK evidence of transition to a well-adapting society? *Global Environmental Change* 20, 627–635.

Trenberth, K. E. (2011) Changes in precipitation with climate change. *Climate Research* 47, 123–138.

Turner, A., and Annamalai, H. (2012) Climate change and the South Asian summer monsoon. *Nature Climate Change* 2, 587–595.

UNFCCC (2012) Approaches to address loss and damage associated with climate change impacts in developing countries that are particularly vulnerable to the adverse effects of climate change to enhance adaptive capacity. Conference of the Parties 18th Session, Doha, 26 November–7 December 2012. Agenda item

3(b) Report of the Subsidiary Body for Implementation. Revised proposal by the President, Draft decision -/CP.18.

U.S. National Academy of Sciences (1977) *Climate, Climatic Change and Water Supply—Overview and Recommendations*, Pp. 1–23, Washington, DC.

Vernon-Kidd, D.C., and Kiem, A.S. (2009) Nature and causes of protracted droughts in southeast Australia: Comparison between the Federation, WWII, and Big Dry droughts. *Geophysical Research Letters* 36, L22707. doi:10.1029/2009GL041067

Vörösmarty, C.J., Green, P., Salisbury, J., and Lammers, R.B. (2000) Global water resources: vulnerability from climate change and population growth. *Science* 289, 284–288.

Weaver, C.P., Lempert, R.J., Brown, C., Hall, J.A., Revell, D., and Sarewitz, D. (2013) Improving the contribution of climate model information to decision making: the value and demands of robust decision frameworks. *WIREs Clim Change* 4, 39–60. doi: 10.1002/wcc.202

Wilby, R.L., Beven, K.J., and Reynard, N.S. (2008) Climate change and fluvial flood risk in the UK: more of the same? *Hydrological Processes* 22, 2511–2523.

Wilby, R.L., Troni, J., Biot, Y., Tedd, L., Hewitson, B.C., Smith, D.M., and Sutton, R.T. (2009) A review of climate risk information for adaptation and development planning. *International Journal of Climatology* 29, 1193–1215.

World Bank (2006) *Ethiopia: Managing Water Resources to Maximize Sustainable Growth*. Country Water Resources Assistance Strategy. The World Bank, Washington, DC.

Zhang, X., Zwiers, F.W., Hegerl, G.C., Lambert, F.H., Gillett, N.P., Solomon, S., Stott, P.A., and Nozawa, T. (2007) Detection of human influence on twentieth-century precipitation trends. *Nature* 448, 461–465.

Ziervogel, G., Johnston, P., Matthew, M., and Mukheiber, P. (2010) Using climate information for supporting climate change adaptation in water resource management in South Africa. *Climatic Change* 103, 537–554.

7   The Role of Cities as Drivers of International Transboundary Water Management Processes

Anton Earle

Introduction

The global population is now predominantly urbanised, the result of a long-term urbanisation process as countries pass through the various stages of economic development (UN-HABITAT, 2010). Some countries in Africa and Asia have yet to attain the 50% urbanisation level; however, most are urbanising rapidly and will attain this level within the next two to three decades. The implication for water management in arid regions is that many of these large urban centres have sourced their water supplies from ever greater distances. The need to secure water, as well as the services related to water, has meant that cities in arid regions have become increasingly involved in international transboundary water management processes. This involvement is a factor usually overlooked in the analysis of transboundary water management processes in arid regions.

A large body of research has sought to understand better the drivers of conflict and cooperation between states in the management of transboundary watercourses, but finding a robust theory of why states choose various approaches has proved elusive (Earle, Jägerskog, & Öjendal, 2010). Prominent researchers in the field of transboundary water management (TWM) conclude that, in situations of water scarcity in international basins, issues of state sovereignty are paramount and that 'relations and engagement takes place in a world of disappeared hydropolitics', making their research by outsiders very difficult (Allan & Mirumachi, 2010). What is needed is an opening up of this 'black box' of interstate interactions around TWM issues.

Hegemonic states at times choose to engage in cooperative processes, while at other times, or indeed in other basins, they do not (Zeitoun & Mirumachi, 2008). Data on national-level water scarcity are not good predictors of the actions taken by states, which can sometimes result in seemingly suboptimal outcomes. In the majority of research efforts on TWM, countries are viewed and analysed as homogenous units, with water resource use and allocation happening at the national level and cascading down to a broad range of users. The flaw of this approach is that it omits subnational actors (state and nonstate) and the role they may play in driving TWM processes at the national level. This chapter argues that the importance of large urban centres in driving water management decisions on international transboundary watercourses is underestimated. Although municipal water use is typically only around 10% of national water use, the greater ability of cities to pay for water and related services translates to a much higher degree of influence in TWM processes than rural users have.

This chapter contends that large urban centres (cities) influence the agenda in TWM in three main ways: (1) their increased capacity to pay for water resources (in comparison with rural water users) means that they can harness large-scale water transfers for their use, (2) via their need for electricity and other services (such as flood protection) where cities are dependent on water resources, and (3) the need for their politicians to secure a political power base in the rural areas. For example, cities such as Johannesburg, Amman, Windhoek, Lusaka, Bangkok, and Cairo have all played a driving role in the development of water resources or the evolution of water management institutions in the international transboundary basins around them.

Cities, as the under-recognised actors in international TWM, are likely to place increasing demand on watercourses and ecosystems further away from them, with implications for the 'donor' region as well as for the recipients of the water. This chapter develops a conceptual framework for the ways in which large urban centres influence international transboundary processes in arid areas, building on discussion of the coexistence of both conflict and cooperation between states. Two cases from southern Africa are discussed, involving the area around Johannesburg in the Orange-Senqu River and the area around Windhoek and the Okavango River. Because this chapter is a short exploratory piece aiming to outline an approach to understanding TWM not previously explored, the cases are chosen based on the authors' detailed knowledge. Additional empirical work is needed to apply the framework to a larger number of cases in a variety of regions, in order for comparisons between cases to be made and to support wider generalisation.

The City: A Consumer as Well as a Manager of Water Resources

The pace of urbanisation reached its peak in the mid-20th century, a point when roughly one-third of the planet's population lived in urban settlements. The rate of urbanisation has slowed since then, but has led to over half the planet's population being urbanised by the close of the first decade of the 21st century (UN-HABITAT, 2010). Some regions of the world, such as North America, Latin America, and Oceania are predicted to be over 80% urbanised by 2050 (UN-DESA, 2009). There is considerable regional variation, with Africa and Asia currently being 40% and 42% urbanised and predicted to reach 62% and 65% by 2050, respectively (UN-DESA, 2009). Despite regional differences, it is clear that the future of humankind is an urbanised one.

These urban centres are almost entirely reliant on food, energy, building materials, and other products from rural areas, some proximate and others imported. Irrigation of food crops accounts for over 70% of water use globally, including surface water as well as groundwater withdrawals (Comprehensive Assessment of Water Management in Agriculture, 2007). Added to this is the water used in the processing and transport of food, resulting in large water footprints for cities. The flow of virtual water into cities is also an important element of their overall water footprints; however, the transfer of physical water into cities may have more of an impact on the surrounding ecosystems (Jenerette, Wu, Goldsmith, Marussich, & Roach, 2006). These water transfers remove water from surrounding ecosystems, with the return flows usually being highly polluted.

The provision of sufficient reliable supplies of fresh water is a core part of ensuring water security for cities. Various definitions of water security have been developed, and the concept is discussed in more detail in the introductory section of this book. However, in the context of cities, three key themes emerge. First is the need to supply sufficient quantities of clean water to residents for their daily needs, second is the provision of electricity linked to hydropower development, and third is the protection from water-related hazards such as floods (GWP, 2010). Municipal governments are directly engaged in securing access to reliable water supplies and the services derived from them, in many cases accessing sources far beyond their geographical jurisdiction (Jenerette, Wu, Goldsmith, Marussich, & Roach, 2006). In some cases, the water resources being accessed are transboundary—that is, shared between countries—thus linking urban water security to basin and regional water security.

Most research on TWM has focused on the state level, investigating how sovereign states cooperate or compete over transboundary watercourses. It has been argued that the 'state' is an abstraction, and when it is deconstructed what emerges is a representative group of individuals deriving their legitimacy from the bureaucratic structures of the state (Kranz & Mostert, 2010). The monolithic view of 'the state' being the prime actor in TWM needs to be challenged; it is important to investigate the range of subnational actors that influence the positions taken up by a state with respect to a specific basin.

There are instances of subnational actors engaging directly with their respective counterparts across international borders. Such 'paradiplomacy,' as it is sometimes called, refers to the international relations conducted by subnational, regional, local, or noncentral governments on their own, with a view to promoting their own interests. An example of this would be the agreements entered into between United States and Canadian provinces around provisions for limiting acid rain in the 1980s (Smith, 1988). Most countries reserve the right to enter into foreign agreements for the national government, usually prohibiting subnational entities such as provinces or states from entering into such agreements. Some countries permit regional governments to enter into international agreements, such as the German

Länder, Canadian provinces, and Belgian regions; however, there are usually limitations placed on these actions (Keating, 1999). Globalisation and the rise of transnational management regimes (around natural resources as well as economic activity and human rights issues), coupled with improvements in communications brought about by the Internet in the 1990s, saw an increase in the activity of such subnational units on the international stage.

In the field of TWM, however, such agreements between subnational units are rare, especially those involving cities. This may be linked to the fact that access to water is viewed as a sensitive security issue best handled by national governments, coupled with the large scale over which basin-wide management institutions are spread—typically encompassing more than one subnational actor in a country. A notable exception is the 1988 agreement between the government of South Africa and the Botswana Water Utility Corporation (WUC)—a parastatal entity tasked with providing water to urban centres in that country. The agreement concerned the construction of the Molatedi Dam on a tributary of the Limpopo River in South Africa and the subsequent transfer of 7.3 million cubic metres of water a year to the city of Gabarone (JTPC, 1988). In return, the WUC pays the South African government a volumetric fee based on covering the construction, operation, and maintenance of the scheme. This arrangement is unusual; usually a national government would enter into an agreement directly with its counterpart organisation (such as a department or ministry) in the other country, and not with a corporation. The likely reason for this setup is that in 1988 South Africa was still run under the Apartheid regime, to which the Botswana government was increasingly hostile.

The more common approach for the interests of cities to be articulated at the international level is through exerting pressure on the national-level negotiators representing the country in the TWM institutions. Thus, cities are one of many stakeholders seeking to highlight their issues at the international level, but they have a much clearer and more direct impact in the outcome of TWM processes than do the other stakeholders. These impacts are discussed below.

Driving the Development of Water Transfers

As cities develop, secure water supplies for domestic needs, industrial, and municipal needs also grow. Urban consumption of water is driven both by increasing population as well as higher per capita use as a direct result of economic development (Jenerette, Wu, Goldsmith, Marussich, & Roach, 2006). With increased levels of urbanisation globally, it is likely that the number of interbasin water transfers will increase, and some will involve international transboundary watercourses (Heyns, 2002).

In most cases, the large-scale transfer of water from distant basins is beyond the budget even of wealthy cities, necessitating contributions from national sources that are motivated by national interest considerations, due to the social and economic importance of cities (Jenerette, Wu, Goldsmith,

Marussich, & Roach, 2006). In order to protect the water security of cities, they usually receive the highest levels of reliable supply in both commitment and delivery. The granting of special rights to cities, due to their national importance, was well articulated by U.S. President Theodore Roosevelt in 1906 when considering the case of the transfer of water from the Owens Valley to the city of Los Angeles, stating 'yet it is a hundred or a thousand fold more important to the state and more valuable to the people as a whole if [this water is] used by the city than if used by the people of the Owens Valley' (Reisner, 1993). Thus, the precedent was set for granting cities special provisions to ensure their water security by accessing resources located in ever more distant basins.

Cities as Drivers of Water-Related Services

Most of the advances made in supplying electricity to households globally have been in urban areas, as the higher population densities provide the economies of scale needed to cover costs (OECD/IEA, 2010). The majority of least-developed countries have rural electrification rates of less than 10%, while urban access is typically more than 50%. Globally, of the estimated 1.3 billion people living without electricity, around 84% live in rural areas (OECD/IEA, 2011). Urban residents enjoy greater access to electricity, and they also use more electricity. For instance, the city of Dar es Salaam in Tanzania accounts for more than 50% of all the electricity used in the country, the bulk of which is generated from hydropower (Odhiambo, 2009). Cities have been extraordinarily good at securing their electricity supplies, frequently relying on sources located far away.

According to the World Commission on Dams, just under one-fifth of global electricity generated comes from hydropower sources and one in three nations depends on hydropower to supply more than half their electricity (WCD, 2000). In some basins, the reliance on hydropower is particularly acute; for instance, the Mekong River basin has an estimated 30,000 MW of potential hydropower currently supplying the bulk of the region's electricity (MRC, 2010). At least four dams have been constructed in the headwaters region by China, providing electricity to fuel the rapid economic growth in its cities, with plans to construct more dams (Lebel, Garden, & Imamura, 2005). In Thailand, the need to provide electricity to the rapidly growing city of Bangkok spurred some of the initial large investments necessary to develop a nationwide grid, eventually incorporating electricity imports from hydropower dams in the Mekong Basin in Laos (Greacen & Palettu, 2007). Two hydropower dams in Laos export electricity to Thailand—the Nam Ngum and the Theun Hinboun—with plans by the Thai Ministry of Energy for another four dams in Laos (Greacen & Palettu, 2007). These developments, along with the construction of dams in the other basin states, will certainly have an impact on the ecology of the Mekong River (MRC, 2010).

In southern Africa, the Cahora Bassa Dam built on the Zambezi River in Mozambique in 1974 supplies hydro-electricity to the Johannesburg area in

South Africa, more than 1,400 kilometres away. This electricity feeds into the South African national grid. It is instructive to note, however, that the long-distance transmission line from Mozambique terminates at the Apollo Inverter Plant in Centurion, a city midway between Johannesburg and Pretoria (ABB, 2012). The Cahora Bassa project was originally conceived as a multipurpose reservoir leading to conjunctive development of large-scale agriculture, light industry, and electricity, akin to the Tennessee Valley in the United States. Due to the unstable political climate in Mozambique in the late 1960s and early 1970s, there was little interest amongst settlers to move to the valley, and thus the project proceeded as a single-purpose hydropower dam with a sole client—South Africa (Isaacman & Sneddon, 2003).

Cities are also heavily involved in water management in relation to flood defence. Cities are protected by flood defences upstream, although water may be allowed to flow into rural areas previously not prone to flooding (Lebel, Garden, & Imamura, 2005). Again, the justification is that the economy and functioning of cities is so important to the national well-being that sacrificing rural land is justified. These interventions include both structural (building of infrastructure to provide flood defence) and nonstructural measures, such as diverting water to farmland. The city of Vienna, Austria, is located on the banks of the Danube River (flowing through or along 10 countries) and suffered regularly from flood damage (ICPDR, 2008). By 1988, a 21-kilometre relief channel called the New Danube had been constructed, protecting the city from future floods (UN-Habitat, 2006). This multipurpose project is now viewed as good practice for integrated flood management in the basin.

Control over Water for Political Purposes

The first two ways that cities influence the agenda on TWM, discussed previously, deal with the development of infrastructure to supply water or services derived from water. However, cities also influence the TWM agenda in a less-tangible, more abstract way through the powerful city-based actors' control of the discourse around TWM processes. States are represented in TWM institutions by central government officials based in the capital city. These officials operate under direction of politicians, ultimately deciding what position to adopt on TWM issues. Frequently, the capital city lies outside the basin being discussed, and at times there can be a disconnect between the aspirations of basin inhabitants and the government officials. By adopting a particular position with regards to the management of a transboundary basin, a city-based politician may gain support amongst a specific constituency. This could be linked to being seen to defend the national sovereignty or 'rights' of the country, or linked to being able to secure more benefits from the watercourse.

This mechanism of influence may be somewhat abstruse, and not many obvious examples are at hand. Possibly the clearest manifestation of this situation is on the Nile River, where the 10 basin states have, over the past two decades, been negotiating to form a basinwide management framework (NBI, 2011). At present, Egypt and the Sudan enjoy rights to the full flow of the river, which were awarded to them in 1959 during the period when most of the other basin states were colonial subjects. Egyptian politicians have characterised the need to maintain this full 'right' to the waters of the Nile as an existential struggle, critical to the future development of their country (Allan, 2002; Cascao, 2009; Cascao & Zeitoun, 2010). They have successfully controlled the discourse around the negotiations to arrive at an equitable allocation of the water from the river. However, there are scenarios whereby Egypt could benefit from allowing some development of the water resources of the Nile by other basin states; for instance, the construction of hydropower dams in the highlands of Ethiopia could provide storage capacity for Egypt and reduce overall evaporative losses (Kalpakian, 2004). Thus, Egypt could allow a renegotiation of the water allocation agreement, allowing more use by the upstream states, with very little loss of water for itself. However, the political discourse internally in Egypt is that the country's Nile-water right needs to be defended; any politician going against this stance risks losing political support. The need to maintain domestic political support means that a specific type of dominant discourse is transmitted from the capital city, Cairo, to the multilateral negotiation process on the river.

Cities Influencing TWM in Practice

Three drivers for cities influencing the TWM agenda are described above. This influence is articulated in the TWM institutional context through projects and on processes involving the basin. Projects would include the development of infrastructure, such as that needed for the transfer of water, generation of electricity, and the construction of flood defences. Table 7.1 lists some of the developments linked to supplying water and derived services to cities from a selection of international transboundary watercourses (the selection was made on the basis of links between the cities and respective basins being evident and is not authoritative or complete). Processes would refer to the overall basin relations. Is there a spirit of cooperation among the basin states or does conflict prevail? Processes can result in agreements (treaties) between states, as well as the formation of basin management organisations—whether basinwide or partial.

The needs of cities and city-based actors influence these projects and processes, whether proactively (as in the municipal government taking the lead in pursuing a specific objective) or reactively (as in national government representatives taking the lead in responding to the needs of a city). Either way, the development needs of cities influence the TWM agenda. This section discusses two cases illustrating some of these processes—the Lesotho Highlands Water Project and the Okavango to Windhoek water transfer.

Table 7.1 Examples of cities receiving services from transboundary basins

Basin	City	Service	Scheme	Status
Orange-Senqu	Johannesburg	Water supply	Lesotho Highlands Water Project	Existing
Limpopo	Gaborone	Water supply	Molatedi Dam transfer	Existing
Okavango	Windhoek	Water supply	Eastern National Water Carrier	Proposed
Zambezi	Bulawayo Gaborone Pretoria	Water supply	Matabeleland Zambezi Project North-South Water Carrier	Proposed
Zambezi	Johannesburg	Electricity	Cahora Bassa HP dam	Existing
Mekong	Bangkok Kunming	Electricity Electricity	Nam Ngum and Theun Hinboun dams in Laos Several planned in Laos A cascade of HP dams in the upper reaches	Existing Planned Existing and planned
Danube	Vienna	Flood defence	New Danube channel	Existing
Jordan	Amman	Water supply	Deir Alla-Amman carrier	Existing
Nile	Cairo	Political support	Maintenance of Nile River flows for Egypt's benefit	Ongoing

Lesotho Highlands Water Project

The development of the city of Johannesburg proceeded rapidly, based on the large gold deposits in the vicinity, and from the start, water scarcity was a problem (Turton, Meissner, Mampane, & Seremo, 2004). The city was founded because of the gold deposits, which happened to be situated on the watershed between two river basins—the Limpopo River draining northeast and the Vaal River (a tributary of the Orange-Senqu River) draining west. The implication is that all water required by the city needs to be pumped in from the outlying areas. Coupled with the naturally dry climate (evapotranspiration potential is more than double the annual average rainfall), this has made the provision of water to the city a constant challenge. The result of the fast-paced growth of Johannesburg over the past century has been an ever-longer water supply network. A public utility, Rand Water, is tasked with the bulk water provision to Johannesburg, as well as to various other municipalities in the Gauteng Province region, an area which generates over 35% of the country's GDP and around 10% of Africa's GDP (StatsSA, 2010).

By the 1970s, it was apparent that the future water needs of Johannesburg would not be satisfied through the flow of the Vaal River, located around 65 kilometres from the city (Turton, Meissner, Mampane, & Seremo, 2004). That river had served as the principal source of water for the city for more than half a century, but had now reached the limit of its development potential. In 1980, the chief engineer of the Rand Water Board (as it was called then) called for a 'total water strategy' for the present-day Gauteng region (Turton, Meissner, Mampane, & Seremo, 2004). His proposal listed several possible interbasin transfer schemes, some within South Africa and two internationally—from the headwaters of the Orange-Senqu River in Lesotho and from the Okavango River in Botswana, respectively. He further noted that for the last two schemes to be feasible, there had to be a political settlement with the countries involved, suggesting that membership of a regional cluster of countries may pave the way. The use of the term *total water strategy* is interesting, as it parallels the total national strategy South Africa was using at the time to fight domestic as well as foreign opposition to its Apartheid policies (Turton, Meissner, Mampane, & Seremo, 2004). In essence, such a strategy blurred the lines between civilian life and the battlefield—a securitisation of all spheres of life in the country. The chief engineer was picking up on this thinking, explicitly linking the water security of the Johannesburg region to that of national security.

The Lesotho Highlands option had been identified as a water supply for Johannesburg in the 1950s and discussions between the two countries proceeded through the 1970s (Robbroeck, 2007). Phase 1 of the project is now complete and supplies 770 million cubic metres of water a year to South Africa, earning Lesotho around $66 million (U.S. dollars) in royalty fees during 2010 (LHWP, 2011). In 2011, the two countries agreed to proceed with the second phase of the project, which will eventually increase the amount of water transferred to South Africa to 2,000 million cubic metres.

As a direct result of the LHWP, a binational management organisation was created by the two countries—the Lesotho Highlands Water Commission (LHWC) (originally called the Joint Permanent Technical Commission)—and it was tasked with oversight of the project (Earle, Malzbender, Turton, & Manzungu, 2005). Due to the concerns of the downstream basin state, Namibia, about possible negative impacts of the LHWP, a process was started in 1995 to establish a basinwide organisation (Kranz & Vidaurre, 2008). This culminated in the formation of the Orange-Senqu River Commission (ORASECOM) in 2000, incorporating all four basin states, including Botswana. The existence of ORASECOM has meant that the hegemonic dominance of South Africa over the other basin states is now challenged, providing the opportunity for integrated management and development of the basin (Kranz & Vidaurre, 2008).

The water resource needs of Johannesburg led directly to the development of a large-scale international transboundary project, as well as to a process to establish a basin management organisation. The city's development needs have strongly influenced the international TWM agenda in the Orange-Senqu Basin. It is more difficult to show whether the city was proactive in this engagement—to what degree did it influence the national-level politicians and leadership? The decision to investigate and ultimately pursue the LHWP was taken by the national Department of Water Affairs (DWA) and endorsed at the political level (Robbroeck, 2007). For assessing the need for such a water transfer, the DWA relies on water-use demand projections from Rand Water, the sole supplier of water to the city of Johannesburg. The municipality of Johannesburg has a seat on the governing board of Rand Water, and in that way can influence their agenda, triggering the 'total water strategy' call. In practice, a constellation of actors was involved in promoting the LHWP—the construction industry, the provincial government, industries, and mines, to name a few. But the linking factor in each case is that the water needs of the city of Johannesburg would be offered as a motivation for the scheme.

Okavango to Windhoek Water Transfer

Windhoek, the capital and largest city of Namibia, lies in the arid centre of the country at an altitude of around 1,650 metres (Turton & Earle, 2004b). As the city developed during the 20th century, so, too, did the need to look at alternative water resources to complement the seasonal rivers in the area. Namibia is the driest country in sub-Saharan Africa, making water management a constant challenge. By the 1960s, the locally available surface and groundwater sources had been fully exploited and a range of alternatives were considered. Due to the high altitude of the city, the distance from the coast (500 km), and the high transport costs, the desalination of seawater was not feasible (Crovello, Davidson, & Keller, 2012).

In the 1970s, the department responsible for water management in the country investigated the possibility of transferring water from the wetter north of the country (Andersson, L., Wilk, J., Todd, M. C., Hughes, D. A., Earle, A., Kniveton, D., Layberry, R. & Savenije, H. H. G. (2006) Impact of climate change and development scenarios on flow patterns in the Okavango River. J. Hydrol. 331(1-2), 43–57). A study carried out in 1972 (and updated in 1993) proposed the extension of the Eastern National Water Carrier to the Okavango River (Andersson, 2006) to supply an estimated 25 million cubic metres of water to Windhoek per year (OKACOM, 2011). At present, the carrier runs from the town of Grootfontein (about 250km south of the Okavango River) and transfers water mainly via an open canal south to the Windhoek region, tapping groundwater sources in the region as well as flow stored in dams. The need to ensure Windhoek has access to a reliable supply of water was identified by the government as one of great importance, as water scarcity would lead to a drop in economic growth of the city and the country (Heyns, 2002).

The situation in Windhoek became particularly acute during the drought of 1996, prompting the Namibian government to publicly announce plans for the transfer. Reaction to the transfer plan was strong and negative amongst the population living downstream in Botswana, as well as from regional and international environmental groups (Turton & Earle, 2004a). Concerns were raised about the impact the transfer would have on the Okavango

Delta—the inland terminus of the river that is considered a pristine wetland of international importance (OKACOM, 2011).

A consequence of the plans to develop the transfer was the formation of a basinwide basin management organisation—the Permanent Okavango River Basin Water Commission, or OKACOM. After Namibia gained independence in 1990, it established bilateral relations on water management issues with Angola and Botswana (Pinheiro, Gabaake, & Heyns, 2003). Namibia invited representatives from these two other basin states to a meeting in Windhoek in 1991, where it introduced the idea of forming a basinwide commission. This was accepted by the other states and resulted in the formation of OKACOM in 1994. At the inaugural meeting of OKACOM in 1994, Namibia officially informed the other two countries of its plans (based on the recommendations of the 1993 study) to develop the final component of the Eastern National Water Carrier and consequently transfer water to Windhoek (Pinheiro, Gabaake, & Heyns, 2003). The desire of Namibia to secure Windhoek's water supply was a driver in the formation of OKACOM.

Although the water transfer plans are on hold at the time of writing, OKACOM has been active in developing a management framework for the basin. It can be argued that OKACOM has greatly improved coordination in the basin between the countries and incorporated the views of local stakeholders in its operations (NBI, 2011). As with the Johannesburg case, it is difficult to show to what degree the city of Windhoek proactively promoted the need for the water transfer. The studies conducted to investigate water supply options were both run under the national water department. Namwater, the public water utility responsible for bulk water supply to the city, is now responsible for future water resource development in the country. Unlike Rand Water, this is a nationwide agency that does not have a space on its board of directors reserved for the city of Windhoek. The Association of Local Authorities (which would include city municipalities) is represented on the Namwater board, providing a possible avenue for the articulation of the interests of Windhoek (Republic of Namibia, 1997). However, it would appear that in this case the initiative was that of the national department responsible for water, but, again, the need to secure the city water supply was and remains a key motivation for the completion of the carrier.

Conclusion

The discussion and cases have explored some of the ways which cities influence the TWM agenda. Recognising that a larger number of cases would need to be investigated in more depth before it is possible to describe the way in which cities exert this influence, it is possible to identify several salient points. One is that large cities are viewed by national politicians and decision makers as 'too big to fail'—their needs are given priority over the needs of rural water users. In some cases, such as Johannesburg, the water supply needs of the city are framed in securitising language, explicitly stating that the national interest would be jeopardised through not responding adequately to supply needs. The second point of interest is that the process leading to the development of water supply or hydro-power infrastructure on international transboundary watercourses is expensive, in terms of money, time, and political capital. The various feasibility and impact studies that need to be carried out are usually more complex than national ones, and contain a wider range of issues, due to the inclusion of more countries. Technical and political representatives of these countries need to devote a lot of time spanning many years to reach agreement. Coupled with the actual construction costs of infrastructure, these projects are beyond the budget of individual cities— they require assistance from national sources. States relinquish some of their sovereignty by entering into cooperative agreements with other states and, by becoming dependent on them for the provision of water resources. This has an impact on the regional balance of power, frequently in a positive way, as a hegemonic state is constrained in its ability to act unilaterally.

This chapter has provided an overview of the impact of the water-security needs of cities at the international TWM level. Cities are influential drivers of infrastructure development and institutional and policy processes such as agreements and river basin organisations. Further investigation is required to determine how these needs are articulated at the national level. Most likely it is not through any one channel or mechanism, but rather a range of actors that reflect the needs of cities and (mostly independent of one another) motivate national governments to take up specific initiatives at the TWM level. Through better understanding of these channels, it should be possible for researchers and practitioners to learn more about the decisions taken by states in TWM processes, and to unpack further the 'black box' that represents international negotiation processes and outcomes.

References

ABB (2012) 'Cahoro Bassa HVDC Transmission', www.abb.com/industries/ap/db0003db004333/ba5c7f1924f82c77c1257743003d666d.aspx, accessed 20 Feb 2012.

Allan, J. A. (2002) *The Middle East Water Question—Hydropolitics and the Global Economy*, IB Tauris, London.

Allan, J. A., and Mirumachi, N. (2010) 'Why negotiate? Asymmetric endowments, asymmetric power and the invisible nexus of water, trade and power that brings apparent water security', in A. Earle, A. Jägerskog, and J. Öjendal (eds) *Transboundary Water Management: Principles and Practise*, Earthscan and SIWI, London and Stockholm.

Cascao, A. E. (2009) 'Changing power relations in the Nile river basin: unilateralism vs. cooperation?', *Water Alternatives*, vol 2, no 2, pp245–268.

Cascao, A. E., and Zeitoun, M. (2010) 'Power, hegemony and critical hydropolitics', in A. Earle, A. Jägerskog, & J. Öjenda (eds) *Transboundary Water Management: Principles and Practice* Earthscan, London.

Comprehensive Assessment of Water Management in Agriculture (2007) *Water for Food, Water for Life: A Comprehensive Assessment of Water Management in Agriculture*, Earthscan and IWMI, London and Colombo.

Crovello, S., Davidson, J., and Keller, A. (2012) *Perception and Communication of Water Reclamation for the Sustainable Future of Windhoek*, Windhoek, City of Windhoek.

Earle, A., Jägerskog, A., and Öjendal, J. (2010) 'Introduction: setting the scene for transboundary water management approaches', in A. Earle, A. Jägerskog, and J. Öjendal (eds) *Transboundary Water Management: Principles and Practise*, Earthscan, London.

Earle, A., Malzbender, D. B., Turton, A. R., and Manzungu, E. (2005) *A Preliminary Basin Profile of the Orange-Senqu River*, University of Pretoria.

Greacen, C., and Palettu, A. (2007) 'Electricity sector planning and hydropower in the Makong Region', in L. Lebel, J. Dore, R. Daniel, and Y. S. Koma (eds) *Democratizing Water Governance in the Mekong Region*, Mekong Press, Chiang Mai.

GWP (2010) 'What is water security?', www.gwp.org/The-Challenge/What-is-water-security, accessed 2 July 2012.

Heyns, P. (2002) 'The interbasin transfer of water between SADC countries: a development challenge for the future', in A. R. Turton and R. Henwood (eds) *Hydropolitics in the Developing World: A Southern African Perspective*, University of Pretoria.

ICPDR (2012) 'Floods on the Danube', www.icpdr.org/main/issues/floods, accessed 20 Feb 2012.

Isaacman, A., and Sneddon, C. (2003) 'Portuguese colonial intervention, regional conflict and post-colonial amnesia: Cahora Bassa dam, Mozambique 1965–2002', *Portuguese Studies Review*, vol 11, no 1, pp207–236.

Jenerette, G. D., Wu, W., Goldsmith, S., Marussich, W. A., and Roach, W. J. (2006) 'Contrasting water footprints of cities in China and the United States', *Ecological Economics*, vol 57, no 3, pp346–358.

JPTC (1988) 'The Supply of Water from the Molatedi Dam on the Marico River', Coordinated by the Joint Permanent Technical Committee (JPTC) and signed at Mmabatho, South Africa.

Kalpakian, J. (2004) *Identity, Conflict and Cooperation in International River Systems*, Ashgate, Aldershot.

Keating, M. (1999) 'Regions and international affairs: motives, opportunities and strategies', *Regional and Federal Studies*, vol 9, no 1, pp1–16.

Kranz, N., and Mostert, E. (2010) 'Governance in transboundary basins—the roles of stakeholders; concepts and approaches in international river basins', in A. Earle, A. Jägerskog, and J. Öjendal (eds) *Transboundary Water Management: Principles and Practise*, SIWI and Earthscan, Stockholm and London.

Kranz, N., and Vidaurre, R. (2008) *Institution-Based Water Regime Analysis Orange-Senqu: Emerging River Basin Organisation for Adaptive Management*, NeWater, Berlin.

Lebel, L., Garden, P., and Imamura, M. (2005) 'The politics of scale, position, and place in the governance of water resources in the Mekong Region', *Ecology and Society*, vol 10, no 2, pp18–38.

LHWP (2011) 'Water and electricity sales historical data', www.lhwp.org.ls/Reports/PDF/Water%20Sales.pdf, accessed 2 September 2012.

MRC (2010) *State of the Basin Report*, Mekong River Commission, Vientiane.

NBI (2011) *Component 3 Report: River Basins Organizations Survey*, Nile Basin Initiative, Entebbe.

Odhiambo, N. M. (2009) 'Energy consumption and economic growth nexus in Tanzania', *Energy Policy*, vol 37, no 2, pp617–622.

OECD/IEA (2010) *Comparative Study on Rural Electrification Policies in Emerging Economies*, OECD and the IEA, Paris.

OECD/IEA (2011) *Energy for All: Financing Access for the Poor*, OECD and the IEA, Paris.

OKACOM (2011) 'OKACOM Annual Report 2011', OKACOM, Maun.

Pinheiro, I., Gabaake, G., and Heyns, P. (2003) 'Cooperation in the Okavango river basin: the OKACOM perspective', in A. R. Turton, P. Ashton, and E. Cloete (eds) *Transboundary Rivers, Sovereignty and Development: Hydropolitical Drivers in the Okavango River Basin*, University of Pretoria.

Reisner, M. (1993) *Cadillac Desert: The American West and its Disappearing Water*, Penguin, New York.

Republic of Namibia (1997) *Namibia Water Corporation Act, No. 12 of 1997*, Windhoek, Republic of Namibia.

Robbroeck, T. (2007) 'Water on the brain', *Civil Engineering*, vol 16, no 6, pp11–19.

Smith, N. P. (1988) 'Paradiplomacy between the U.S. and Canadian Provinces: the case of acid rain memoranda of understanding', *Journal of Borderlands Studies*, vol 3, no 1, pp13–38.

StatsSA (2010) 'National statistics', www.statssa.gov.za/publications/publication-search.asp, accessed 2 August 2012.

Turton, A. R., and Earle, A. (2004a) *An Assessment of the Parallel National Action Model as a Possible Approach for the Integrated Management of the Okavango River Basin*, University of Pretoria.

Turton, A. R., and Earle, A. (2004b) 'Managing change in the Okavango Basin', *IHDP Update*, pp4–5.

Turton, A. R., Meissner, R., Mampane, P.M., and Seremo, O. (2004) *A Hydropolitical History of South Africa's International River Basins*, Water Research Commission, Pretoria.

UN-DESA (2009) *2009 Revision of World Urbanization Prospects*, UN-DESA, New York.

UN-Habitat (2006) 'Best practises database', www.unhabitat.org/bp/bp.list.details.aspx?bp_id = 1771, accessed 2 August 2012.

UN-Habitat (2010) *State of the World's Cities 2010–2011 Bridging the Urban Divide*, UN-Habitat, Nairobi.

WCD (2000) *Dams and Development: A New Framework for Decision-Making*, World Commission on Dams, Cape Town.

Zeitoun, M., and Mirumachi, N. (2008) 'Transboundary water interaction I: reconsidering conflict and cooperation', *International Environmental Agreements*, vol 8, pp297–316.

8 The Water–Energy Nexus
Meeting Growing Demand in a Resource-Constrained World

Antony Froggatt

The Water–Energy Nexus and Its Main Drivers

The production of energy and delivery of usable water are inextricably linked. Energy is often required to access water and it is essential for treating and transporting it, while most of the main global energy sources are dependent on availability of water for production or cooling. In the coming decades, sustaining the efficient use of shared water and energy resources is likely to be central to peace and prosperity in many regions. Figure 8.1 shows the current per capita consumption of energy and water in countries of the G-20. As can be seen, there are significant differences in per capita energy consumption, for example between the United States and India. If global per capita consumption of energy continues to approach the levels of Europe or the United States—which is an aspiration for those in the developing world—total consumption worldwide will increase and potentially exceed sustainable limits.

Rapid change in the patterns and magnitude of global trade is also accelerating the use of water and energy, as consumption is increasingly distant to sources of production, as recognised through concepts such as footprinting, virtual water, and embedded carbon. Around one-quarter of China's energy consumption and CO_2 emissions are used in the manufacture of goods for export (Carbon Trust, 2011). The energy sector is also a major user of water; therefore, production of exported goods is driving competition with other major water users, such as the agricultural sector. Partly as a result of this competition for water, as well as the growing international trade of soy and growth in demand for animal feed, China's virtual water imports doubled between 2001 and 2007. As the availability of currently exploited domestic resources decreases and the impacts of climate change become more extreme, the significance of trade in products with embedded resources is likely to increase, especially from areas with relatively high water availability, such as Argentina and Brazil (Dalin et al., 2012).

Nowhere is the link between water and energy more apparent than in climate change. The concentration of CO_2 emissions in the atmosphere continues to increase and, according to the World Meteorological Organisation, they had reached 390.9 parts per million (ppm) in 2011 (WMO, 2012). Because CO_2 is the most significant long-lived greenhouse gas, it will

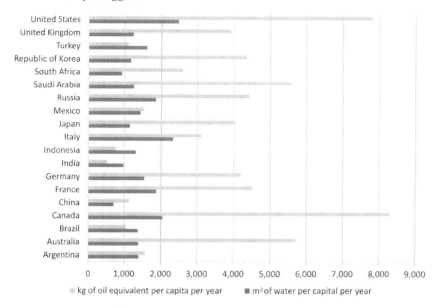

Figure 8.1 Energy and water consumption per capita in countries of the G-20 in 2010
Source: Earthtrends (2012)

be the main driver of climate change. The World Bank assumes that without significant emissions reductions, the world's average temperature could increase by 4° C by 2060. Such an increase has huge potential consequences, especially on precipitation patterns, because in a warmer world generally dry areas will become drier and wet areas wetter (World Bank, 2012), and is, therefore, likely to generate new threats to the infrastructure that underpins traditional energy and water production and delivery systems. The CO_2 from the burning of fossil fuels in the energy sector is responsible for 57% to the total greenhouse gases (UNFCCC, 2009).

This chapter highlights the resource implication of the current consumption levels of both water and energy, and how under current policies these will increase significantly. It further explores the supply implications of this predicted increase in demand and addresses the consequences of greater reliance on unconventional sources. Finally, it argues that a transformation of the sectors is urgently needed, but at the heart of which must be an understanding of the complexities of the energy–water nexus, and therefore policies are needed that simultaneously address their resource security concerns.

Water for Energy

Global energy use and demand projections show rapid growth during the coming decades as population and per capita consumption continue to grow (GEA, 2012; BP, 2012). Further demand will result from the delivery of modern energy services to the 2.7 billion people who still rely on traditional biomass fuels for cooking and heating and the 1.3 billion people who do not have access to electricity (IEA, 2012). Many parts of the world are still undergoing rapid industrialisation, urbanization, and rising consumer demand, which will drive unprecedented power capacity additions and demand for transportation fuel. Based on the assumption that current policies deliver the energy services and savings anticipated, the International Energy Agency (IEA) forecast a 1.3% annual growth rate for energy in the coming decades, leading to a 40% increase in energy consumption by 2035, relative to current levels (IEA, 2012). This increase is equivalent to the total global consumption in 1970. The energy sector is a major contributor to global CO_2 emissions. Developing along a business-as-usual energy use pathway is expected to result in atmospheric CO_2 concentrations leading to global average warming of up to 6°C (IEA, 2009).

Water availability is vital for the energy sector; without it, little energy can be produced or consumed, be it in the extraction of fossil fuels and their processing, water for cooling in power stations, or driving the generators for hydropower. Water is also an important element of new renewable energy sources, such as biofuels and biomass, for the manufacture of equipment and for Concentrated Solar Power (CSPs). The energy sector is therefore a significant user of water; the World Energy Council (WEC) estimated that in 2005, total water use associated with energy production (including biofuels) was in the order of 1,600 km³ (1.6 trillion m³). More than 80% of water use by the energy sector is for the production of traditional biomass, with 'only' 257 km³ associated with commercial energy and electricity production (WEC, 2010). The IEA's 2012 World Energy Outlook (WEO), for the first time, devoted a chapter to water use in the energy sector and estimated that global water withdrawals for energy production in 2010 were 583 billion m³ (not including traditional biomass) (IEA, 2012). However, estimates of water use for energy vary considerably, for example, due to differences in system boundaries, lack of consistent data, and methodologies (Rothausen and Conway, 2011).

The massive recent expansion in production of biofuels has become a major source of demand for water (IEA, 2012; IPIECA, 2012). However, there are large variations in relative consumption volumes of each source, due to differences in climatic conditions, soil types, species, and agricultural systems. The WEC study demonstrates this using data on the water requirements of biofuels across a range of countries, including: the Netherlands, 24.16 m³/GJ; the United States, 58.16 m³/GJ; Brazil, 61.20 m³/GJ; and Zimbabwe 142.62 m³/GJ (WEC, 2010). On average, production of a single litre of biofuel requires about 2,500 litres of crop evapotranspiration (the amount of water taken by the plants from the soil and expelled through transpiration; for more information, see Allen et al., 1998) and 820 litres of irrigation water (de Fraiture et al., 2008). Globally, irrigation water allocated to biofuel production is estimated at 44 km³—roughly 2% of all irrigation water—while the full implementation of current national biofuel

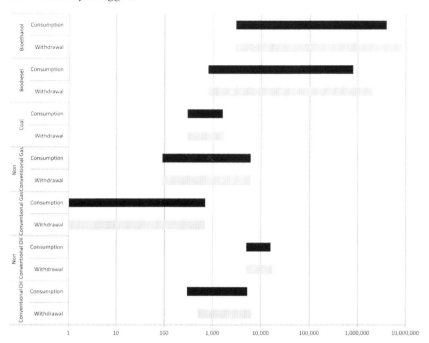

Figure 8.2 Water use of primary energy carriers (litres/tonne of oil equivalent)
Source: IEA (2012)

policies would require 30 million hectares of cropland and 180 km^3 of additional irrigation water (UNESCO, 2010).

Figure 8.2 gives a range of water use for various primary energy producers and shows the extent to which they vary across the sources. This is both a reflection of the water used in their extraction and transformation and the energy intensity of each resource. Oil is the largest contributor to global energy demand and therefore the total volume of water required in oil production is very large, with one analysis suggesting that the total water use is around 166 km^3 per year (WEC, 2010). The high level of water consumption in crude oil production is due to the volume of water being used during extraction, which accounts for around 90% of the average water use in the process as a whole. This aggregate number masks significant local differences and impacts; for example, offshore oil extraction does not face the same water resource issues as many on-land operations. The amount of oil used and its impact on the environment varies significantly depending on geography, geology, recovery techniques, and reservoir depletion, but on average, one barrel of oil requires seven barrels of water (ETIPRG, 2011).

Coal is the most carbon-intensive of all conventional fuels and has the greatest volume of reserves (BP, 2012a). Furthermore, despite recognition of the need to reduce greenhouse gas emissions, the use of coal is increasing faster than any other fuel. Water is required for coal cutting, dust suppression during mining, and coal washing. While coal's water use per unit of energy is lower than oil, the growing importance of coal in the energy sector makes its water use an area of particular global concern (IEA, 2012). Already-limited water availability is affecting production efficiency and future plans, such as in the countries with the largest proposed growth in coal, China and India (Bloomberg, 2012). For example, in China, by far the world's largest coal user and producer, coal mining, processing, combustion, and coal-to-chemical industries are responsible for 22% of the nation's total water consumption, second only to agriculture (State of the Planet, 2011), with suggestions that the total amount of water required is 120 km^3 annually (Circle of Blue, 2011). By 2020, the coal sector will be responsible for 27% of China's total water consumption, with an estimated 34 billion m^3 of water used by coal-fired power plants alone each year. The problem is compounded by the fact that the vast coal reserves are in the arid northern and western provinces, where there is greater competition for water resources (Circle of Blue, 2011).

Attention has been given to large-scale accidents, such as the Deepwater Horizon in 2011, where following an explosion and fire on the Transocean's Deepwater Horizon rig, hydrocarbons escaped from the Macondo well for 87 days (BP, 2010). The U.S. federal government officials estimated that the well ultimately released more than 200 million gallons (or 4.9 million barrels) of crude oil into the ocean (CRS, 2010). However, continual and sometimes routine impacts on water bodies are equally, if not more, damaging. Even in cases of careful environmental management, the extraction of fossil fuels creates liquid and solid wastes that can contaminate nearby water bodies and groundwater. Mining and drilling for fossil fuels also brings to the surface materials including water, and generate large quantities of waste materials or by-products, creating large-scale waste-disposal challenges (Allen et al., 2011).

Water is both consumed in the process of electricity production and it is withdrawn for nonconsumptive use from existing water bodies (such as the sea, rivers, or lakes) for cooling and returned to the water bodies. Not all power systems require similar volumes of water for the same energy outputs. Older electricity-generating plants generally use once-through cooling systems, while more modern plants can also use recirculating air or water systems. The once-through systems have a much higher rate of water withdrawal (around 30 times greater than recirculation systems) but much less total water use (about a tenth) (WEC, 2010). Figure 8.3 highlights the range of water footprints with various electricity-generation sources and technologies. Of note is how small the volume of water is, in relation to the whole system, that is required for the production of the fuel. In addition, large differences exist in the volume of water required between and within the fuel sources and sectors. Coal, in particular, shows large variations, with Integrated Gasification Combined Cycle (IGCC) plants the most efficient. In IGCC power plants, the coal is gasified and then burned in a gas turbine. However, in the more traditional coal stations, the fuel is burned directly

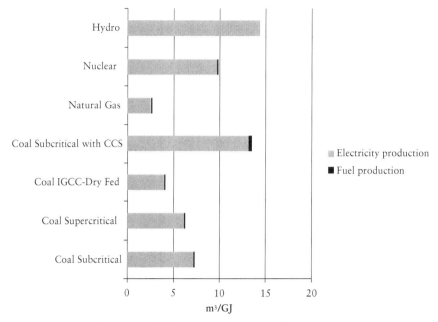

Figure 8.3 Water footprint of electricity generation (m³/GJ)
Source: WEC (2010)

to create the heat to drive the turbines; these are subcritical pulverised coal combustion (PCC) systems. Efficiency improvements have been achieved by operating these at higher temperatures and pressures; these are Supercritical and Ultrasupercritial PCC units (World Coal, 2012). The more modern and efficient power stations improve both energy efficiency and volume of water required.

An important potential factor in the water use of fossil fuel plants may be whether carbon capture and storage (CCS) is developed and implemented. This would significantly increase the volume of water required for electricity generation, not only because the use of CCS requires additional energy and reduces overall efficiency, but also because the process itself requires water. Overall water use in pulverised combustion coal stations could increase by 90% and by 76% in gas stations operating CCS (WEC, 2010).

While no fuel is burned in hydro-power, water is used to drive the turbines, and the need to alter the water course can have a measurable impact on the rate of water evaporation. This is due to the damming of rivers to create reservoirs, with much greater surface areas for evaporation affecting consumptive water use (WEC, 2010).

The IEA's WEO 2012 includes an estimation of water demand according to projections of growth and sources of energy. Figure 8.4 shows growth in the expected water withdrawals and consumption in OECD and non-OECD countries under the IEA's new energy policies scenario. This scenario assumes the effective implementation of existing policy commitments, which

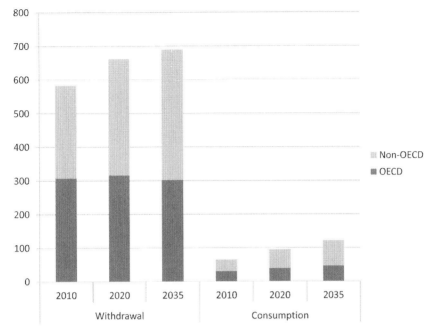

Figure 8.4 Global water use for energy production in the International Energy Agency's New Policy Scenario (billion m^3)

Source: IEA (2012)

is projected to lead to a global increase in water withdrawals of more than 20% compared to 2010, with all of this growth occurring in the non-OECD countries and Asia accounting for about 70%. The IEA also assesses the water requirements for an energy sector scenario in which GHG emissions are reduced in line to ensure that global mean temperatures do not exceed two degrees above preindustrial levels. Under this scenario, there could only be a 4% increase in global in water withdrawals by 2035 associated with the energy sector (IEA, 2012).

Energy for Water

About 7% of commercial energy production is used globally for managing the world's fresh water supply (Brazilian et al., 2011). However, this figure hides large regional variations; for example, in California, water-related use of electricity is about 19% of the total, due to the very high intensity and large area of irrigated agriculture (Krebs, 2007). In Spain, the water sector uses around 6% of the electricity, with 25% of this used for water for food production (De Stefano and Ramon Llamas, 2013). In the UK, the water industry uses about 3% of the total national electricity consumption (Rothausen and Conway, 2011).

Energy use in the water sector can be split into two parts: construction and operation. The energy associated with the sector's infrastructure requirements, dams, piping, and treatment plants is itself significant. However, much larger is the energy associated with operation, where it is used for a variety of processes, including the lifting of groundwater, pumping water through pipes, treatment, and increasingly for desalination. The overall energy use intensity varies considerably between processes, such as pumping and treatment as a result of differences in the energy requirements of the extraction and treatment stages (Hussey, 2010). It the United States alone, the California Energy Commission in 2005 estimated that energy intensity ranges from a lower-bound estimate of 0.61 kWh/m^3 (2,300 kWh/million gallons [MG] in original) to an upper-bound estimate of 9.88 kWh/m^3 (37,400 kWh/MG in original) (Griffiths-Sattenspiel and Wilson, 2009).

As with the energy sector, demand is likely to increase significantly. Based on an average economic growth scenario and assuming no major efficiency gains, the 2030 Water Resources Group (WRG) anticipates that global water requirements will grow from 4,500 km^3 in 2009 to 6,900 km^3 in 2030. The bulk of this increase is anticipated to be from emerging economies, such as India, with a predicted increase by 760 km^3 (double current levels) due to rising populations and changing diets. Meeting this predicted increase will require a move away from business as usual, according to the WRG scenario, as efficiency improvements in supply productivity and the expected increase in supply cannot meet the forecasted demand. Consequently, unless these trends are altered, fossil water reserves would be further depleted, water reserved for environmental needs would be drained, or—more simply—some of the demand would not be met (WRG, 2009).

Internationally, growing energy production and demand is affecting water availability because increasing demand for energy requires more water and further supply enables more exploitation of water resources. However, in some areas, increasing demand and higher standards of water quality, combined with growing energy scarcity, threaten economic growth, such as in China and India. Indeed, China exemplifies many of the challenges facing water and energy management. In 2005, roughly half its cities faced some form of water shortage (World Bank, 2006), and the energy intensity of water provision is also increasing rapidly, particularly as the transportation of water increases in the face of growing demand and physical water scarcity and as energy-intensive water treatment becomes more widely used. Major transfer projects include the South–North transfer, a $62 billion, 50-year project to divert 44.8 km^3 of water per year from the Yangtze River in southern China to the Yellow River Basin in arid northern China (Water-Technology.net, 2012). Kahrl and Roland-Holst (2009) calculated that for China, the average energy intensity of water use increased from 7.94 Watthours (Wh)/m^3 in 2002 to 9.31 Wh/m^3 in 2004. China's water demand is expected to increase from 618 km^3 to 818 billion m^3 in 2030, with just over 50% for agricultural use (almost 25% for rice alone); 32% for industrial use, with a large share required for additional demand from the coal sector; and 18% for domestic use (WRG, 2009).

Land irrigation is a vital part of the global food production system and it provides around 40% of the world's food from 20% of the cultivated land (Turral et al., 2011). The amount and type of irrigation varies considerably from country to country and region to region. However, it has been estimated that in 1995, the energy associated with irrigation of crops was about 1.5 EJ. This relates to about 10% of total energy use in the food sector, with the largest share (31%) attributed to field machinery. For some countries, the figures are significantly higher, such as in the United States, where irrigation accounts for 13% of energy used in agriculture (Pfeiffer, 2004), and in India, where 15 to 20% of total electricity consumption is in the agricultural sector (Hoff, 2011). In China, emissions from groundwater abstraction for irrigation are roughly 0.5% of national total emissions and groundwater use grew from 10 km^3 to over 100 km^3 between the 1950s and 2000s, whilst remaining largely unregulated (Wang et al., 2012).

Due to growing competition for land and rising populations, depleting water resources, demand for more water-intensive food sources (particularly meat), and changing weather patterns, the amount of irrigated land will need to increase significantly (Turral et al., 2011), as will the energy requirement.

The Increasing Role of Unconventional Sources

The growing global demand for both water and energy, coupled with the depletion of existing reserves, is leading to the exploitation and development of different resource types and the utilisation of new technologies. Some of these require previously unexploited or little-used materials, such as rare earth metals, while in other cases, previously unextractable resources are now being utilised. The latter case is the rise in the use of unconventional, often fossil fuel, resources, and their new extraction or processing requirements are usually resource intensive, and therefore new resource constraints and shortages for other sectors are rising.

This drive to the unconventional sources highlights the policy conflicts at the heart of the resource and, in particular, energy demand. While addressing climate change should lead to an increase in low-carbon energy sources, concerns over security of supply and higher fossil fuel prices have also driven the development of unconventional fossil fuels, which often have higher carbon and are usually more water intensive than conventional fossil fuels. This highlights the conflicts that are increasingly prevalent in energy policy between energy security and climate change concerns. At the heart of this is coal, which is abundant and is located in areas that don't always have oil. The transport sector is dominated by oil (92%) and, therefore, access to oil is a key security of supply concern. Converting coal to liquids (CTL) was developed and widely deployed in South Africa, with its large coal reserves, during the Apartheid-era sanctions against the country. Sasol, a South African-based company, is a world leader in CTL, and around 30% of the country's gasoline and diesel needs are produced from indigenous coal. Other countries considering development of CTL include China and India. However, studies have shown that CTL fuels could increase lifecycle greenhouse gas emissions by up to 2.5 times those of conventional diesel, and increase water consumption by an order of 10 (Mielke et al., 2010).

The speed of development of unconventional gas or shale gas has surprised many. In the United States, its contribution to supply rose from 1% to 20% by 2010, while the IEA anticipates that its global contribution to gas supply could increase from 12% in 2008 to 24% in 2035, with other major developments in Canada, China, Russia, and Australia (IEA, 2011). Shale gas exploitation uses a process called *fracking,* which involves the injection of large quantities of water into rock to fracture it and release the encapsulated natural gas. The average water use for fracking is considerably larger than that of conventional gas, although it varies considerably between projects, and it can be larger than coal (IEA, 2012) (see Figure 8.2). Developments in the United States have also highlighted operational and technological problems associated with fracking, including contamination of water bodies. Analysis of the Marcellus and Utica shale formations has shown that in active gas extraction areas, the average methane concentration in drinking-water wells was 19.2 mg/litre, whilst background concentration in neighbouring non-gas-extracting regions of similar geological structure was 1.1 mg/litre (European Parliament, 2011).

Exploitation of Canadian oil sands, an unconventional oil source, has also become a significant contributor to North America's liquid fuel use. Globally, the IEA anticipates that with current policies, the use of unconventional oil will increase 300%, and that by 2035 it could provide about 10% of global supply (IEA, 2012). Water is used in both mining and *in situ* production of oil sands, with processing requiring between 2.5 and 4 times more water than conventional oil (WEC, 2010), although other sources (see Figure 8.2) (IEA, 2012) quote much larger volumes of water.

Similar resource problems occur for water, because growing demand is driving use of unconventional and harder-to-reach sources, some of which require much higher energy intensity, such as desalination. The global contracted capacity of desalination plants was 71.7 million m^3 daily in 2010, including those under construction. This reflects a three-fold increase in cumulative desalinated water capacity over the last decade, and has resulted in more than 15,000 desalination plants globally. Among the top 10 countries by total installed capacity are Saudi Arabia, United Arab Emirates (UAE), Spain, the United States, China, and Algeria, each having installed desalination capacity of more than 2 million cubic meters per day since 2003 (IDA, 2011). The percentage of desalinated water (or recovered wastewater) varies considerably between countries, but can be as high as 55% in Kuwait, 50% in Qatar, and 30% in the UAE. The energy intensity of the desalination processes also varies, but is significant in many Middle Eastern states, accounting for roughly 13% to 22% of total electricity demand in the UAE, 8% to 13% in Qatar, and 5% to 8%

in Kuwait (Siddiqi and Anadon, 2011). The energy efficiency of the desalination process is crucial to the overall costs, as in the United States it is estimated that electricity expenses represent from one-third to one-half of the operating cost of desalination. Currently, the typical energy intensity for seawater desalination is 3 to 7 kilowatt-hours of electricity per cubic meter of water (kWh/m^3) (CRS, 2013).

The most important example is Saudi Arabia, which is producing nearly half of the world's desalinated water—about 24 million m^3 per day. While there are moves towards the use of solar power, at the world's largest plant under construction (at Al-Khafji), the current equipment is extremely fossil-fuel intensive. Of the current desalination capacity, 90% runs on gas or oil, requiring a daily consumption equivalent to about 1.5 million barrels of oil (Lee, 2010). If the full capital, energy, and operating costs were taken into account, the delivered cost of water in Riyadh is likely to be in the region of US$6.00/$m^3$, whereas water from the current system is being sold for around US$0.03/$m^3$ (Gasson, 2011).

Desalination is also being used to meet growing water demand outside the Middle East. In Australia, the use of desalination is expected to increase from 45 Giga litres (GL)/year in 2008 (0.12 million m^3/day) to 450 GL/yr (1.2 million m^3/day) by 2013, providing 0.57% of the total in 2008 and 4.3% in 2013 (Hoang et al., 2009). In the United States, as of 2005, the total capacity of desalination plants was approximately 1,600 million gallons per day (6 million m^3/day), which represents more than 2.4% of total supply (CRS, 2013).

Another form of energy intensive water is that found in bottles, and it therefore could be considered as an 'unconventional source'. More than 200 billion litres of bottled water were sold globally in 2007. The energy use associated with the distribution of a bottle of water is significantly greater than municipally produced tap water. In the United States, the energy input equivalent to between 32 and 54 million barrels of oil were required for the annual consumption of bottled water in 2007. The vast majority of the energy consumed in the bottling of water is associated with the production of polyethylene terephthalate (PET) plastic bottles. However, long-distance travel can be a significant energy user. Overall, energy use for bottled water is between 1,000 and 5,000 times that of piped water per unit of energy (Gleick and Cooley, 2009).

Conclusion

The sustainable use of energy and water are inextricably linked at the local, national, and regional levels. Water is critical for many processes of energy production, and water use can be energy intensive. Climate change is also likely to have far-reaching consequences for the hydrological cycle and water availability. However, different resources, extractions, and processes have different resource-efficiency levels, and so not all energy or water source options have the same impact on the other.

In the coming decades, global energy demand is expected to increase by 40% by 2035 (IEA, 2012) and water demand by 50% by 2030 (WRG, 2009). Meeting these anticipated demand requirements is leading to the exploration for and exploitation of both existing and new resource types. Although these help to reduce the predicted supply gap in one resource, they are often accelerating demand in another. In the extreme cases, some newer energy sources, such as biofuels or oil sands, require large amounts of water, while some new water processes, such as desalination, are very energy intensive. This rise of the unconventional sources can result in a vicious cycle of resource depletion, and further undermines the sustainability of current consumption patterns.

In this context of rapid change and uncertainty, the demand for resources is increasing at a faster rate than at any other time in recorded history and is already causing energy and water security problems. The resultant insecurities of water and energy are already increasing the likelihood of conflicts within and between countries. Water access has been identified as a contributing factor to tribal conflict and violent civil unrest, in particular in India and Pakistan (Pacific Institute, 2011). Access to energy can also be a significant indicator and/or cause of regional insecurity, such as in the Ukraine–Russia gas transit dispute in 2005 and the development of the Iranian nuclear power programme. It is also already possible to point to areas in which conflict may occur in the future as resource demand continues to rise. These possible 'flashpoints' are where conflicts are not inevitable, but in which new disputes may occur and control of water or energy resources are used for political advantage. One example of this is in the Nile Basin, where both Ethiopia and Sudan have expressed a desire to build hydroelectric dams, but Egypt has stated that an interference with the Nile waters would be regarded as an act of aggression (Lee et al., 2012).

Understanding the extent and complexity of the interrelationship between water, energy, and other resources, as well as climate change, is an essential first step to reducing geopolitical risks and creating future resource supply security. Without this, it will be impossible to measure the actual or likely total demand of each resource, the range of technical options available, the right incentives for rapid and transformative change, and the investment priorities needed to engage the private sector in what are often public sector projects.

While many of the linkages between energy and water are clearly demonstrated, experiences have shown that 'energy and water issues are rarely integrated into policy' (Gleick, 2008), be it on the national or international level. On the global level, combination of changing weather patterns and trends in resource production and consumption are moving the water and energy security up their separate domestic and international political agendas. However, all too often, they are being considered in isolation of each other. The 2011 G-20 summit declaration calls for the multilateral development banks to finalise their 'joint action plans on water, food and agriculture', without mentioning the links to energy production (G20, 2012). Therefore international bodies, which were often established to promote or oversee a particular sector or industry, need to expand their remits to address the complexities of multiple resource impacts.

In an increasingly globalised world, there is a need to understand the complexities of production and supply chains. In the framework of the international negotiations around climate change, there have been ongoing discussions on the importance of producer versus consumer responsibility for greenhouse gas emissions. Within the climate negotiations, this issue is far from settled, but the process itself is an important mechanism for recognising the extent of the problem. However, much more needs to be done on understanding the resource and emissions implications of global trade and, in particular, in relation to the embedded content of water and energy in traded goods.

References

Allen, L., Cohen, M., Abelson, D. and Miller, B. (2011) 'Fossil fuels and water quality', in P.H. Gleick (ed) *The World's Water*, vol 7, Pacific Institute 2011.

Allen, R., Pereria, L., Raes, D. and Smith, M. (1998) *Crop Evapotranspiration— Guidelines for Computing Crop Water Requirements—FAO Irrigation and Drainage*, Paper 56, Food and Agriculture Organisation of the United Nations.

Bloomberg (2012) 'China, India lack water for coal plant plans, GE Director says', *Bloomberg*, 8 June.

BP (2010) Deepwater Horizon, Accident Investigation Report, Executive Summary, 8 September.

BP (2012) *BP Energy Outlook 2030*, January.

BP (2012a) *BP Statistical Review of World Energy*, June.

Brazilian, M., Rogner, H., Howells, M., Hermann, S., Artent, D., Gielen, D., Studuto, P., Mueller, A., Komor, P. and Tol, R. (2011) 'Considering the energy, water and food nexus: towards an integrated modelling approach', *Energy Policy*, vol 39, no 12, pp7896–7906.

Carbon Trust (2011) *Global Flows—International Carbon Flows* (CTC795), May.

Circle of Blue (2011) 'Double choke point: demand for energy tests water supply and economic stability in China and the U.S', *Circle of Blue*, 22 June.

CRS (2010) *Deepwater Horizon Oil Spill: The Fate of the Oil*, Congressional Research Service, December.

CRS (2013) *Desalination and Membrane Technologies: Federal Research and Adoption Issues*, Congressional Research Service, January.

Dalin, C., Konar, M., Hanasaki, N., Rinaldo, A. and Rodriguez-Iturbe, I. (2012) 'Evolution of the global virtual water trade network', *Proceedings of the National Academy of Science of the United States*, 23 February.

De Stefano, L. and Ramon Llamas, M. (2013) *Water Agriculture and the Environment in Spain: Can We Square the Circle*, CRC Press/Balkema, Netherlands.

Earthtrends (2012) EarthTrends Environmental Information Data-set, World Resources Institute.

ETIPRC (2010) *Water Consumption of Energy Resource Extraction, Processing and Conversion*, by Erik Mielke, Laura Diaz Anadon, Vekatesh Narayanamurti", Energy Technology Innovation Policy Research Group, Harvard Kennedy School, Belfer Center of Science and International Affairs, October 2010.

European Parliament (2011) *Policy Impacts of Shale Gas and Shale Oil Extraction on the Environment and on Human Health*, Directorate General for Internal Policies Policy Department A: Economic and Scientific, June.

De Fraiture, C., Giordano, M. and Liao, Y. (2008) 'Biofuels and implications for agricultural water use: blue impacts of green energy', *Water Policy*, vol 10, supplement 1, pp67–81.

G20 (2012) 'Building our common future: renewed collective action for the benefit of all', *Cannes Summit Final Declaration*, 4 November.

Gasson, C. (2011) 'Fixing Saudi's subsidised leaks', *Global Water Intelligence*, vol 12, no 12, www.globalwaterintel.com/archive/12/12/analysis/fixing-saudis-subsidised-leaks.html.

GEA (2012) *Global Energy Assessment—Toward a Sustainable Future*, Cambridge University Press, Cambridge and New York and the International Institute for Applied Systems Analysis, Laxenburg, Austria.

Gleick, P. (2008) *Water and Energy (and Climate): Critical Links*, Pacific Institute, Oakland, USA.

Gleick, P. and Cooley, H. (2009) 'Energy implications of bottled water', *Environmental Research Letters*, vol 4, no 014009, pp1–6.

Griffiths-Sattenspiel, B. and Wilson, W. (2009) *The Carbon Footprint of Water*, The River Network, Oregon, USA.

Hoang, M., Bolto, B., Haskard, C., Barron, O., Gray, S. and Leslie, G. (2009) *Desalination in Australia*, CSIRO, National Research Flagship, Water for a Healthy Country, February.

Hoff, H. (2011) 'Understanding the nexus', Background Paper for the *Bonn 2011 Conference: The Water, Energy and Food Security Nexus*, Stockholm Environment Institute, Stockholm.

Hussey, K. (2010) *Interconnecting the water and energy cycles, Identifying and exploiting the synergies*, Vice Chancellor's Representative in Europe, Karen Kussey, Research Fellow, Crawford School of Economics and Government, The Australian National University, 11 May.

IDA (2011) *Desalination Yearbook 2010–11*, International Desalination Association Massachusetts, USA.

IEA (2009) *World Energy Outlook 2009*, International Energy Agency, Paris.

IEA (2011) *Are We Entering a Golden Age of Gas, Special Report of the World Energy Outlook 2011*, International Energy Agency, Paris.

IEA (2012) *World Energy Outlook 2012*, International Energy Agency, Paris.

IPIECCA (2012) *The Biofuel and Water Nexus, Guidance Document for the Oil and Gas Industry*, International Petroleum Industry Environmental Conservation Association, London, UK.

Kahrl, F. and Roland-Holst, D. (2009) 'China's water-energy nexus', *Water Policy*, vol 10, supplement 1, pp1–16.

Krebs, M. (2007) *Water Related Energy Use in California*, Assembly Committee on Water, Parks and Wildlife, California Energy Commission, www.energy.ca.gov/2007publications/CEC-999–2007–008/CEC-999–2007–008.PDF.

Lee B., Preston F., Kooroshy J., Bailey R. and Lahn G., (2012) *Resources Future*, Chatham House, December.

Lee, E. (2010) 'Saudi Arabia and desalination', *Harvard Review*, 23 December.

Mielke E., Anadon, L. and Narayanamurti V. (2010) *Water Consumption of Energy Resource Extraction, Processing and Conversion*, Energy Technology Innovation Policy Research Group, Harvard Kennedy School, Belfer Center of Science and International Affairs, October.

Pacific Institute (2011) *Water Conflict Chronology*, updated September 2011, www.worldwater.org/conflict.html.

Pfeiffer, D.A. (2004) *Eating Fossil Fuels*, From the Wilderness Publications, www.fromthewilderness.com/free/ww3/100303_eating_oil.html.

Rothausen, S. and Conway, D. (2011) 'Greenhouse-gas emissions from energy use in the water sector', *Nature Climate Change*, vol 1, pp210–219.

Siddiqi, A. and Anadon, L., (2011) 'The water–energy nexus in Middle East and North Africa', *Energy Policy*, vol 39, no 8, pp4529–4540.

State of the Planet (2011) *How China is Dealing with Its Water Crisis*, May, http://blogs.ei.columbia.edu/2011/05/05/how-china-is-dealing-with-its-water-crisis.

Turral, H., Burke, J. and Faures, J.-M. (2011) *Climate Change, Water and Food Security*, Food and Agriculture Organisation of the United Nations, Rome, Italy.

UNESCO (2010) *Water and Biofuels*, United Nations Educational, Scientific and Cultural Organisation, www.unesco.org/new/en/natural-sciences/environment/water/wwap/facts-and-figures/all-facts-wwdr3/fact-22-water-biofuels, accessed February 2012.

UNFCCC (2009) *Fact sheet: the need for mitigation*, United Nations Framework Convention on Climate Change, November.

Wang, J., Rothausen, S., Conway, D., Lijuan, Z., Wei, X., Holman, I. and Li., Y. (2012) 'China's water-energy nexus: greenhouse-gas emissions from groundwater use for agriculture', *Environmental Research Letters*, vol 7.

Water-Technology.net (2012) *South-to-North Water Diversion Project, China*, www.water-technology.net/projects/south_north, accessed December 2012.

WEC (2010) *Water for Energy*, World Energy Council.

WMO (2012) *Provisional Statement on the State of Global Climate in 2012*, World Meteorological Organisation, 28 November.

World Bank (2006) *Water Quality Management: Policy and Institutional Considerations*, World Bank, Washington, DC.

World Bank (2012) *Turn Down the Heat. Why a 4°C Warmer World Must Be Avoided*, A Report for the World Bank by the Potsdam Institute for Climate Impact Research and Climate Analytics.

World Coal (2012) *Improving Efficiencies*, www.worldcoal.org/coal-the-environment/coal-use-the-environment/improving-efficiencies, accessed December 2012.

WRG (2009) *Charting Out Water Future, Economic Frameworks to Inform Decision Making, 2030*, Water Resources Group, London, UK.

9 Water Security for Ecosystems, Ecosystems for Water Security

David Tickner and Mike Acreman

Introduction

The recent discourse on water security has emerged against a backdrop of increasing human demand for, and struggle over, scarce water resources across the globe (e.g., UN Development Programme, 2006; World Water Assessment Programme, 2009). The socioeconomic dimensions of water scarcity and water security have been extensively documented (see, for instance, most of the other chapters in this book). In contrast, the consequences of water policy and management decisions for freshwater ecosystems, and the potential knock-on effects on society from changes to those ecosystems, have often either been implicit in academic and policy debates about water security or they have been overlooked. This is despite fact that dams, over-abstraction of water, and other pressures have led to high-profile failures of freshwater ecosystems, such as the Murray-Darling River and the Aral Sea, with very substantial socioeconomic and political impacts. In fact, rivers and wetlands are now among the most threatened of all ecosystems globally (Finlayson et al., 2005; Vörösmarty et al., 2010; WWF, 2012). More broadly, the magnified political and media profile of environmental sustainability issues, dating back to the 1992 UN Conference on Environment and Development, suggests that sustainable management of natural resources requires specific attention to environmental, as well as socioeconomic, concerns.

Reasons for the lack of attention to freshwater ecosystems in the water security discourse include the historical gap in concepts and language, and perceived gaps in priorities, between those primarily concerned with conserving river, lake, and wetland ecosystems and those whose job it has been to develop and manage water resources for human use. There are signs that these hitherto distinct specialisms are now converging, at least in some key aspects. Recognising the need to optimize multiple uses of freshwater ecosystems in a world of more than 7 billion people, conservation NGOs such as WWF and The Nature Conservancy have begun to engage constructively in policy debates and basin-scale strategic planning. Similarly, many in the water resource management sector have reflected on lessons from decades of reliance on supply-side solutions, expensive built infrastructure, and end-of-pipe technical fixes and have become open to a wider range of possible interventions, including those that maintain or restore key ecosystem functions or embrace natural infrastructure—such as wetlands—as an effective and sustainable solution to many water management challenges. Importantly, concepts and tools that highlight the value—in dollar terms or otherwise—of natural capital and ecosystem services (de Groot, 1992; Barbier, 2009) have gained traction with both groups. This chapter describes the evolution of approaches to both freshwater conservation and water resource management and points to evidence that new approaches, which place the maintenance of strategically important ecosystem services at the heart of water policy, offer significant scope for the pursuit of a range of complementary social, environmental, and economic outcomes. A specific focus is on natural infrastructure and environmental flows as tools for securing long-term benefits for society and for ecosystems.

The next section summarizes the evolution of freshwater ecosystem conservation, especially the development of ecosystem service approaches. A brief perspective is also given of the historical development of water resource management, particularly as it has affected ecosystems. Then there is a short exploration of the growing consensus around the importance of environmental flows as a means of safeguarding natural capital, ensuring the continuation of ecosystem services and providing long-term water security at the basin scale. The chapter concludes with reflections on the challenges of, and opportunities for, implementing environmental flows as a foundation for basin-scale water security.

A Short History of Freshwater Ecosystem Conservation

From Birds to Basins

The genesis of freshwater conservation lay in the desire of interest groups within society to protect individual aquatic species or specific wetland habitats because of their intrinsic value. In the UK, the establishment of the Royal Society for the Protection of Birds in 1889 stemmed from concern over decline in avian populations, including those of wetland species, such as grebes and egrets (RSPB, 2009). Similarly, the foundations of what would become The Nature Conservancy were laid in the 1920s and 1930s by scientists who wanted to catalogue and 'save' the United States' remaining wilderness areas (The Nature Conservancy, undated). This focus on species and habitats remains a powerful motivation for conservation efforts in developed countries and emerging economies alike.

Towards the end of the 20th century, the language and aims of the growing environmental movement subtly shifted. There was a realisation that, for reasons of ethics and effectiveness, conservation efforts would seldom bear fruit if they pitted the needs of wildlife against those of society and an acceptance that many cherished landscapes—including some wetlands—were, in part, the result of historical human intervention. Conservationists also began to understand that decisions affecting habitats and species were often taken by politicians and captains of industry for whom biodiversity was a luxury, at best, and a hindrance, at worst. It became clear that environmental organisations had relatively little political and economic influence in such arenas. Thus, in order to achieve their goals, conservationists started to work in partnership with other interest groups to develop ideas and language that were relevant to broader political debates and that resonated with those closer to power. The concept of sustainable development was set out by the influential Brundtland Report (World Commission on Environment and Development, 1987) and in 1992 the UN Conference on Environment and Development in Rio de Janeiro had the effect of magnifying the political profile of interlinked environmental, social, and economic sustainability issues. The apparent acceptance by political and economic decision makers of the main principles of sustainable development represented a change for the conservation movement in that it held the promise, so far largely unfulfilled, of a level playing field in which the short-term needs of people would be better balanced with long-term care of ecosystems.

Freshwater conservationists were early adopters of the principles of sustainable development. The Ramsar Convention, signed by representatives from 18 countries in 1971 (and now with more than 160 contracting parties) included as a central concept the 'wise use' of wetlands, defined now as the 'maintenance of [wetlands'] ecological character, achieved through the implementation of ecosystem approaches, within the context of sustainable development' (Finlayson et al., 2011; Finlayson, 2012). Thus, wise use explicitly allowed for exploitation of wetland resources within sustainable limits. Approaches such as Integrated River Basin Management (IRBM) were also developed by conservationists in the 1980s and 1990s, based on the Ecosystem Approach (Maltby et al., 1999). The principles underpinning IRBM had much in common with those on which Integrated Water Resource Management was based (IWRM, see below), but included additional emphasis on important ecological processes that underpinned delivery of ecosystem services.

Recent Developments: Natural Infrastructure and Ecosystem Services

The central tenets of both the wise use concept and the IRBM approach included an acceptance that freshwater ecosystems could be valued not only for their biodiversity but also for the benefits they could bring to sections of society. Ecosystem services—including the supply of clean water, flood attenuation, groundwater recharge, and freshwater fisheries—stem from the processes by which the natural environment produces resources (sometimes called *natural capital* or *natural infrastructure*) that provide benefits to humans (Barbier, 2009; Fischer et al., 2009). The state of, and threats to, ecosystem services globally were synthesized in the Millennium Ecosystem Assessment (Hassan et al., 2005). This authoritative assessment showed that future provision of freshwater ecosystem services was profoundly threat-ened by human impacts on natural infrastructure, such as rivers, lakes, wet-lands, and aquifers. Specific threats varied from place to place, but globally the major impacts were due to:

- Habitat fragmentation and loss of ecosystem connectivity following the construction of dams for water storage, flood management, and hydro-power generation (Nilsson et al., 2005)
- Over-abstraction of water from rivers, lakes, and aquifers, mostly for agricultural use (Comprehensive Assessment of Water Management in Agriculture, 2007)
- Rapid industrial and agricultural development which, combined with frequent use of rivers as sewers of first resort, has led to huge pollution loads in large systems such as the Yangtze (Müller et al., 2008)

Tools for analysing, quantifying, and valuing ecosystem services had been developed by the academic community in previous decades (Maltby and Acreman, 2011), and the new millennium saw an increase in the pro-file of these approaches among applied researchers and policymakers. For instance, the UN Environment Programme commissioned a landmark report on The Economics of Ecosystems and Biodiversity (TEEB, 2010). Building on this, and based on the model of the Intergovernmental Panel on Climate Change, an Intergovernmental Science-Policy Platform on Biodiversity and Ecosystem Services (IPBES) was established in 2012 as a mechanism to build capacity for using ecosystem science in decision mak-ing. These tools showed more clearly than before that healthy freshwater ecosystems can provide human communities with important elements of economic security (e.g., natural resources such as water, fish, medicines, timber), social security (e.g., protection from natural hazards, such as floods), and ethical security (e.g., upholding the rights of people and other species to water) (Acreman, 2001). This, in turn, demonstrated that allo-cations of water for 'the environment' were not necessarily at the expense of human security (Acreman, 1998). Data about ecosystem services could therefore enhance decisions about tradeoffs in water management. In the Mekong Basin, for example, analysis has highlighted the need for consid-eration of risk to freshwater fisheries and associated food security for tens of millions of people at the downstream end of the system in any decisions about hydropower development (Ziv et al., 2012). In the United States, Europe, and China, assessments have demonstrated that maintenance or restoration of floodplains can provide substantial ecosystem services and may be economically attractive alternatives to costly engineered flood con-trol infrastructure (Opperman et al., 2009; Ebert et al., 2009).

Although frameworks continue to be refined, a typology of freshwater ecosystem services has become well established (Table 9.1), and significant policy decisions by governments have begun to acknowledge the need to optimize continued delivery of these services where they are strategically

Table 9.1 A typology of freshwater ecosystem services

Ecosystem service types	Examples	Permanent/ temporary rivers and streams	Permanent lakes, reservoirs	Seasonal lakes, marshes, and swamps, including floodplains
Provisioning services				
Food	Production of fish, wild game, fruits, grains, etc.	xxx	xxx	xxx
Fresh water	Storage and retention of water; provision of water for irrigation and for drinking	xxx	xxx	xx
Fibre and fuel	Production of timber, fuelwood, peat, fodder, aggregates	xx	xx	x
Regulating services				
Climate regulation	Regulation of greenhouse gases, temperature, precipitation, and other climatic processes; chemical composition of the atmosphere	x	xxx	x
Hydrological regimes	Groundwater recharge and discharge; storage of water for agriculture or industry	xxx	xxx	xx
Pollution control and detoxification	Retention, recovery, and removal of excess nutrients and pollutants	xxx	xx	x
Erosion protection	Retention of soils and prevention of structural changes	xx	x	x
Natural hazards	Flood control, storm protection	xx	xxx	xxx
Cultural services				
Spiritual and inspirational	Personal feelings and well-being; religious significance	xxx	xxx	xx
Recreational	Opportunities for tourism and recreational activities	xxx	xxx	xx
Supporting services				
Biodiversity	Habitats for resident or transient species	xxx	xxx	xx
Soil formation	Sediment retention and accumulation of organic matter	xxx	x	xx
Nutrient cycling	Storage, recycling, processing, and acquisition of nutrients	xxx	xxx	xxx

Note: Relative magnitude of selected services per unit area for selected ecosystem types, from low (x) to medium (xx) to high (xxx) (adapted from Finlayson et al., 2005).

important. For instance, the Australian federal government has developed a Murray-Darling Basin Plan that aims to 'save the river system' on the basis that it is 'the largest environmental asset on the continent and [Australia's] most important production asset' (Cullen, 2012). Even more recently, conservationists have articulated ecosystem service concepts through tools that aid understanding of water footprints and water-related business risks. Such approaches have resonated strongly with powerful economic audiences, such as multinational food and beverage companies that depend on reliable clean water supplies (Chapagain and Tickner, 2012).

Water Resource Management and Ecosystems

Technical Water Resource Planning: The Exploitation Paradigm

Pegram et al. (2013) identified two major phases in the history of river basin planning: technical water resources planning, and, more recently, 'strategic' approaches to river basin management (see Figure 9.1).

Technical water resources planning had its roots in the earliest human attempts to manage the hydrological cycle. In places such as ancient China and Mesopotamia, local water management schemes were normally developed for a specific purpose, such as irrigation. Although engineering technology advanced considerably, this small-scale, single-issue approach was dominant until the 1800s, when the first attempts at basin-scale planning were undertaken in the United States and Europe. The Tennessee Valley Authority, established by the U.S. Congress in 1933, and perhaps the first true basin-wide authority, focused on the building and operation of water and energy supply facilities. It inspired similar institutional development elsewhere and helped to trigger a massive expansion of water infrastructure construction during the middle of the 20th century. Throughout this period, the emphasis of water resource management was on harnessing and

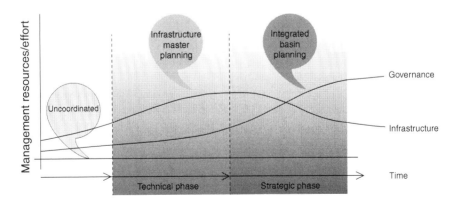

Figure 9.1 The two phases of river basin planning
Source: Pegram et al. (2013)

exploiting nature through engineering (an emphasis that remains in many parts of the world). Consideration of the impacts on ecosystems and biodiversity, or of the knock-on effects on downstream sections of society, was largely absent. When the human population was relatively low, this caused little concern, and it has been argued that the environmental costs were often outweighed by the social and economic benefits of water infrastructure development (Grey and Sadoff, 2007).

Toward the end of the 20th century, concern grew about anthropogenic impacts on river ecosystems and potential knock-on effects on society, including the impacts on human health from poor water quality. Public debate about river and coastal pollution in the United States and Europe gave rise to, respectively, the Clean Water Act in 1972 and the EU Urban Wastewater Treatment Directive in 1991. These instruments led to the investment of billions of dollars in improved wastewater treatment infrastructure on both sides of the Atlantic Ocean. Another shift in policy emerged from mounting awareness of the substantial negative impacts that dams could have on freshwater biodiversity and on people, either directly through displacement, or indirectly as a consequence of effects on ecosystem services to communities, such as downstream river flows, sediment replenishment, and fisheries. The World Commission on Dams report (World Commission on Dams, 2000) concluded that, although large dams had made an important and significant contribution to human development by providing stable water resources and flood alleviation, the social and environmental costs had, in too many cases, been unacceptable and often unnecessary. It estimated that between 40 and 80 million people around the world had been displaced as a result of the construction of large dams. Moreover, the livelihoods of at least 472 million people are thought to have been adversely impacted by dams (Richter et al., 2011). Largely because of these impacts, and after vociferous campaigns against specific initiatives, such as the Narmada River project, spearheaded by high-profile figures such as Arundhati Roy (Roy, 1999), many decision makers and investors became nervous of further involvement in dam building. Following from the World Commission on Dams report, the International Hydropower Association produced sustainability guidelines (IHA, 2004) to promote greater consideration of environmental, social, and economic aspects in the scoping of new hydropower projects and the management and operation of existing hydropower schemes. The guidelines provided criteria for assessing energy options (including energy needs against supply-side and demand-side efficiency measures), resource depletion, poverty reduction, carbon intensity and greenhouse gas emissions, impacts on ecosystems, and waste products. The subsequent Hydropower Sustainability Assessment Protocol (IHA, 2010) was an attempt to provide an enhanced sustainability assessment tool to measure and guide performance in the hydropower sector, to provide more consistency in the approach to assessment of hydropower project sustainability. The protocol was produced during a process that involved representatives of organisations from a diversity of sectors, with differing views and policies on sustainability issues related to hydropower development and operation. The 14 forum members included representatives of governments of developed and developing countries, commercial and development banks, social and environmental NGOs, and the hydropower sector.

Strategic Basin Planning: IWRM and Beyond

The Integrated Water Resources Management (IWRM) paradigm emerged toward the end of the 20th century, partly as a response to such controversies. The key principles of IWRM were captured in the report of the 1992 Dublin Conference (International Conference on Water and the Environment, 1992) and political momentum for the new approach was reflected in the inclusion of a call for all countries to develop national-level IWRM plans in the World Summit on Sustainable Development Plan of Implementation in 2002. The principles clearly acknowledged the need for a holistic approach to water management 'for the development of human societies and economies, and for the protection of natural ecosystems on which the survival of humanity ultimately depends'. More recently, a critique of IWRM has emerged in light of increasing experience of implementing national policies for water management, which were developed in the wake of the Dublin Conference. This critique was summarized by Pegram et al. (2013) and includes:

- The need for water management and river basin planning to be based more on local experience and needs than on a universal set of principles
- The fact that IWRM principles were largely developed in the European context where institutional and infrastructure development was already advanced, and that this may not be the case in emerging economies where the 21st-century challenge of balancing social, economic, and environmental water needs is greatest
- The recent emergence of a more strategic approach that moves beyond the view that natural ecosystems should be 'protected' by compliance with minimal environmental safeguards and instead emphasizes the key ecosystem functions and services that are required to attain priority human development goals

Underpinning the new strategic approaches to river basin planning set out by Pegram et al. (2013) is 'the greater recognition and understanding [by water policy makers and water resource managers] of the reliance of human social and economic systems on the goods and services provided by natural and ecological water systems'. This twin focus on achieving development goals and on viewing natural infrastructure and ecosystem services as tools for achieving these goals is a marked departure from the emphasis on harnessing and controlling nature, which was central to the Tennessee Valley Authority and other 20th-century engineering-led initiatives.

Environmental Flows as Tools for Water Security

The discussion above suggests that, as a result of hard experience, paradigms have begun to shift in both conservation and water management communities of practice. These movements seem to have been based on an increased acceptance by both communities that functioning ecosystems can be tools for social and economic development. The extent to which this apparently convergent evolution will become the basis for sustainable approaches to water resource management policy and practice remains uncertain and will vary from place to place, depending on cultural, political, and socioeconomic circumstances. The specific set of demands on local water resource managers will be critical, be they related to pollution control and water quality, energy generation and infrastructure development, food security and irrigation, or other issues. However, one particular challenge is likely to be important in many parts of the world as policymakers and water resource managers strive to bring about basin-scale water security: that of maintaining sufficient flows of water through river systems in order to sustain a range of ecosystem services that in turn contribute to the attainment of strategic social and economic development priorities (see Figure 9.2).

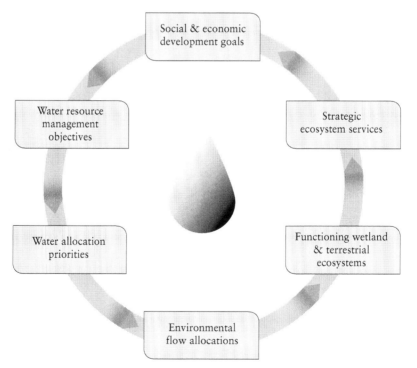

Figure 9.2 The link between development goals, water resource management, environmental flows, and ecosystem services

River Flows as a Test of Basin-Scale Water Security

There is much discussion in the academic and grey literature (including in this book) about the meaning of water security. Regardless of formal definitions, it is clear that 'security' will look and feel different for each water user, depending on context. In order to help decision makers assess whether policy decisions have led to improvements in water security for multiple users in a given river basin, a suite of indicators will be necessary, including a combination of social, economic, and environmental metrics. Among these measures, the state of the basin's hydrological resources is key. For surface water systems, the flow of water through river, lake, or wetland systems is arguably the critical determinant of ecosystem services and the single most important hydrological indicator (aquifer levels are the equivalent for groundwater systems). For many water resource managers, the continued flow of water to large cities, coastal communities, power stations, or farmers at the downstream end of a river system will be essential. Equally, ecosystem services such as sediment transport or freshwater fisheries—both highly dependent on river flows—can be strategically important for broader development objectives relating to land management and food security. From a freshwater conservation perspective, researchers have cited the 'environmental flows' of water through a river system as the master variable influencing ecosystem health (Poff et al., 1997). Thus, the maintenance of defined quantities of river flow over a given period (or defined aquifer levels) can be regarded as the hydrological litmus test of basin-scale water security from both socioeconomic and environmental perspectives.

Environmental flows are often defined in terms of the quantity, timing, and elements of the quality of water flows and levels in a river system required to sustain freshwater and estuarine ecosystems and the human livelihoods and well-being that depend on these ecosystems (see the Brisbane Declaration, 2007, for a widely accepted definition). The basic premise that underpins any environmental flow assessment is that, while it is important to maintain defined river flows over a period of time, in many rivers some water can be used by people without causing unacceptable harm to ecosystems and the services they provide. The science of environmental flows has progressed significantly in the past 20 years, with rapid methodological development (Acreman et al., in press). More than 200 different methods have been developed for assessing environmental flows, varying from quick modelling approaches that require little work beyond obtaining and processing existing data, to intensive efforts that demand several years of dedicated fieldwork (O'Keeffe and Le Quesne, 2009). Each has advantages and disadvantages, and each is more useful in some situations than in others.

A common feature of many environmental flow assessment methods is the use of a combination of hydrological, ecological, and socioeconomic criteria to determine the most appropriate flows over a period of time for a given river system (or stretch of a river system). Thus, environmental flow assessments commonly require a combination of technical assessments and dialogue with local water users, riparian communities, and decision makers. The emphasis is on facilitating stakeholder decisions about appropriate tradeoffs between rival allocations of water, with input about potential ecological implications provided by scientists. The point of such processes is to ensure stakeholder buy-in to decisions about water allocations for environmental flows. Conclusions about the optimal flow for any given river stretch at any given point in time therefore depend significantly on what society wants from the river, whether it is in a protected national park, a strategically important irrigation district, or an urban area (or a combination of these). An illustration of this can be found in an assessment of the environmental flow requirements of the Upper Ganga River in India, which took account of the cultural and spiritual importance of the Ganga to the Hindu faith (O'Keeffe et al., 2012). This was reflected practically in the expectation of the local community in Bithoor, just upstream of Kanpur, that the monsoon season river flow should be sufficient to inundate the temple and, in so doing, wash the feet of Brahma at least once a year. One notable exception to this multidisciplinary approach to setting objectives can be found in the EU, where the Water Framework Directive dictates a scientifically defined primary target for all water bodies, Good Ecological Status, with stakeholder participation focusing on measures to achieve this (Acreman and Ferguson, 2010).

The Implementation Challenge

Failure to safeguard environmental flows in key rivers has already had significant socioeconomic and political, as well as environmental, repercussions. As discussed, there has been a high-profile debate in Australia about the Murray-Darling River, and various state and federal government initiatives have been instigated, often accompanied by very substantial investment of public funding for institutional and infrastructural measures. A system of tradeable permits also allows former users to sell their water holdings, which provides flexibility in the system and, in theory at least, enables economically efficient allocation of water. In China, steady increases in water use since 1949 led to reduced flow in the Yellow River and to the drying up of the main stem of the river for long periods, with consequent impairment to transport of the river's high sediment load to the sea. This situation stimulated the Chinese authorities to invest substantially in new monitoring infrastructure and inter-provincial water allocation and regulation schemes, so that a minimum annual volume of 21 billion m^3, out of the average annual runoff of 58 billion m^3, could be set aside in order to maintain sediment transport functions (Shen and Speed, 2009). Challenges remain in terms of implementation, but the relative success of this scheme still stands as the world's largest reallocation of water for the maintenance of ecosystem services (Le Quesne et al, 2010).

China and Australia have by no means been the only countries to introduce policy mechanisms in order to maintain or restore environmental flows. The first in a new wave of national water legislation that recognized the importance of flows was enacted in Mexico in 1992 as a consequence of rapid agricultural development, growing pressure on surface water resources, and overexploitation of groundwater. The Mexican Water Law established a clear institutional framework with a central National Commission on Water (Comisíon Nacional del Agua, or CONAGUA) to oversee its implementation. Subsequently, CONAGUA worked to develop a *norma,* or technical regulatory standard, for setting environmental flows for the country's water bodies. This *norma* was agreed on in 2012 and provided a legal basis for ensuring that environmental flows are included in basin-water calculations and water resource availability assessments across the country (Le Quesne et al., 2010; Eugenio Barrios Ordonez, WWF Mexico, personal communication). Elsewhere, the South African National Water Act was passed in 1998. The act recognized the requirement for sustainable utilisation of the water resource through the definition of an environmental reserve (i.e., the quantity and quality of water required to protect aquatic ecosystems). In theory, once it was agreed to by the Minister, this environmental reserve, along with a reserve for basic human needs, was binding on any institution involved in water resource management (Le Quesne et al., 2007). National laws, policies, and standards with similar stipulations on environmental flows as a priority for water use followed in places as diverse as Tanzania, Pakistan, Australia, Costa Rica, Brazil, the EU, and various U.S. states. Environmental flows have also become mainstreamed into the policies of major institutions, including the World Bank (Hirji and Davis, 2009).

As with the Yellow and Murray-Darling rivers, these legislative and policy requirements for the maintenance of environmental reserves or environmental flows were predicated largely on the need to ensure water security for downstream users and to ensure that ecosystem services were secured in the long term. In this sense, they reflected convergence in conservation and water management approaches. Despite this, implementation of the laws and policies has been slow. Le Quesne et al. (2010), building on the work of Moore (2004), surveyed the implementation status in a number of countries and documented challenges perceived to be most pertinent to the actual establishment and maintenance of environmental flows (see Figure 9.3). The most significant obstacles related to:

- Lack of political will and stakeholder support for setting strategic direction, securing planning resources, and enforcing implementation
- Insufficient resources and capacity, especially for water management institutions charged with enforcing water allocations
- Presence of institutional barriers and conflicts of interest, for example, between agencies that plan and manage hydropower, agriculture, land use, urban development, industrial planning, and natural resources

Fundamentally, the challenge of securing environmental flows for basin-scale water security is a challenge of water allocation. Given the increasing pressures on water resources in many parts of the world from growing populations, economic development, and the uncertainties caused by climate change, there will be no universal silver bullet for this. Solutions must also be

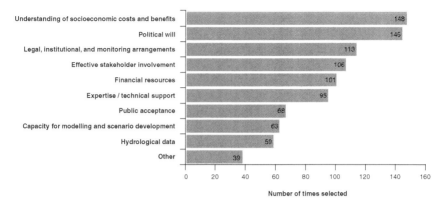

Figure 9.3 Difficulties and obstacles to understanding and implementing environmental flows

Source: Moore (2004) and Le Quesne et al. (2010)

carefully tailored to local social, economic, and environmental contexts. Nevertheless, Speed et al. (2013) drew on a global review of lessons from strategic water allocation policy and practice to derive a set 'golden rules' that could guide water managers in their efforts (see Table 9.2).

One potential reason for optimism is the increasing attention being paid to global water challenges, often from surprising sources. The involvement of significant business sectors in water policy debates has been discussed earlier in this chapter. Water security is also now an issue for the defence and foreign policy communities in some countries (e.g., Osikena and Tickner, 2010; Intelligence Community Assessment, 2012). This new attention may mean that political stakes in the water debate will rise. It may make life more complicated for freshwater conservationists and water managers, bringing even more perspectives and institutional priorities to the table. However, the increased profile of water security issues among politically and economically powerful audiences may also present opportunities. Conservationists and water managers will need to understand broader social and economic goals to which sustainable, efficient, and equitable water management (incorporating good understanding of natural infrastructure and ecosystem services) can contribute. The involvement of new, and significant, political players also brings potential for extra support to 'soft path' solutions (Gleick, 2002) that address long-standing obstacles, including the need for stronger, better-resourced water management institutions that can define locally appropriate environmental flows, and implement water allocation decisions accordingly. As water security dialogues become more complex and more contested, policy entrepreneurs (Huitema and Meijerink, 2010) can play a key role in convening stakeholders and facilitating policy change. Scientists will also continue to be important in ensuring that good information is available so that planning and management decisions are well informed, and they will need to forge new partnerships with stakeholders in order to support the

Table 9.2 Ten 'golden rules' for strategic water allocation

1	In basins where water is becoming stressed, it is important to **link allocation planning to broader social, environmental, and economic development planning.** Where interbasin transfers are proposed, allocation planning also needs to link to plans related to that development.
2	Successful basin allocation processes depend on the existence of **adequate institutional capacity.**
3	The degree of complexity in an allocation plan should **reflect the complexity and challenges in the basin.**
4	Considerable care is required in defining the amount of water available for allocation. **Once water has been (over) allocated, it is economically, financially, socially, and politically difficult to reduce allocations.**
5	**Environmental water needs provide a foundation** on which basin allocation planning should be built.
6	The water needs of certain **priority purposes should be met before water is allocated among other users.** This can include social, environmental, and strategic priorities.
7	In stressed basins, **water efficiency assessments and objectives should be developed within or alongside the allocation plan.** In water-scarce situations, allocations should be based on an understanding of the relative efficiency of different water users.
8	Allocation plans need to have a **clear and equitable approach for addressing variability between years and seasons.**
9	Allocation plans need to incorporate **flexibility in recognition of uncertainty over the medium to long term** in respect of changing climate and economic and social circumstances.
10	**A clear process is required for converting regional water shares into local and individual water entitlements,** and for clearly defining annual allocations.

Source: Speed et al. (2013)

Note: Text in bold is the authors' emphasis.

development of shared goals (Poff et al., 2003). Increased involvement of a wider set of stakeholders will make it more important than ever that common concepts and a common lexicon are established.

Conclusions

There is increasing recognition that identifying and maintaining strategically important ecosystem services is an integral element of 21st-century water management. The flow of a river regulates functions of, and services from, freshwater ecosystems. It is, therefore, the key hydrological variable—the prime natural capital—of which water managers should take account as they seek to establish water security for multiple users at the basin scale. Although other pressures take their ecological toll, river flow is also often the most important factor determining the state of freshwater biodiversity. Defining environmental flows, and allocating water to maintain defined flows, should therefore be a point of cohesion between freshwater conservationists and water resource managers.

Implementing environmental flows has proven to be a challenge in most parts of the world, largely because of significant increases in demand for water allocations for human uses, especially irrigated agriculture. Decision makers face a tough future as they make tradeoffs between upstream allocations and downstream ecosystem services. One particular challenge relates to an absence of political will in some countries for making such difficult decisions. Even with greater political buy-in, tradeoffs in water allocation will be necessary. For the foreseeable future, human water use will need to increase in some river basins. In others, especially closed basins in which over-allocation of water is an impediment to long-term water security for multiple stakeholders, steps to increase allocation of water to maintain critical environmental flows will become ever more urgent. It would be an especially fervent conservationist who believed that water security for ecosystems and biodiversity should be the priority everywhere, at all costs. Equally, only the most regressive engineer would believe that ecosystems are simply a luxury that must be sacrificed for human water security. Our ability to define and implement locally appropriate environmental flows, which safeguard ecosystem services, contribute to development goals, and support freshwater biodiversity, will be the litmus test of whether we can achieve sustainable water security at the basin scale.

References

Acreman, M.C. (1998) 'Principles of water management for people and the environment', in A. de Shirbinin and V. Dompka (eds) *Water and Population Dynamics*, American Association for the Advancement of Science, Washington DC.

Acreman, M.C. (2001) 'Ethical aspects of water and ecosystems', *Water Policy Journal*, vol 3, no 3, pp257–265.

Acreman, M.C. and Ferguson, A. (2010) 'Environmental flows and the European Water Framework Directive', *Freshwater Biology*, vol 55, pp32–48.

Acreman, M.C., Overton, I., King, J., Wood, P., Cowx, I., Dunbar, M.J., Kendy, E. and Young, W. (in press) 'The changing role of science in environmental flows', *Hydrological Sciences Journal*.

Barbier, E.B. (2009) 'Ecosystems as natural assets', *Foundations and Trends in Microeconomics*, vol 4, no 8, pp611–681.

Brisbane Declaration (2007) *Environmental Flows Are Essential for Freshwater Ecosystem Health and Human Well-Being*, 10th International Riversymposium and International Environmental Flows Conference, www.eflownet.org/download_documents/brisbane-declaration-english.pdf.

Chapagain, A.K. and Tickner, D. (2012) 'Water footprint: help or hindrance', *Water Alternatives*, vol5, no 3, pp563–581.

Comprehensive Assessment of Water Management in Agriculture (2007) *Water for Food, Water for Life: A Comprehensive Assessment of Water Management in Agriculture*, Earthscan, London and International Water Management Institute, Colombo.

Cullen, S. (2012) 'Burke unveils final Murray-Darling plan', *ABC News*, www. abc.net.au/news/2012-11-22/burke-unveils-final-murray-darling-plan/4386298, accessed 29 November 2012.

Ebert, S., Hulea, O. and Strobel, D., (2009) 'Floodplain restoration along the Lower Danube: a climate change adaptation case study', *Climate and Development*, vol 1, no 3, pp212-219.

Finlayson, C.M. (2012) 'Forty years of wetland conservation and wise use', *Aquatic Conservation: Marine and Freshwater Ecosystems*, vol 22, pp139-143.

Finlayson, C.M., Davidson, N., Pritchard, D. Milton, G.R. and Mackay, H. (2011) 'The Ramsar Convention and ecosystem-based approaches to the wise use and sustainable development of wetlands', *Journal of International Wildlife Law and Policy*, vol 14, pp176-198.

Finlayson, C.M., D'Cruz, R. and Davidson, N.C. (2005) *Ecosystems and Human Well-Being: Wetlands and Water Synthesis*, World Resources Institute, Washington, DC.

Fischer, B., Turner, R.K. and Morling, P. (2009) 'Defining and classifying ecosystem services for decision making', *Ecological Economics*, vol 68, pp643-653.

Gleick, P.H. (2002). 'Water management: soft water paths' *Nature*, vol 418, no 6896, pp373-373.

Grey, D. and Sadoff, C.W. (2007) 'Sink or swim: water security for growth and development', *Water Policy*, vol 9, no 6, pp545-571.

de Groot, R.S. (1992) *Functions of Nature*, Wolters-Noordhoff, Groningen, The Netherlands.

Hassan, R., Scholes, R. and Ash, N. (eds) (2005) *Ecosystems and human well-being: current state and trends, Findings of the Condition and Trends Working Group*, Millennium Ecosystem Assessment, Island Press, Washington, Covelo, and London.

Hirji, R. and Davis, R. (2009) *Environmental Flows in Water Resources Policies, Plans, and Projects Findings and Recommendations*, World Bank, Washington, DC.

Huitema, D. and S. Meijerink (2010) 'Realizing water transitions: the role of policy entrepreneurs in water policy change', *Ecology and Society*, vol 15, no 2, article 26.

Intelligence Community Assessment (2012) *Global Water Security*, Intelligence Community report for the U.S. State Department.

International Conference on Water and the Environment (1992) *International Conference on Water and the Environment: Development Issues for the 21st Century*, Report of the Conference, Dublin.

International Hydropower Association (IHA) (2004) *Sustainability guidelines*, International Hydropower Association, London.

International Hydropwer Association (IHA) (2010) *Hydropower Sustainability Assessment Protocol*, International Hydropower Association, London.

Le Quesne, T., Kendy, E. and Weston, D. (2010) *The Implementation Challenge: Taking Stock of Government Policies to Protect and Restore Environmental Flows*, WWF-UK, Godalming, UK.

Le Quesne, T., Pegram, G. and von der Heyden, C. (2007) *Allocating Scarce Water: A Primer on Water Allocation, Water Rights and Water Markets*, WWF-UK, Godalming, UK.

Maltby, E. and Acreman, M.C. (2011) 'Ecosystem services of wetlands: pathfinder for a new paradigm', *Hydrological Sciences Journal*, vol 56, no 8, pp1-19.

Maltby, E., Holdgate, M., Acreman, M.C. and Weir, A. (eds) (1999) *Ecosystem Management; Questions for Science and Society*, Sibthorp Trust, Nantwich, UK.

Moore, M. (2004) 'Perceptions and interpretations of environmental flows and implications for future water resource management: a survey study', Master's degree thesis, Department of Water and Environmental Studies, Linköping University, Sweden.

Müller B., Berg M., Yao, Z.P., Zhang, X.F., Wang, D. and Pfluger, A. (2008) 'How polluted is the Yangtze River? Water quality downstream from the Three Gorges Dam', *Science of the Total Environment*, vol 402, pp232–247.

The Nature Conservancy (undated) 'Our history: history and milestones of The Nature Conservancy', www.nature.org/about-us/vision-mission/history/index. htm, accessed 21 November 2012.

Nilsson C., Reidy C., Dynesius, M. and Revenga C. (2005) 'Fragmentation and flow regulation of the world's large river systems' *Science*, vol 308, pp405–408.

O'Keeffe, J. and Le Quesne, T, (2009) *Keeping Rivers Alive: A Primer on Environmental Flows and Their Assessment*, WWF Water Security Series No. 2, WWF-UK, Godalming, UK.

O'Keeffe, J., Kaushal, N., Smakhtin, V. and Bharati, L. (2012) *Assessment o f Environmental Flows for the Upper Ganga Basin*, WWF India, New Delhi.

Opperman, J.J., Galloway, G.E., Fargione, J., Mount, J.F., Richter, B.D. and Secchi, S. (2009) 'Sustainable floodplains through large-scale reconnection to rivers', *Science*, vol 326, no 5959, pp1487–1488.

Osikena, J. and Tickner, D. (2010) *Tackling the World Water Crisis: Reshaping the Future of Foreign Policy*, Foreign Policy Centre, London.

Pegram, G., Le Quesne, T., Yuanyuan, L., Speed, R. and Li, J. (2013) *River Basin Planning: Principles, Procedures and Methods for Strategic River Basin Planning*, UNESCO, Paris and WWF-UK, Godalming, UK.

Poff, N.L., Allan, J.D., Bain, M.B., Karr, J.R., Prestegaard, K.L., Richter, B.D., Sparks, R.E. and Stromberg, J.C. (1997) 'The natural flow regime: A paradigm for river conservation and restoration', *Bioscience*, vol 47, no 11, pp769–784.

Poff, N.L., Allan, J.D., Palmer, M.A., Hart, D.D., Richter, B.D., Arthington, A.H., Rogers, K.H., Meyer, J.L. and Stanford, J.A. (2003) 'River flows and water wars: emerging science for environmental decision making', *Frontiers in Ecology and the Environment*, vol 1, no 6, pp298–306.

Richter, B.D., Postel, S., Revenga, C., Scudder, T., Lehner, B., Churchill, A. and Chow, M. (2011) 'Lost in development's shadow: The downstream human consequences of dams', *Water Alternatives*, vol 3, no 2, pp14–42.

Roy, A. (1999) *The Cost of Living*, Flamingo, London.

RSPB (2009) 'History of the RSPB', www.rspb.org.uk/about/history, accessed 21 November 2012.

Shen, D. and Speed, R. (2009) 'Water resources allocation in the People's Republic of China', *Water Resources Development*, vol 25, no 2, pp209–225.

Speed, R., Le Quesne, T., Pegram, G., Li, Y. and Zhiwei, Z. (2013) *Basin Water Allocation: Principles, Procedures, and Methods for Strategic Water Allocation*, UNESCO, Paris and WWF-UK, Godalming, UK.

TEEB (2010) *The Economics of Ecosystems and Biodiversity: Mainstreaming the Economics of Nature: A Synthesis of the Approach, Conclusions and Recommendations of TEEB*, Nagoya, Japan.

UN Development Programme (2006) *UN Human Development Report: Beyond Scarcity: Power, Poverty and the Global Water Crisis*, Palgrave MacMillan, Basingstoke and New York.

Vörösmarty, C.J., McIntyre, P.B., Gessner, M.O., Dudgeon, D., Prusevich, A., Green, P., Glidden, S., Bunn, S.E., Sullivan, C.A., Reidy Liermann, C. and Davies, P.M. (2010) 'Global threats to human water security and river biodiversity', *Nature*, vol 467, pp555–561.

World Commission on Dams (2000) *Dams and Development: A New Framework for Decision-Making*, Report of the World Commission on Dams, Earthscan, London.

World Commission on Environment and Development (1987) *Our Common Future*, Oxford University Press, London.

World Water Assessment Programme (2009) *The United Nations World Water Development Report 3: Water in a Changing World*, UNESCO, Paris and Earthscan, London.

WWF (2012) *Living Planet Report: Biodiversity, Biocapacity and Better Choices*, WWF International, Gland, Switzerland; Zoological Society of London, London; Global Footprint Network, Oakland, California; European Space Agency, Paris.

Ziv, G., Baran, E., Rodríguez-Iturbe, I. and Levin, S.A. (2012) 'Trading-off fish biodiversity, food security, and hydropower in the Mekong River Basin', *Proceedings of the National Academy of Sciences*, vol 109, no 15, pp5609–5609.

10 From Water Productivity to Water Security

A Paradigm Shift?

Floriane Clement

A Paradigm Shift?

The concept of 'water security' has gained considerable prominence in international and national policy arenas since the early 2000s. That popularity was notably exemplified by the Ministerial Declaration of The Hague on Water Security in the 21st Century in 2000. The declaration, adopted by around 120 ministers of water, stated that the international community had one common goal: to 'provide water security' (World Water Forum, 2000). This was reaffirmed by the Food and Agricultural Organization for the United Nations (FAO) in 2002, who declared that 'water security is the main goal inspiring the international community's emerging agenda for the 21st century' (FAO Legal Office, 2002).

Water security is often associated to food security debates. This is perhaps not surprising, since water is a critical input to agricultural output, and since the agricultural sector accounts for some 70% of global water use. The connection between water and food security is commonly based on the following narrative: Growing populations require more food, the water needed to produce that food is already scarce, and urgent action is needed to protect water resources if we are to avoid both food and water crises (Falkenmark and Lundqvist, 1998; CAWMA, 2007).

The food and water crises rationale is shared by another influential concept in the water–food nexus debates, that of water productivity. Water productivity has also found a central place in the international agenda, as exemplified by this quote from Kofi Annan, former Secretary-General of the United Nations: 'We need a "Blue Revolution" in agriculture, focused on increasing productivity per unit of water, or "more crop per drop"' (Annan, 2000), and in the World Water Day 2007 report 'Increasing water productivity holds the key to future water scarcity challenges' (UN-Water and FAO, 2007, p21).

Both water productivity and water security have been heralded as paradigms to guide policies and interventions addressing global and local water and food issues. Whereas the concept of water productivity was particularly influential in the 1990s to early 2000s, it has in recent years lost ground to water security discourses. Because discourses constitute the world as much as they express multiple visions, ideologies, and interests, this study explores whether this discursive change from productivity to security reflects a paradigm shift, an evolution in thinking that will affect agricultural water policies and interventions across the globe, or no more than a repackaging of similar arguments under a different label.

At first glance, water security discourses seem to have opened to new narrative themes and enlarged the room for maneuver for institutional change. Yet a text-based analysis of the discourses of water productivity and water security reveals that the latter mark an expansion of the water productivity debates but not a real move towards integrating power and political issues in water management. Food-water debates have remained largely apolitical and the causes and solutions to enhance water security are still largely framed in economic and techno-managerial terms. Furthermore, water security discourses envision a very limited role for local people and communities to play in decision-making processes. Findings suggest that overcoming the existing limitations of the water productivity paradigm will require a more substantial discursive change through, for instance, an integration of political chains of explanation into mainstream discourses.

Definitions

A brief overview of the definition and meanings of *water productivity* and *water security* offers a preliminary insight of how both concepts have produced misunderstandings through 'discursive closure' (Hajer, 1995). *Discursive closure* can be described as the process of reducing complexity and diversity into a simplified representation of the reality without acknowledging the simplification. Knowledge simplification is a natural, and often necessary, process. What is problematic is when simplified knowledge is promoted as unquestionable and universally true.

Water productivity evolved from a measure of how efficiently a system converts water into goods and services (Molden, 1997), to one of the necessary steps to reach the overall goals of food security and poverty alleviation. Water productivity is part of a 'web' of productivity of resources, including agricultural productivity, land/soil productivity, energy productivity, labor and capital productivity, and so forth. Its trajectory is associated with that of productivity, which was introduced as an economic ratio measuring the efficiency of production to a normative concept: Every agricultural and industrial system should be productive. Similarly, water productivity has been heralded as a necessary and indisputable objective to overcome water and food crises. It has become a paradigm for water management, with a particular set of assumptions, research questions, and methodologies, which has rallied a specific community of practices.

Discursive closure has been induced in several ways. First, water productivity is presented as a single methodical measure of efficiency, but there are five possible choices for the denominator: the volume of water used, diverted, beneficially and non-beneficially consumed, beneficially consumed, and net irrigation requirements (Playan and Mateos, 2006). The numerator has also been expressed in two ways:[1] (1) in the 'physical water productivity', it is the weight/volume of agricultural or livestock outputs produced; (2) in the 'economic water productivity', it represents the economic value derived from the outputs. What water productivity means also depends on the nature of the processes (physiological, biophysical, managerial, economic) considered and the scale of analysis (plant, field, farm, or basin) (Molden et al., 2003). Which ratio, process, and scale are considered holds implications for the type of interventions recommended to increase water productivity. Yet policy debates commonly advocate to 'increase water productivity' as if there was one single ratio.

Second, the subjective nature of water productivity has not been acknowledged. Yet the perceived value of the outputs varies depending on cultural and social determinants. For instance, crops might be esteemed not only for their economic or nutritional value but also for their taste or their ritual value (e.g., taro in the northern Salomon Islands; cf. Connell, 1978). Such values might differ not only from one social group or one community but also even from one individual to another. This leaves to the question of who is to define the value of outputs and what is considered a desirable level of productivity. There is also substantial disagreement on how much *more* productive a system *can* be (Molden et al., 2010).

Lastly, water productivity discourses represent inputs and outputs as resources naturally flowing into and outside of the system, but have hardly considered issues of access and control over the resources or their benefits. This supposed neutrality neatly conceals the politics of water management.

These issues do not invalidate water productivity as a concept. What is problematic is that its subjective nature is hardly acknowledged and discussed (with a few recent exceptions: Boelens and Vos, 2012; Lankford, 2012) and the central role of scientists in the production of knowledge is rarely questioned.

Water Security

Multiple narratives have contributed to the rise of the concept of water security. Initially rooted in military, strategic, and geopolitical concerns related to water as a source of international conflict, the securitization of water is part of a larger trend of securitization of the environment which has narrowed public policy debates (Brock, 1997).

Water security has emerged as part of a 'web' of securities (Zeitoun, 2011), long after food security was introduced at the World Food Conference and institutionalized by the creation of the World Food Security Committee in 1974.

Forms of discursive closure include the following. First, despite its increasing prominence, water security has rarely been defined, either in academic and nonacademic publications. In some instances, 'water security' is used as a rallying flag in the publication title, but is not or hardly mentioned in the text (especially in the media, but also in Falkenmark and Lundqvist,

1998; Brichieri-Colombi, 2004; Perry, 2010). More often, water security is affirmed as an indisputable goal to achieve, but what it entails is not explicit (e.g., in Clarke, 1991).

Brauch (2007) notes that 'security' holds an objective sense (a measure of the absence of a threat), a subjective one (the absence of fear for a potential threat), and an intersubjective meaning, where security is a social construction of the reality. Definitions of water security use objective terms, varying from notions of water quantity, water availability, water access, or a combination of availability and access (see Appendix).

To these objective terms are often attached more subjective determinants, such as 'sufficient' or 'acceptable'. Obviously, what is sufficient or acceptable in one place or for one community might not be so in another setting. Yet water security discourses rarely debate how these terms are defined and, as importantly, who should define them. For Mekonnen (2010), what is problematic is that water security is a normative concept, setting a desirable goal, for example, 'the availability of an acceptable quality and quantity of water . . . and acceptable level of water-related risks' (2010, p438), without considering if it is achievable. Fixing an unattainable goal as a non-negotiable policy objective is likely to lead to a political impasse (Mekonnen, 2010).

Discursive closure is a common trait of environmental debates. The next section sheds further light on the nature of whose voices and views are allowed to participate and influence water and food debates through a more in-depth text analysis.

Discourses and Narratives

Methodology

Water issues, as environmental issues, are inherently discursive, characterized by 'a complex and continuous struggle over the definition of the meaning of the environmental problem itself' (Hajer, 1995). Discourses are not only an expression of how social actors envision the world, but also significantly contribute to shape and constitute the world (Fairclough and Vodak, 1997). A discourse is defined as 'an institutionalised way of talking that regulates and reinforces action and thereby exerts power' (Link, 1983, p60, in Wodak and Meyer, 2010). Of particular interest for this analysis are the narratives, a particular form of discourse that can be defined as a story line describing an issue or event according to a certain logical form of explanation. Narratives are essential political instruments. On the one hand, they reduce the fragmentation of discourses by offering a common storyline through which different actors' position themselves, and, on the other hand, they largely contribute to discursive closure by offering simplified forms of explanation of complex problems.

Narratives are typically formed of a logical series of explanatory arguments. The following narrative components were selected in this chapter the rationale, or why the considered issue is important; the identified causes of the problems; and the potential solutions. The role assigned to different groups of actors in deciding on the design and implementation of solutions, and the scale at which problems are considered and solutions implemented, were also examined. The rationale and causes of problems inform the analyst on the current and ideal dominant vision of the world, and are the elements of the narrative that most significantly contribute to discursive closure, by tacitly dictating the required direction of institutional change. The types of solutions proposed, together with the scale of action and role of actors, directly reflect the interests and ideologies of actors and define, more or less explicitly, whose voices are legitimate in policy and development processes.

The analysis of the narratives associated with water productivity and water security was primarily based on a review of academic papers, but also included reports from international intergovernmental organizations, newspaper articles, and so forth. Academic papers were searched in the following databases: Water Resources Abstracts, EconLit, Sociological Abstracts, Science Direct, and CAB Direct. A qualitative detailed analysis was conducted on a sample of 34 articles for water productivity and 34 articles for water security and a quantitative one on a larger sample: 59 articles and 51 articles for water productivity and water security, respectively. Purely technical papers on water productivity (e.g., calculation of the ratio for different crops) and water security (e.g., on desalination) were not included, and only water security papers related with food security were considered.

Comparing Narratives

Table 10.1 offers a comparative overview of the key elements of the narratives defined above. In order to acknowledge discursive diversity, while at the same time distinguishing the most representative views associated with each concept, only the views most commonly found in the literature reviewed appear in the table. The argumentation following on the water security and water productivity narratives discusses these dominant views, but readers should keep in mind that discourses are not homogenous and counternarratives exist. As Cook and Bakker observed (2012), discourses on water security vary considerably across academic disciplines.

Rationale

In water-productivity debates, water and food scarcity is seen as escalating due to population and economic growth (Table 10.1). Because the agricultural sector is the largest water consumer and other sectors' water needs are increasing, the resulting rationale is to improve water productivity: grow more food with less water, so that more water will be available for other natural and human uses (Molden et al., 2003). In this view, increasing water productivity is often indisputably presented as the best solution to achieve food security and even alleviate poverty: 'For many regions of the world,

Table 10.1 Elements of dominant narratives on water security and water productivity

	Water productivity	*Water security*
Rationale	*Population and economic growth require using water more efficiently.* If we can grow more food with less water, more water will be available for other natural and human uses. *Water productivity increase is key to achieve food security, prevent environmental degradation, reduce poverty, and foster economic growth.*	The social, economic, and environmental future depends on how *efficiently* and equitably is water used. *Water security is key to achieve food security*, social and political stability, *halt environmental degradation, reduce poverty, and foster economic growth.*
Causes of the problem when specified (from most to least recurrent)	1. *Poor management of water and land* 2. Poor infrastructures and institutions, *lack of knowledge* on how to increase water productivity, and lack of incentives to do so 3. *Low water prices, climatic conditions, and climate change*	1. *Population and economic growth* 2. *Climatic and biophysical conditions, climate change, poor policies* 3. *Poor management of water, low water prices, lack of knowledge* on biophysical processes 4. Unsustainable consumption patterns
Solutions proposed to address the problem (from most to least recurrent)	1. *Techno-managerial and economic options for demand management* 2. *Better science:* evaluation tools, analytical methods, understanding of biophysical processes 3. Institutional change: participation of users in water management, *IWRM*	1. Increase water productivity and other *techno-managerial and economic options for supply management* (infrastructures, allocation mechanisms) 2. Improved policies (coordination, *IWRM*) and planning (mapping, assessing water resources and demand) 3. Dialogue 4. Adaptation, *additional knowledge*, and decision-making tools
Role of actors involved (from most to least recurrent)	1. Researchers: *provide technical solutions*, develop data set and institutional networks	1. National policymakers: coordinate intersectoral policies, allocate water and arbitrate conflicts, design 'good' water policies, provide strong institutions

(Continued)

Table 10.1 (Continued)

	Water productivity	Water security
	2. Farmers: adopt good practices, new technologies, and seed varieties with the support of government and research organisations; *participate in water management*	2. Researchers: provide tools for decision-making, provide and disseminate knowledge, *design technical solutions*
		3. Donors: provide investment capacity
	3. National policymakers: create a favourable institutional and economic environment for farmers to adopt technologies and involve users in water management	4. Communities: *have a say in water management*
Scale of issue	Primarily plant, farm, and basin level	Primarily global, regional, and national level
Scale of action	Primarily plant, *farm/field level* (adopt adequate technologies and practices), irrigation system level (deliver and distribute water), and basin level (IWRM, allocate water among sectors)	Primarily national and *farm level*

In italics: common characteristics of narratives between water security and productivity.

increasing water productivity ... holds the greatest potential to improve food security and reduce poverty' (Giordano et al., 2006, p4); or it is seen as the only solution to save the world from a water crisis: 'In most countries suffering water shortages, at the heart is the question of whether a water crisis can be adverted or whether water can be made more productive' (Hamdy et al., 2003). Water is primarily represented as a physical and economic resource that should be managed efficiently. Such problem framing has implicitly given a prominent place to engineering, natural sciences, and economics to not only address agricultural water issues, but also prevent environmental degradation, foster economic growth, and reduce poverty.

Water security is also presented as a guiding concept to achieve multiple objectives, but aims at broader societal goals than water productivity, among which are ensuring social and political stability. The water crisis and food gap is a recurrent theme in water security debates, as well. However, the latter present water crisis not only as a food security problem but also as a national and international security issue, as suggested by the frequent use of alarmist expressions, for instance: 'All these issues [water security problems], which call for urgent attention are generating intense discussion at present' (Xia et al., 2006), or 'this article is not intended to be alarmist but its message is urgent' (Tandon, 2007).

The urgency of action is further reinforced by the high frequency of terms such as 'conflict', 'threat', and 'crisis' (see Table 10.2). Such terms are also present in food security debates, as indicated in this statement from Jacques Diouf, then-president of the United Nations Food and Agricultural Organization (FAO), in an international meeting held in Rome in July 2011: 'If we

Table 10.2 Number of occurrences of keywords in water security and productivity literatures

Theme	Key term	Number of occurrences of key term in the literature on	
		Water productivity	*Water security*
Crisis narrative	Crisis	33 *(17)*	150 *(34)*
	Threat	35 *(20)*	213 *(37)*
	Conflict	74 *(21)*	**546** *(35)*
	Stress	186 *(42)*	273 *(39)*
	Scarcity	210 *(45)*	**618** *(41)*
Guiding principles	Equit*	292 *(26)*	219 *(30)*
	Efficien*	**1189** *(55)*	579 *(38)*
Context: Politics and culture	Power*	31 *(20)*	200 *(26)*
	Politic*	56 *(21)*	**919** *(39)*
	Cultur*	10 *(6)*	68 *(17)*
Scale	Global	370 *(42)*	**1165** *(44)*
	National	81 *(16)*	409 *(27)*
	Household	278 *(31)*	280 *(27)*
	Field/farm	**1576** *(55)*	213 *(19)*
Institutions	Access	250 *(34)*	**655** *(42)*
	Capabilit*	2 *(2)*	39 *(13)*
	Rights	105 *(28)*	**330** *(29)*
	Governance	20 *(10)*	**430** *(30)*
	Participat*	91 *(26)*	217 *(31)*
	Voice	4 *(5)*	29 *(13)*
Related concerns	Food	**889** *(54)*	**2066** *(46)*
	Health	120 *(25)*	346 *(40)*
	Energy	173 *(32)*	**519** *(37)*

number of occurrences of the key term in the publications on water productivity and water security reviewed by the author. In italics font and parenthesis: number of publications in which the key term was found. In bold font: number of occurrences higher than the sum of the average number of occurrences and the standard deviation (indicating very intensive use of the key term)

Source: For water security: 50 journal papers, book chapters, and reports, total of 577,539 words; for water productivity: 59 journal papers, book chapters, and reports, total of 506,109 words. Non-relevant terms were removed (e.g., coefficient for 'efficien*' and words related with energy for 'power')

*means the suffix was removed for the search

want to avoid future famine and food insecurity crises in the region, countries and the international community urgently need to bolster the agricultural sector and accelerate investments in rural development' (FAO, 2011). As illustrated by Diouf's quote the securitization of water discourses often has an instrumental purpose: propelling water and food to the top of international policy and aid agendas (Brock, 1997; Allouche, 2011). It has, for instance, contributed to justify the huge investment of development banks to build large-scale water storage structures in sub-Saharan Africa as a guarantor of water security (Grey and Sadoff, 2007; Merrey, 2009).

Causes of Low Water Productivity and Water Insecurity

Low water productivity is usually described as resulting primarily from farmers' inadequate water and land management practices, poor irrigation infrastructure, and the low performance of the irrigation water bureaucracy (Table 10.1). Biophysical conditions, lack of economic incentives, and a lack of knowledge among policymakers and farmers are some of the causes identified to increase crop water productivity. Institutions are perceived as inadequate when they constrain the adoption of appropriate technologies (Hamdy et al., 2003). All causal factors are apolitical and acultural, which is further suggested by Table 10.2 (see 'Context'). Problem framing indicates a clear direction for institutional change and policy interventions: Support production and dissemination of science and provide adequate economic incentives for farmers and bureaucrats to adopt a 'rational' behavior.

Many of the roots of water insecurity are shared with the water productivity narratives, for example, population and economic growth, climatic conditions, poor economic policies, poor management of land and water, lack of knowledge, and biased ways of thinking in the policy-making arena. Yet a marked difference is that 'politics' are clearly a dominant theme in water security discourses (Table 10.2). Furthermore, the broader range of causal factors leaves more room for maneuvering, in terms of institutional and political change.

Solutions to Water Productivity and Security Challenges

Apolitical problems call for apolitical solutions. In dominant discourses, the solutions needed to increase water productivity are presented as clearly identified and incontestable, ranging from improved management practices, use of appropriate and new technologies, rational utilization of water, change of cropping patterns, engineering programs for improved infrastructures, and so forth. Institutional change is in most cases represented as a rational process such as 'modernization' and 'applying improved administrative principles' (Kijne, 2003) administered by an invisible hand, free of political, social, or cultural constraints.

Appropriate policies and planning and dialogue among water users occupy a central place to achieve water security. Yet, despite the resurgence of the 'politics' theme in water security discourses, the view that simple technical and economic fixes (e.g., water storage) can solve complex water issues still dominates water security for food security debates (see Table 10.1, and Cook and Bakker, 2012). This trust in techno–managerial solutions is also noticeable in the prominence of the term 'efficiency/efficient' in the literature on water security reviewed (Table 10.2). In the food–water debates, increasing water productivity is indeed at the top of the basket of good actions to enhance water security recommended by intergovernmental organizations such as the FAO, the Global Water Partnership (GWP), and the World Economic Forum Water Initiative (WEFWI) (GWP, 2000; FAO Media Centre, 2011; WEFWI, 2011). The way issues and their solutions are framed in agricultural water security discourses indicates that the latter have subsumed water productivity debates but have not marked a real shift towards power-centered approaches.

Scale and Interlinkages with Other State/Livelihood Concerns

Water productivity aims at responding to global water scarcity, but the scale of action is circumscribed to well-delineated system boundaries, often the field or farm, sometimes the irrigation system or the river basin. Scale has been an issue hotly debated by water productivity scholars: The water considered 'lost' at one scale can be 'recovered' at another (Bluemling et al., 2007; Molden et al., 2010). The appropriate scale of analysis and intervention is framed by purely scientific considerations. The neglect of nonhydrological scales might lead to discarding a range of political, social, and economic interventions (Brichieri-Colombi, 2004) and might reduce the categories of actors perceived as legitimate to take part in decision-making processes, as preference might be given to those who hold scientific knowledge.

On the contrary, the rationale for water security is articulated primarily at national and global scales (Tables 10.1 and 10.2). The national scale is a favored scale of action (Tables 10.1 and 10.2) for planning, coordination among water sectors, and dialogue among water users. The global scale reaffirms the significance of water security to humanity's future but is not fully utilized for a critical analysis of the global factors for water (in)security. For instance, only a few authors (Tandon, 2007; Zeitoun, 2011) question the consumption patterns of developed countries and their responsibility for water insecurity in other parts of the globe. Mostly, global security is thought of as 'transboundary', and refers to issues of water sharing and water conflicts among states.

The difference of scale between water productivity and water security discourses suggests a substantial change in thinking, indicating as well a significant evolution in the role of different groups of actors, such as farmers, scientists, and national policymakers. The increasing consideration of water security at the household level (e.g., Calow, 2010; Asian Development Bank, 2011; Chenoweth et al., Chapter 19) signals an evolution of the concept.

Actors' Roles

In the water productivity literature, farmers are identified as the active implementers of practices and technologies, but largely according to the guidance and specifications of recognized knowledge holders, such as scientists or extensionists. Farmers are the primary responsible for increasing water productivity, but scientists hold knowledge—and power (Lankford, 2012). The state is to support science by creating a supportive economic and institutional environment for farmers to adopt practices and technologies and bureaucrats to work diligently and efficiently. Although these are absent from academic and international discourses, multinational agrochemical companies have also played a central role by supplying the inputs necessary to increase productivity. Private interests have usually colluded with those of intergovernmental organizations, funding agencies, and state governments to design national programs for the 'modernization' of agricultural systems and increase of agricultural productivity (Keeley and Scoones, 2003).

Water security discourses advocate for a greater involvement of multiple stakeholders in decision making, as testified by a strong connection between water security and 'good governance' (Table 10.2; Sojamo et al., 2012). The state is to hold a primary role in the water security literature—probably because it has usually been considered as the responsibility of the state to ensure the security of its citizens. Researchers are to support policymakers to analyze the complex reality and make the right decisions. As discussed in the rationale subsection, water security represents a strong justification for donors to provide a package of funds, technology, and policy assistance. Given the dominance of techno-managerial solutions in water–food security discourses, we can expect that the collusion of multiple interests to modernize agriculture and dominate food and virtual water trades still guarantees a high role to multinational agribusiness players, as argued by Sojamo et al. (2012). Individuals and communities are, as a whole, thought of as either the passive victims of insecurity or the recipients of security. A prominent argument is that they need to be protected and provided services meeting their needs. To conclude, the role of several actors has been transformed to some extent with an increased role of the state, shifting from a knowledge intermediary (between scientists and farmers) to a planner and leader in charge of coordinating water uses and facilitating dialogue among water users.

New Narratives, Better World?

Both water productivity and water security have the attributes of the buzzwords commonly found in development discourses: reconciling different interests under one broad goal, which one cannot contest without being suspected of being anti-moral (Cornwall, 2010). The problem is not as much that the necessity to be productive is not questioned, but rather that there is hardly any discussion on the tradeoffs between productivity and other potentially desirable objectives. Many canal systems, 'upgraded' to concrete and metal structures in the name of higher productivity, have overall not met their objectives, because interventions have overlooked issues of equity, access, resilience, and vulnerability (Lankford, 2009; Ostrom et al., 2011). Furthermore, empirical evidence suggests that farmers might have other priorities than productivity and that highly productive systems can be unsustainable and detrimental to livelihoods (Boelens and Vos, 2012). There are numerous examples of communities showing a higher preference to social acceptability and sustainability than to efficiency, an argument particularly developed by anthropologists (e.g., Mosse, 1997; Cleaver, 2000).

It is worth examining whether the narrative of security opens discursive and political space conducive to improvement of well-being. Well-being is understood here according to the capability approach as the 'freedom people have to achieve the various "beings" and "doings" they have reason to value' (Sen, 1999). We consider, on the one hand, the capability to access capitals and, on the other hand, the capability to influence decision making.

Distribution, Access, and Equity

A famous example of the productivity and efficiency paradigm applied to farming systems is the green revolution. The impact of the green revolution on poverty and well-being is the topic of polarized debates outside of the scope of this chapter, but two major insights are of direct relevance here. First, most studies agree that the distribution of the potential benefits from productivity gains has been unequal among different social groups (Beck, 1995; Das, 2002). Scholars have indeed evidenced potential conflicts between water productivity gains and equity. For instance, encouraging higher water productivity through the use of high-value crops can undermine access to water for other uses, notably women's water access for domestic use (Ahlers and Zwarteveen, 2009). Calder et al. also found that, to achieve higher productivity, watershed management programs in India have often favored the replacement of grazing land (often labeled 'wasteland') by privately owned tree plantations. This has been detrimental to the poorest section of communities, who most used grazing land, whereas the tree plantations primarily benefited the local elite or better-off in the communities (Calder et al., 2008). Increase in productivity does not always aggravate inequity, but as mentioned earlier, it is important to acknowledge at the onset of any intervention the different capabilities of farmers to increase water productivity and their different capabilities to benefit from any technology (e.g., for risks associated with sprinklers, see Lankford, 2009).

The discourse analysis outlined in this chapter suggests that in food–water debates, the techno-managerial theme that has prevailed in water productivity narratives still dominates in water security narratives, with 'productivity' and 'efficiency' as a chief guiding principle (Table 10.2). Definitions of water security have considered access and distributive issues (Zeitoun, 2011; cf. also Table 10.2), which is likely to support a move beyond the 'food gap' narrative. Yet the notion of equity still receives little attention (Table 10.2).

Discourses of water security tend to apply to populations rather than to individuals and tend to project a gross representation of 'citizens' as aggregated actors who pursue a common goal and have common perceptions and interests. Notably, discourses of water security have remained relatively gender blind (with a few exceptions: Brauch 2007; Tandon, 2007).

Participation and Voice in Decision Making

A second lesson from studies on the impacts of the green revolution on poverty is that the distribution of benefits depends on the 'room for maneuver' that farmers hold to influence social, economic, and political structures (cf. Franks and Cleaver, 2007). Water security discourses venture into some of the politics of access, use, and allocation of water resources. They have opened to new narrative themes, which present water issues as political, social, and economic problems, and of water as a resource interconnected with other resources (food, energy; see Table 10.2). However, dominant water–food discourses remain largely apolitical (see also Allouche, 2011). New narratives have opened up political space with a greater involvement of the state, but give a limited role to play to local people apart from a form of participation that is organized and controlled by the state. As for water productivity, they are asked to participate but are not really given a 'voice' (Table 10.2).

Conclusion

As scholars, development practitioners, and policymakers, we can use and advocate particular concepts because they are popular in our epistemic community, or we can choose to critically examine those concepts, and to some extent reappropriate and reframe them. This chapter proposed a reflection on the recent prominence of the water security concept in food security debates and its gradual yet growing dominance over the concept of water productivity, based on discourse analysis.

The concept of water security offers new forms of narratives which can overcome some of the limitations of the concept of water productivity. Those narratives have, for instance, integrated the notion of access, which was absent from water productivity discourses. They have also enlarged the scale of action from the field to the national and global level, thus opening more space for political change and institutional pluralism. Yet issues related with power distribution, such as equity, gender, and justice, have received limited attention in the water security debates. When apolitically linked with the food security narrative, water security remains largely guided by the pursuit of productivity. Therefore, this discourse analysis indicates an enlargement of debates but not a paradigm shift. The way agricultural water management issues are envisioned and conceptualized has not radically changed and has remained largely framed in apolitical and techno-managerial terms.

To move to more political debates, one could propose to refine/replace water security with more politically engaged terms such as *water justice* and *water capability*. However, there is a high probability that these terms progressively lose their political content, to become buzzwords. Cornwall and Brock (2005) rather propose to use chains of equivalence that link buzzwords with more political terms. A greater entry and attention to anthropological studies of water security in international debates could help building such chains and emphasize the political, cultural, and symbolic aspects of water, moving beyond the consideration of water as a resource to the perception of water as a 'total social fact' (Orlove and Caton, 2010). An anthropological approach would also support a finer analysis of actors' different interests and the meanings and values given to water resources and water management options according to class, gender, ethnicity, religion, caste, and so forth. More particularly, it would be worth exploring the subjective and intersubjective meanings of water security and give a greater voice to those who are water insecure.

Note

1. There is even a third but far-less-used definition, which is that of 'nutritional water productivity', whereby the numerator is converted in a nutritional value (Renault and Wallender, 2000).

References

Ahlers, R. and Zwarteveen, M.Z. (2009) 'The water question in feminism: water control and gender inequities in a neo-liberal era', *Gender, Place & Culture*, vol 16, no 4, pp409–426.

Allouche, J. (2011) 'The sustainability and resilience of global water and food systems: political analysis of the interplay between security, resource scarcity, political systems and global trade', *Food Policy*, vol 36, supplement 1, ppS3–S8.

Annan, Kofi A. (2000) 'We the Peoples - the role of the United Nations', Millenium report of the Secretary-General of the United Nations in the 21st Century, Chapter 4 "Sustainaing our Future", available at: http://www.un.org/millennium/sg/report/ch4.pdf.

Asian Development Bank (2011) *Asian Water Development Outlook 2011. A preview*, ADB, Manilla.

Beck, T. (1995) 'The Green Revolution and poverty in India: a case study of West Bengal', *Applied Geography*, vol 15, no 2, pp161–181.

Bluemling, B., Yang, H. and Pahl-Wostl, C. (2007) 'Making water productivity operational—a concept of agricultural water productivity exemplified at a wheat-maize cropping pattern in the North China plain', *Agricultural Water Management*, vol 91, nos 1–3, pp11–23.

Boelens, R. and Vos, J. (2012) 'The danger of naturalizing water policy concepts: water productivity and efficiency discourses from field irrigation to virtual water trade', *Agricultural Water Management*, vol 108, pp16–26.

Brauch, H.G. (ed) (2007) *Environment and Security in the Middle East: Conceptualizing Environmental, Human, Water, Food, Health and Gender Security*, Springer, Dordrecht.

Brichieri-Colombi, S.J. (2004) 'Hydrocentricity: a limited approach to achieving food and water security', *Water International*, vol 29, no 3, pp318–32.8

Brock, L. (1997) 'The environment and security: conceptual and theoretical issues', in N.P. Gleditsch (ed) *Conflict and the Environment*, Kluwer, Boston.

Calow, R.C., MacDonald A. M., Nicol, A.L. and Robins N.S. (2010) 'Ground Water Security and Drought in Africa: Linking Availability, Access, and Demand.' *Ground Water*, vol 48, no 2, pp246–256.

Calder, I.R., Gosain, A., Rama Mohan Rao, M.S., Batchelor, C., Snehalata, M. and Bishop, E. (2008) 'Watershed development in India. 1. Biophysical and societal impacts', *Environment, Development and Sustainability*, vol 10, no 4, pp537–557.

Clarke, R. (1991) *Water: The International Crisis*, Earthscan, London.

Cleaver, F. (2000) 'Moral ecological rationality, institutions and the management of common property resources', *Development and Change*, vol 31, no 2, pp361–383.

Comprehensive Assessment of Water Management in Agriculture (CAWMA) (2007) *Water for Food, Water for Life: A Comprehensive Assessment of Water Management in Agriculture*, Earthscan, London and International Water Management Institute, Colombo.

Connell, J. (1978) 'The death of taro: local response to a change of subsistence crops in the Northern Solomon Islands', *Mankind*, vol 11, no 4, pp445–452.

Cook, C. and Bakker, K. (2012) 'Water security: debating an emerging paradigm', *Global Environmental Change*, vol 22, no 1, pp94–102.

Cornwall, A. (2010) 'Introductory overview—buzzwords and fuzzwords: deconstructing development discourse', in A. Cornwall and D. Eade (eds) *Deconstructing Development Discourse. Buzzwords and Fuzzwords*, Practical Action Publishing in association with Oxfam GB, Rugby, UK.

Cornwall, A. and Brock, K. (2005) 'What do buzzwords do for development policy? A critical look at "participation", "empowerment" and "poverty reduction"', *Third World Quarterly*, vol 26, no 7, pp1043–1060.

Das, R.J. (2002) 'The Green Revolution and poverty: a theoretical and empirical examination of the relation between technology and society', *Geoforum*, vol 33, no 1, pp55–72.

Fairclough, N. and Vodak, R. (1997) 'Critical discourse analysis', in T.A. van Dijk (ed), *Discourse Studies: A Multidisciplinary Introduction*, Sage, London.

Falkenmark, M. and Lundqvist, J. (1998) 'Towards water security: political determination and human adaptation crucial', *Natural Resources Forum*, vol 22, no 1, pp37–51.

FAO (2011) 'Rome emergency meeting rallies to aid Horn of Africa', www.fao.org/news/story/en/item/82543/icode, accessed 31 January 2012.

FAO Legal Office (2002) *Law and Sustainable Development since Rio—Legal Trends in Agriculture and Natural Resource Management*, FAO, Rome.

FAO Media Centre (2011) 'Water is key to food security', Q&A with FAO Assistant Director-General for Natural Resources, Alexander Mueller, www.fao.org/news/story/en/item/86991/icode, accessed 31 January 2011.

Franks, T. and Cleaver, F. (2007) 'Water governance and poverty: a framework for analysis', *Progress in Development Studies*, vol 7, no 4, pp291–306.

Giordano, M., Rijsberman, F.R. and Saleth, R.M. (2006) *More Crop Per Drop: Revisiting a Research Paradigm: Results and Synthesis of IWMI's Research 1996–2005*, IWA Publishing, London.

Global Water Partnership (GWP) (2000) *Towards Water Security: A Framework for Action*, GWP, Stockholm and London.

Grey, D. and Sadoff, C.W. (2007) 'Sink or swim? Water security for growth and development', *Water Policy*, vol 9, no 6, pp545–571.

Hajer, M.J. (1995) *The Politics of Environmental Discourse: Ecological Modernization and the Policy Process*, Oxford University Press, Oxford.

Hamdy, A., Ragab, R. and Scarascia-Mugnozza, E. (2003) 'Coping with water scarcity: water saving and increasing water productivity', *Irrigation and Drainage*, vol 52, no 1, pp3–20.

Keeley, J. and Scoones, I. (2003) *Understanding Environmental Policy Processes: Cases from Africa*, Earthscan, London.

Kijne, J.W. (2003) *Unlocking the Water Potential of Agriculture*, FAO, Rome

Lankford, B. (2009) 'Viewpoint—the right irrigation? Policy directions for agricultural water management in sub-Saharan Africa', *Water Alternatives*, vol 2, no 3, pp476–480.

Lankford, B. (2012) 'Fictions, fractions, factorials and fractures; on the framing of irrigation efficiency', *Agricultural Water Management*, vol 108, pp27–38.

Mekonnen, D.Z. (2010) 'The Nile Basin Cooperative Framework Agreement negotiations and the adoption of a water security paradigm: flight into obscurity or a logical cul-de-sac?' *European Journal of International Law*, vol 21, no 2, pp421–440.

Merrey, D.J. (2009) 'Will future water professionals sink under received wisdom, or swim to a new Paradigm?' *Irrigation and Drainage*, vol 58, no S2, ppS168–S176.

Molden, D. (1997) *Accounting for Water Use and Productivity*, International Irrigation Management Institute, Colombo.

Molden, D., Murray-Rust, H., Sakthivadivel, R. and Makin, I. (2003) 'A water productivity framework for understanding and action', in J. W. Kijne, R. Barker, and D. Molden (eds) *Water Productivity in Agriculture: Limits and Opportunities for Improvement*, CABI, Wallingford, UK.

Molden, D., Oweis, T., Steduto, P., Bindraban, P., Hanjra, M.A. and Kijne, J. (2010) 'Improving agricultural water productivity: between optimism and caution', *Agricultural Water Management*, vol 97, no 4, pp528–535.

Mosse, D. (1997) 'The symbolic making of a common property resource: history, ecology and locality in a tank-irrigated landscape in South India', *Development and Change*, vol 28, no 3, pp467–504.

Orlove, B. and Caton, S.C. (2010) 'Water sustainability: anthropological approaches and prospects', *Annual Review of Anthropology*, vol 39, pp401–415.

Ostrom, E., Lam, W.F., Pradhan, P. and Shivakoti, G. (2011) *Improving Irrigation in Asia: Sustainable Performance of an Innovative Intervention in Nepal*, Edward Elgar Publishing, Cheltenham, UK.

Perry, C. (2010) 'Water security—what are the priorities for engineers?' *Outlook on Agriculture*, vol 39, no 4, pp285–289.

Playan, E. and Mateos, L. (2006) 'Modernization and optimization of irrigation systems to increase water productivity', *Agricultural Water Management*, vol 80, no 1–3, pp100–116.

Renault, D. and Wallender, W.W. (2000) 'Nutritional water productivity and diets', *Agricultural Water Management*, vol 45, no 3, pp275–296.

Sen, A. (1999) *Development As Freedom*, Oxford University Press, Oxford.

Sojamo, S., Keulertz, M., Warner, J. and Allan, J.A. (2012) 'Virtual water hegemony: the role of agribusiness in global water governance', *Water International*, vol 37, no 2, pp169–182.

Tandon, N. (2007) 'Biopolitics, climate change and water security: impact, vulnerability and adaptation issues for women', *Agenda*, vol 21, no 73, pp4–17.

UN-Water and FAO (2007) 'Coping with water scarcity. Challenge of the 21st century', www.fao.org/nr/water/docs/escarcity.pdf.

Wodak, R. and Meyer, M. (eds) (2010) *Methods of Critical Discourse Analysis*, Sage, London.

World Economic Forum Water Initiative (WEFWI) (2011) *Water Security. The Water-Food-Energy-Climate-Nexus*, Island Press, Washington, DC.

World Health Organisation (2003) *WHD Brochure, Part IV: The priorities and solutions for creating healthy places*, WHO, Geneva.

World Water Forum (2000) Ministerial Declaration on Water Security in the 21st Century, The Hague, 22 March 2000, available at:

Xia, J., Zhang, L., Liu, C. and Yu, J. (2006) 'Towards better water security in North China', *Water Resources Management*, vol 21, no 1, pp233–247.

Zeitoun, M. (2011) 'The global web of national water security', *Global Policy*, vol 2, no 3, pp286–296.

Appendix
Selected definition of water security and references

Definition of water security	Reference
'A situation of reliable and secure access to water over time. It does not equate to constant quantity of supply as much as predictability, which enables measures to be taken in times of scarcity to avoid stress.'	Appelgren, B. (1997) 'Keynote paper—Management of water scarcity: national water policy reform in relation to regional development cooperation', *Second Expert Consultation on National Water Policy Reform in the Near East*, Cairo, 24–25 November 1997. FAO, Rome
'Societies can enjoy water security when they successfully manage their water resources and services to: ** satisfy household water and sanitation needs in all communities, ** support productive economies in agriculture and industry, ** develop vibrant, lovable cities and towns, ** restore healthy rivers and ecosystems, and ** build resilient communities that can adapt to change.'	Asian Development Bank (2011) *Asian Water Development Outlook 2011. A preview*, ADB, Manila
'The availability of, and access to, water sufficient in quantity and quality to meet the livelihood needs of all households throughout the year, without prejudicing the needs of other users.'	Calow, R.C., MacDonald, A.M., Nicol, A.L. and Robins, N.S. (2010) '*Ground water* security and drought in Africa: linking availability, access, and demand', Ground Water, vol 48, no 2, pp246–256
'Every person has access to enough safe water at affordable cost to lead a clean, healthy and productive life, while ensuring that the natural environment is protected and enhanced.'	Global Water Partnership (2000) *Towards Water Security: A Framework for Action*, GWP,Stockholm and London

(*Continued*)

Definition of water security	Reference
'The availability of an acceptable quantity and quality of water for health, livelihoods, ecosystems and production, coupled with an acceptable level of water-related risks to people, environments and economies.'	Grey, D. and Sadoff, C. W. (2007) 'Sink or swim? Water security for growth and development', *Water Policy*, vol 9, no 6, pp545–571
'Sustainable access, on a watershed basis, to adequate quantities of water, of acceptable quality, to ensure human and ecosystem health.'	Norman, E.S., Bakker, K. and Dunn, G. (2011) 'Recent developments in Canadian water policy: an emerging water security paradigm', *Canadian Water Resources Journal*, vol 36, no 1, pp53–66
Household water security is 'the reliable availability of safe water in the home for all domestic purposes.'	World Health Organisation (WHO) (2003) *WHD Brochure, Part IV: The priorities and solutions for creating healthy places*, WHO, Geneva
'A condition in which there is a sufficient quantity of water, at a fair price, and at a quality necessary to meet short and long term human needs to protect their health, safety, welfare, and productive capacity at the local, regional, state, and national levels.'	Witter and Whiteford (1999) in Kaplowitz, M.D. and Witter, S.G. (2002) 'Identifying water security issues at the local level', *Water International*, vol 27, no 3, pp379–386
'(a) population-wide security, that is, everyone can obtain secure water for domestic use; (b) economic security, namely water resources can satisfy the normal requirements of economic development; (c) ecological security, namely water resources can meet the lowest water demands of ecosystems without causing damage.'	Xia, J., Zhang, L., Liu, C. and Yu, J. (2006) 'Towards better water security in North China', *Water Resources Management*, vol 21, no 1, pp233–247

11 Transboundary Water Security

Reviewing the Importance of National Regulatory and Accountability Capacities in International Transboundary River Basins

Naho Mirumachi

Introduction

While the use of the term 'water security' is diffuse across disciplines and contexts (Cook and Bakker, 2012), it is increasingly acknowledged that water security cannot be examined from the water sector alone (WEF, 2011; Zeitoun, 2011). Water issues are tied with problems of food production and land development, increasing global population and, importantly, lifestyle changes that place demands on water resources through consumption of both food and energy. Climate change also has implications on water availability and use with mitigation and adaptation measures reviewing practices within the water sector. In addition, the broad notion of water security is deeply associated with concepts such as poverty reduction, sustainable development, and human security. As represented by the Millennium Development Goals, global agendas have long emphasized the crucial link between water and poverty (Mount and Bielak, 2011). The Ministerial Declaration of the 2nd World Water Forum in 2000 emphasized that water security contributes to sustainable development. Water security has been defined in the context of conflict prevention, based on geopolitical concerns over water availability and its implications to human security (GTZ, 2010). Water use and management through agricultural, climate, and energy policies and practices operate at local, national, international, and global scales. These policies and practices are also influenced by global and international agendas on development and geopolitics. This general description of water security provides some broad contours of the relationship between sectors and scales, and between related concepts.

In order to provide some more detail on the assumptions and implications of water security, this chapter focuses specifically on the international scale, using the lens of transboundary river basins. The chapter builds upon analysis of domestic factors, such as political instability, playing a role in water security at the international level (Mirumachi, 2008). By exploring how national and international scales are interconnected, the chapter argues that transboundary water security reflects the national capacities of basin states to allocate, reallocate, and regulate water resources between different water users and stakeholders. Here, transboundary water security is characterized by the collective capacity to harmonize multisectoral policies within an international transboundary river basin. It is argued that the process to build up collective capacity and to identify ways in which policy is made compatible across sectors is fundamentally political. This process reflects stakeholders' views on the relationship between society and nature, on notions of threats and risks by and to water resources, and on concerns for equity in resource allocation and cost burdens. A focus on the active role of nonstate actors and on the state-bound authority of river basin organizations helps explain the scales of regulatory frameworks and measures of environmental accountability necessary.

The chapter first examines characteristics of transboundary water security. The concept of water security poses questions about water allocation and reallocation. Using a brief example of the Ganges River Basin, we then analyse the role private sector actors are playing in the allocation and reallocation of water resources. This example serves to show how decision making for transboundary water security needs to consider activities of both state and nonstate actors. The next section is an initial examination of river basin organizations as a mechanism for transboundary water security, using the context of the lower Mekong River Basin. Based on analysis from the two previous sections, the next section argues that there is a need for policy harmonization not just between basin states but also between sectors within the nation-state. The chapter concludes by highlighting the importance of national capacities to achieve transboundary water security.

Unpacking Transboundary Water Security
Beyond Interstate Cooperation

Much of the existing literature on water security in international transboundary river basins underscores the importance of interstate cooperation. For example, in the policy literature, cooperative water resources management is the means to achieve water security according to the From Potential Conflict to Cooperation Potential programme of the United Nations Educational, Scientific and Cultural Organization. It is argued that institutions to govern shared waters should consider 'effective transboundary water management' and 'preventive hydro-diplomacy' (Cosgrove, 2002, p75). Effective transboundary water management is seen to facilitate efficient and equitable water allocation between states. Mechanisms for dispute resolution ease tension between states and foster cooperation (Cosgrove, 2002). Similarly, the report by the Royal Academy of Engineering on global water security pointed out how international treaties are the touchstone for ensuring water security between states sharing waters. The report argued the importance of establishing mechanisms for 'international coordination' that could guide national responses (The Royal Academy of Engineering, 2010, p19). In the academic literature, Tarlock and Wouters (2010) argued that the concept of 'hydro-commons' is useful to address global water security from a legal perspective. The idea of 'hydro-commons' disassociates water scarcity from interstate competition and instead encourages the peaceful management of shared waters using legal principles, such as equitable and reasonable utilization. Petersen-Perlman et al. (2012) pose water security as a negative concern for basin states and thus argue the importance of fostering cooperation. The existing literature treats water security as both something to achieve or avoid in transboundary river basins and the need of cooperation is associated with this concept.

Lacking supranational authorities, cooperation is indeed needed for the current practice of putting in place negotiated agreements between basin states. However, empirical studies have shown that cooperation does not necessarily guarantee improved water allocation. In the Aral Sea region, where environmental degradation is a serious issue, interstate cooperation is considered indispensible (e.g., Teasley and McKinney, 2011; Granit et al., 2012; Libert and Lipponen, 2012). In the Chu and Talas Rivers shared by Kazakhstan and Kyrgyzstan, a landmark agreement on water use was signed in 2000. The agreement between the governments of the Republic of Kazakhstan and Kyrgyz Republic on the Use of Interstate Waterworks Facilities on the Chu and Talas Rivers defines the responsibilities of Kazakhstan to share costs of operating and maintaining dams and reservoirs in Kyrgyz territory, and of Kyrgyzstan to provide water to its downstream counterpart. The Chu-Talas Rivers Commission is regarded as a 'success', based on the implementation of cost sharing and expansion of commission activities (Libert, 2011). However, Wegerich (2008) cautioned that such evaluations of success may be overstated when looking at the specific details of water release, which advantage water resource control by Kyrgyzstan. The reality of interstate relations is one of coexisting conflict and cooperation, rather than a unidirectional change from conflict to cooperation (Zeitoun and Mirumachi, 2008; Mirumachi, 2010). Cooperation on its own cannot be an accurate indicator of water security at the transboundary level.

The plural interpretation of the 'best' way to allocate and manage water is one reason why political interactions between basin states are one of coexisting conflict and cooperation. Many basins explicitly and implicitly use the principle of equitable and reasonable utilization from the 1997 United Nations Convention on the Law of the Non-Navigational Uses of International Watercourses to guide the negotiation and establishment of shared river institutions (Salman, 2007). While an understanding of equitable and reasonable utilization is a step towards achieving water security in the basin, questions on reallocation of water from one state to another, from one region to another, and from one sector to another need to be considered, as well. In the Jordan River Basin, reallocating existing and new water sources according to per capita calculations has been proposed as a 'positive-sum' cooperative way of managing shared waters (Phillips et al., 2007, 2009). Desalination would be one way of creating new, additional water sources that would be shared between Israel, Palestine, and Jordan (Phillips et al., 2007, 2009). However, the Israeli position views desalination as a national issue, not within the scope of shared natural resource issues. Consequently, should desalinated water be used to secure base flow of the aquifers and lakes in the region, costs to supply desalinated water need to be negotiated between states (Feitelson and Rosenthal, 2012). Progress on changing the status quo of water allocation has been slow and it shows the heavily politicized nature of shared waters in this region (Phillips, 2012). Importantly, while the Jordan River Basin has long been viewed as a hotspot for water conflict, the contested nature of water allocation and reallocation is common to transboundary basins around the world. The Nile Basin states were faced with similar dilemmas of reallocation when negotiating the Cooperative Framework Agreement. This agreement included concerns for 'water security' and had a specific article that could override existing water allocation measures put in during the colonial times. As basin states that could be most adversely affected in terms of water quantity, Egyptian and Sudanese negotiators resisted the inclusion of this article (Mekonnen, 2010; Nicol and Cascão, 2011).

Reallocation of water for human needs to ecological needs of the water environment is also challenging. 'Food water' issues, or issues of water supply and management for food production (Allan, Chapter 20; see also Allan, 2011), are increasingly being discussed beyond national scales of analysis and at the global scale as the role of global virtual water trade is acknowledged (Aldaya et al., 2010) and as new issues of 'land-grabbing' by foreign investors emerge (Allan et al., 2012). Non-food water issues, or issues relating to water for industrial and individual uses (Allan, Chapter 20; Allan, 2011) continue to be raised on global agendas, most notably in efforts to improve access to water and sanitation. If these demands of both food and non-food water are to be met, how would water for the natural environment be impacted? A good example to explore the issue of reallocating water between human and environmental needs is South Africa. South Africa established a unique water policy where water for the environment is accounted for. The National Water Act recognizes water for basic human needs and water for ecological reserves. The former is often described as ensuring 'some water for all, forever'. The latter refers to the right to provide both sufficient quantity and quality of water for the environment. The legal recognition of these two basic rights makes this national act a progressive one in the region (UNDP, 2006).

While South Africa has, at least legally, embraced the idea of water for both society and nature, regional efforts are not as explicit on such thinking. The member states of the Southern African Development Community (SADC), including South Africa, have signed the Revised Protocol on Shared Watercourses in 2000. This regional initiative is significant, because it embeds international legal principles of water utilization and development following the 1997 UN Convention on the Law of the Non-Navigational Uses of International Watercourses. However, water for the environment is only mentioned in a general manner, in relation to sustainable development. Challenges of information sharing between states exist and differences in financial and human resources are not negligible

(Raadgever et al., 2008; Heyns et al., 2008). Moreover, much of sub-Saharan Africa is dominated by the notion of securing water through large-scale infrastructure to avert intrastate conflict and to alleviate poverty (Swatuk, 2012). 'Difficult hydrology made more so by climate change, combined with watercourses shared by two or more states' drives this imperative and the main contention over allocation is between the agricultural, industrial and urban sectors (Swatuk, 2012, p88). Consequently, considering water for nature and ecosystems is not sufficient and the revised protocol within SADC is not being used to critically address transboundary water security.

Transboundary water security poses deeply political questions about allocation and reallocation of water. The implications extend to how existing socioeconomic mechanisms and institutions related to water need to be readdressed. The answers to these political questions are often guided by geopolitical and development concerns, which in turn reflect fundamental perspectives of society's relationship with water resources and nature. As such, the concept of transboundary water security entails more than international cooperation to sign agreements or to establish multilateral cooperative initiatives. The concept enables us to revisit often anthropocentric assumptions on the relationship between society and nature, which have guided agreements, policies, and management practices.

Actors and Scales in Transboundary River Basins

Much of the formal decision making over international transboundary waters is done between state governments, especially as issues of water allocation may have implications on territoriality and sovereignty. Nonetheless, international financial institutions (IFIs), donor agencies, UN organizations, and global water organizations also facilitate formal decision making by supporting regional initiatives. Organizations like the Global Water Partnership are explicitly incorporating the concept of water security in their transboundary activities, advocating better transboundary cooperation as a way to achieve water security (GWP, 2009). Development aid agencies, like the German donor agency Deutsche Gesellschaft für Internationale Zusammenarbeit (GIZ), have financially supported programs and activities of major basins in the African continent, such as the Nile, Niger, and Orange-Senqu river basins, and cite the importance of conflict prevention for water security (GTZ, 2007, 2010). The relationship between national governments, IFIs, donor agencies, and UN agencies has been notable in the planning and development of large-scale infrastructure projects, facilitating large investments in iconic projects of the hydraulic mission—a phase of intensive water capture through infrastructure development to centrally manage and control river flows (Molle et al., 2009). While infrastructure development is still a key feature of the hydraulic mission phase, the actor landscape has become complex with the rise of nonstate actors investing in transboundary water development projects. These changes highlight the linkage between the water and energy sectors, and between various spatial scales. A brief example from the Ganges River Basin illustrates this point.

In South Asia, the shared Ganges waters have been the stage for both large- and small-scale infrastructure development. Transboundary water development has been characterized by multipurpose dam projects. In particular, between Nepal and India, hydraulic infrastructure can provide flood protection against seasonal water variability but also take advantage of such variability and provide water for irrigation and hydropower. A significant development in recent years is private sector investments in hydropower projects with the introduction of Independent Power Producers (IPPs). IPPs are business organizations that develop and distribute hydropower energy. These private sector actors have gained prominence as national energy reforms and privatization have occurred in South Asia, in particular India (Dubash and Rajan, 2003). In 2008, a consortium of IPPs led by the Indian infrastructure company GMR signed a Memorandum of Understanding with the Nepali government to develop the Upper Karnali Hydroelectric Project, a run of the river hydropower project. Located on the Karnali River, one of the major Ganges tributaries flowing from Nepal to India, this project would have a minimum of 300 MW capacity to generate electricity. This has sparked interest amongst IPPs for further investment opportunities (IPPAN and CII, 2006). Faced with energy shortage in India, Indian IPPs are seeking new sites for investment, including foreign projects. It is said that Nepal has the capacity to develop 83,000 MW and, even if economic viability were taken into account, 43,000MW would be possible. However, currently only a mere fraction has been developed with total capacity at around 500 MW (ADB, 2007). From an Indian perspective, Nepal provides rich opportunities for business expansion (Kawale, 2009).

Backed by IPPs, the Upper Karnali Hydroelectric Project breaks the mould of state-led hydropower development. The governments of India and Nepal have executed river development projects based on bilateral agreements, with notable multipurpose projects in the Kosi and Gandak tributaries of the Ganges River in place since the 1950s. The Kosi and Gandak agreements have set the template for project-based bilateral agreements, but there has been persistent contention over equitable benefit sharing (Dhungel, 2009). Compared to these agreements, projects supported by the IPPs do not involve diplomatic negotiations based on national interests. However, foreign investment in national waters has been a highly divisive issue in Nepal. In 2011, the project office at the Upper Karnali Hydroelectric Project site was burned down, causing the Nepali government to order military presence close to the site (Gautam, 2011). While no connections to this particular incidence have been made, the Maoist political party, the Unified Communist Party of Nepal, have lambasted the project for selling out to Indian interests, arguing that a resort to violent conflict may even be necessary to stop construction plans (Adhikari, 2011). This project shows that energy trade is subject to the domestic approval of water resources development. Because IPPs are private sector actors, they are not bound by precedents of water and benefit sharing in previous international agreements between basin states. Instead, these projects will be subject to environmental impact assessment and other environmental standards of the host nation. Cumulative benefits and threats from altered river flow and ecosystems need to be examined. National capacity will be challenged to anticipate impacts to transboundary water flows and to identify threats to changed water flow from domestic projects. Importantly, public acceptance needs to be gained. The emergence of IPPs and governance mechanisms for such projects can facilitate or hinder basin-wide considerations of water availability and allocation, linking national issues with transboundary water security.

Mechanisms for Transboundary Water Security

The ways in which actors and governance mechanisms matter to achieving water security can also be explored through the following example of recent hydropower development in the Mekong River Basin. Similar to the Ganges River example, private investors are increasingly interested in developing hydropower projects, both on the mainstream and tributaries of the river. In the upper basin, shared by Yunnan Province of China and Myanmar, there are eight mainstream projects planned, in construction, or in operation. In the lower basin, shared by Laos, Thailand, Cambodia, and Vietnam, there are up to 12 mainstream projects that are being investigated. Interest in hydropower is significant in the lower basin for both large- and small-scale projects; over 130 projects are in operation, in construction, or planned between the four states (Haas, 2010; Räsänen et al., 2012). One partial reason suggested for the increased attention to hydropower development involving the private sector is the reduction in public funds (NVE, 2010). The investment and construction of dams along the river show a very complex political economy made up of national governments, quasigovernmental utilities, international financial institutions, and private investors (Matthews, 2012). Vested interests of these actors are interwoven through the planning, financing, and operational norms and practices. The nontransparency of this complex web is often challenged by international and local NGOs demanding more information and deliberation on these dams. The campaign, Save the Mekong, is a good example of a coalition of civil society organizations formed to advocate the socioeconomic and environmental concerns of dam development. The development of hydropower is not confined to government, but it is driven by a growing variety of actors with vested interests.

The lower basin states have planned the development of the river basin through multilateral river basin organizations. The current Mekong River Commission (MRC) is mandated to provide a platform for decision making on principles of sustainable development under the Agreement on the Cooperation for the Sustainable Development of the Mekong River Basin signed in 1995. In the last few years, the MRC has reinvigorated its role in the hydropower debate, establishing the Initiative on Sustainable Hydropower.

The initiative established preliminary guidelines on 'good practice' of hydropower development on the mainstream (MRC, 2009b). In addition, working with WWF and the Asian Development Bank, the initiative developed a policy tool, the Rapid Basin-Wide Hydropower Sustainability Assessment Tool, which would identify 'hydropower sustainability risk and opportunity' (MRC, 2010, p14). The MRC has taken up Strategic Environmental Assessment (SEA) as a way to ensure that mainstream dam development fits well with its Basin Development Plan. By conducting the SEA, benefits, costs, risks, and opportunities are to be highlighted at a basin level (MRC, 2009a). From a policy framework perspective, the lower Mekong Basin has invested in tools to assess basin-wide water security concerns.

However, the issue of hydropower development has pitted contrasting views on the benefits and adverse impacts of these proposed projects and dams under construction. The MRC has developed a macroeconomic perspective of poverty reduction and development, thereby framing dam development as a viable option for economic development. In order to ensure maximum benefit from the dams at minimal adverse impact to socioeconomic and environmental conditions, the above-mentioned initiatives and policy tools have been implemented. This perspective has been critiqued as a gross oversimplification of hydropower development as tradeoffs between economic development and the environment. Dam construction will have a negative impact on fish habitats and affect fish migration. This is not only an important biodiversity concern but also a livelihood problem, as large rural populations living along the river rely on food and income from these fish (Dungan et al., 2010; Friend et al., 2010). The dam debate obscures the complex political economy of hydropower development and risks overlooking these rural communities (Kuenzer et al., 2012). The predominant hydropower discourse is criticized for valuing hydropower projects as 'more important' or 'more valuable' than ecosystems and rural livelihoods, and for dismissing alternative development options (Friend and Blake, 2009; Friend et al., 2010).

The critique of oversimplified development options is useful and necessary in discussing transboundary water development in the Mekong. The MRC provides a focal point to improve transboundary water governance (Grumbine et al., 2012). However, analysis should also extend to the ways in which domestic rules govern private actors. Private investors in hydropower development in the Mekong will be subject to the domestic licensing competition and national legal requirements. Because the MRC is not a supranational authority, it is up to the individual states to regulate these actors in accordance to domestic norms and protocols. In transboundary river basins, the river basin organization is likely to put forth regional economic development goals that all states can agree on, despite differences in national interests. The practice of achieving these goals relates to transboundary water security in that water is allocated and reallocated through an ever-complex set of actors.

River basin organizations will require consistent capacity to deal with surmounting tasks to identify and assess multiple water resource use by various stakeholders. This may not be easily achieved in developing regions, as financial support by member states tends to be small and largely reliant on the support from international financial institutions and donor agencies. Moreover, such support may not be sufficient or sustainable. Empirically, an indicative survey of African transboundary river basins showed that such support has been patchy across basins and time, with preference to support large basins (GTZ, 2007). It should also be pointed out that limitations of river basin organizations notwithstanding, the current set up of transboundary water governance mechanisms often utilizes these organizations and states drive the governance process. However, these international river basin organizations do not replace sovereign basin states. Consequently, transboundary accountability of environmental degradation and compensation for inadequate mitigation measures of dam building cannot to be placed on the international river basin organization alone. These responsibilities need to be taken up at the national governmental level.

Focus on National Capacity to Harmonize Policy across Sectors

To achieve water security, Appelgren and Klohn (1997) suggested that national policies should be harmonized across states within a basin, rather than to engage in lengthy and uncertain processes of establishing binding legal principles at the international level. This bottom-up approach underlines the importance of embedding transboundary water issues in the national agendas. Policy recommendations suggest implementing Integrated Water Resources Management (IWRM) at the transboundary level as a way to harmonization. It is argued that transboundary IWRM would facilitate the production and use of scientific knowledge for relevant and useful decision making (Gooch et al. 2006). The multisectoral approach of IWRM has been cited as a strength to address fragmentation and overlap of institutional responsibilities (GWP, 2010). IWRM is increasingly being mentioned in the context of water security (GWP, 2010; Bogardi et al., 2012, see also Cook and Bakker, 2012) perhaps indicating the benefit of IWRM to establish ownership of efforts for water security.

However, experience in the Mekong River Basin shows that practicing transboundary IWRM is challenging with numerous line agencies of multiple governments dealing with water issues. Bureaucratic silos exist between various departments and ministries relating to water. Competition between bureaucracies for institutional survival also perpetuates particular modes of water resources management and impedes others (Molle et al., 2009). Crucially, there needs to be strong buy-in by these governments to use the river basin organization as a platform to advance transboundary IWRM (Mirumachi, 2012). It goes without saying that using IWRM as an approach for harmonizing policy is a political process. Integrated decisions about water allocation are not merely technical but also political (Allan, 2003). One of the key political questions implicit in IWRM is the balance between water resources management and socioeconomic development (e.g., GWP, 2000,

2009). Harmonizing policy across states will require negotiating different framings and interpretations of 'resource efficiency' and 'equity' that are principles guiding IWRM.

There is certainly scope for institutional development at the regional scale through programs and policies of river basin organizations. However, when private investors potentially have the effect of changing water allocation and environmental management in transboundary rivers, national regulatory capacities are challenged. The implication of the changing actor landscape of transboundary water security is that policy harmonization between sectors within states, and not only between states, becomes all the more important. Put differently, it is up to the basin states to develop both national and basin-wide capacities to regulate river development and achieve water security. Thus, transboundary water security is not limited to an international regulatory or accountability concern; it involves national regulatory and accountability measures as well. This is not to say that basin-wide initiatives or river basin organizations should be abandoned all together. Rather, the limitations of existing state-led governance frameworks need to be pointed out. In addition, overlooking the improvement of individual, national regulatory capacities should be cautioned.

Conclusion

The chapter set out to unpack the concept of transboundary water security. A close examination of water security in international transboundary river basins brought to the fore implications on national capacity to regulate, manage, and govern waters. It was pointed out that large-scale infrastructure development is now not exclusive to state actors, and an increasing number of private sector actors are financing and investing in hydropower projects. Because international river basin organizations are voluntary units comprised of sovereign nation-states, it was argued that national capacities to assess and regulate water use and allocation become crucial. Of course, there is a large role that transboundary agreements and frameworks need to play to ensure equitable allocation and sustainable water use. Focus on national capacity, in addition to basin-wide capacity, deserves merit because it highlights how domestic policy changes may strengthen or weaken transboundary water security. If the whole is to be greater than the sum of its parts, then transboundary water security needs to consider both international and national capacity to harmonize policy between sectors.

Acknowledgments

Elements of this paper were discussed in presentations at the Water Security Research Centre Seminar Series, University of East Anglia, October 2009, and UEA–ICID seminar on Water Security: Progress in Theory and

Practice, November 2011. The author would like to thank the participants of these seminars, Tony Allan, and two anonymous reviewers for constructive comments.

References

Adhikari, P. (2011) '900-MW Upper Karnali Project: Maoist leaders warn govt over contract', *The Kathmandu Post,* 13 June.

Aldaya, M.M., Allan, J.A. and Hoekstra, A.Y. (2010) 'Strategic importance of green water in international crop trade', *Ecological Economics,* vol 69, pp887–894.

Allan, J.A. (2003) 'Integrated Water Resources Management is more a political than a technical challenge', *Developments in Water Science,* vol 50, pp9–23.

Allan, J.A. (2011) *Virtual Water: Tackling the Threat to our Planet's Most Precious Resource,* I.B. Tauris, London.

Allan, J.A., Keulertz, M., Sojamo, S. and Warner, J. (eds) (2012) *Handbook of Land and Water Grabs in Africa: Foreign Direct Investment and Food and Water Security,* Routledge, London.

Appelgren, B. and Klohn, W. (1997) 'Management of transboundary water resources for water security: principles, approaches and State practice', *Natural Resource Forum,* vol 21, no 2, pp91–100.

Asian Development Bank (ADB) (2007) *Summary Environmental Impact Assessment, Nepal: West Seti Hydroelectric Project,* ADB, Manila.

Bogardi, J. J., Dudgeon, D., Lawford, R., Flinkerbusch, E., Meyn, A., Pahl-Wost,C., Vielhauer, K. and Vörösmarty, C. (2012) 'Water security for a planet under pressure: interconnected challenges of a changing world call for sustainable solutions', *Current Opinion in Environmental Sustainability,* vol 4, pp35–43.

Cook, C. and Bakker, K. (2012) 'Water security: debating an emerging paradigm', *Global Environmental Change,* vol 22, pp94–102.

Cosgrove, W.J. (2002) *Water Security and Peace: A Synthesis of Studies Prepared under the PCCP–Water for Peace Process,* UNESCO-IHP, Paris.

Dhungel, D.N. (2009) 'Historical eye view', in D.N. Dhungel and S.B. Pun (eds) *The Nepal–India Water Relationship: Challenges,* Springer, Dordrecht.

Dubash, N.K. and Rajan, S. (2003) 'Electricity reforms in India: political economy and implications for social and environmental outcomes', in N. Wamukonya (ed) *Electricity Reform: Social and Environmental Challenges,* UNEP, Roskilde.

Dungan, P.J., Barlow, C., Agostinho, A.A. , Baran, E., Cada, G.F., Chen, D., Cowx, I.G., Ferguson, J.W., Jutagate, T., Mallen-Cooper, M., Marmulla, G., Nestler, J., Petrere, M., Welcomme, R.L. and Winemiller, K.O. (2010) 'Fish migration, dams, and loss of ecosystem services in the Mekong basin', *Ambio,* vol 39, pp344–348.

Feitelson, E., and Rosenthal, G. (2012) 'Desalination, space and power: The ramifications of Israel's changing water geography', *Geoforum,* vol 43, no 2, pp272–284.

Friend, R. and Blake, D. (2009) 'Negotiating trade-offs in water resources development in the Mekong Basin: implications for fisheries and fishery-based livelihoods', *Water International,* vol 11, supplement 1, pp13–30.

Friend, R., Robert, A. and Keskinen, M. (2010) 'Songs of the doomed: the continuing neglect of capture fisheries in hydropower development in the Mekong', in F. Molle, T. Foran, and M. Käkönen (eds) *Contested Waterscapes in the Mekong Region: Hydropower, Livelihoods and Governance,* Earthscan, London.

Gautam, S. (2011) 'Arsonists target Upper-Karnali project', *The Himalayan Times,* 23 May.

Gesellschaft für Technische Zusammenarbeit (GTZ) (2007) *Donor Activity in Transboundary Water Cooperation in Africa: Results of a G8-Initiated Survey 2004–2007,* GTZ, Eschborn.

Gesellschaft für Technische Zusammenarbeit (GTZ) (2010) *The Water Security Nexus: Challenges and Opportunities for Development Cooperation*, GTZ, Eschborn.

Global Water Partnership (GWP) (2000) *Integrated Water Resources Management*, GWP, Stockholm.

Global Water Partnership (GWP) (2009) *Improving Africa's Water Security: Progress in Integrated Water Resources Management in Eastern and Southern Africa*, GWP, Stockholm.

Global Water Partnership (GWP) (2010) *Water Security for Development: Insights from African Partnerships in Action*, GWP, Stockholm.

Gooch, G.D., Stålnacke, P. and Roll, G. (2006) 'The way ahead for transboundary integrated water management', in G.D. Gooch and P. Stålnacke (eds) *Integrated Transboundary Water Management in Theory and Practice*, IWA Publishing, London.

Granit, J., Jägerskog, A., Lindström, A., Björklund,G., Bullock, A., Löfgren,R., de Gooijer, G. and Pettigrew, S. (2012) 'Regional options for addressing the water, energy and food nexus in Central Asia and the Aral Sea basin', *International Journal of Water Resources Development*, vol 28, no 3, pp419–432.

Grumbine, E., Dore, J. and Xu, J. (2012) 'Mekong hydropower: drivers of change and governance challenges', *Frontiers in Ecology and the Environment*, vol 10, no 2, pp91–98.

Haas, L. (2010) 'Overview of hydropower development in the Mekong', presented at the Regional Workshop for Coordination of Research on Hydropower Development in the Lower Mekong Basin, Vientiane, Lao PDR, 14–15 September 2010.

Heyns, P. S. V. H., Patrick, M. J. and Turton, A. R. (2008) 'Transboundary water resource management in Southern Africa: meeting the challenge of joint planning and management in the Orange River basin', *International Journal of Water Resources Development*, vol 24, pp371–383.

Independent Power Producers' Association Nepal (IPPAN) and Confederation of Indian Industry (CII) (2006) *Research on Nepal India Cooperation on Hydropower*, IPPAN, Nepal and CII, Kathmandu, New Delhi.

Kawale, J. (2009) *Speech at 4th International Hydropower Convention: Hydropower for Progress of Nepal*, Kathmandu, 25–26 April.

Kuenzer, C., Campbell, I., Roch, M., Leinenkugel, P., Tuan, V. Q., Dech, S. (2012) 'Understanding the impact of hydropower developments in the context of upstream–downstream relations in the Mekong river basin', *Sustainability Science*, DOI 10.1007/s11625-012-0195-z.

Libert, B. (2011) 'Shared benefit or mutual barrier' in UNECE (ed) *Technical Cooperation: Success Stories*, United Nations Publication, Geneva, pp26–34.

Libert, B. and Lipponen, A. (2012) 'Challenges and opportunities for transboundary water cooperation in Central Asia: findings from UNECE's regional assessment and project work', *International Journal of Water Resources Development*, vol 28, no 3, pp 565–576.

Matthews, N. (2012) 'Water grabbing in the Mekong basin: an analysis of the winners and losers of Thailands' hydropower development in the Lao PDR', *Water Alternatives*, vol 5, no 2, pp392–411.

Mekonnen, D. Z. (2010) 'The Nile Basin Cooperative Framework Agreement negotiations and the adoption of a "water security" paradigm', *The European Journal of International Law*, vol 21, no 2, pp421–440.

Mekong River Commission (MRC) (2009a) *Inception Report: MRC SEA for Hydropower on the Mekong Mainstream*, MRC, Vientiane.

Mekong River Commission (MRC) (2009b) *Preliminary Design Guidance for Proposed Mainstream Dams in the Lower Mekong Basin*, MRC, Vientiane.

Mekong River Commission (MRC) (2010) *Rapid Basin-Wide Hydropower Sustainability Assessment Tool (RSAT)*, MRC, Vientiane.

Mirumachi, N. (2008) 'Domestic issues in developing international waters in Lesotho: Ensuring water security amidst political instability', in N.I. Pachova, M. Nakayama and L. Jansky (eds) *International Water Security: Domestic Threats and Opportunities*, United Nations University Press, Tokyo.

Mirumachi, N. (2010) *Study of Conflict and Cooperation in International Transboundary River Basins: The TWINS Framework*, unpublished PhD thesis, University of London.

Mirumachi, N. (2012) 'How domestic water policies influence international transboundary water development: a case study of Thailand', in J. Öjendal, S. Hansson and S. Hellberg (eds) *Politics and Development in a Transboundary Watershed— the Case of the Lower Mekong Basin*, Springer Verlag, Dordrecht.

Molle, F., Mollinga, P.P. and Wester, P. (2009) 'Hydraulic bureaucracies and the hydraulic mission: Flows of water, flows of power', *Water Alternatives*, vol 2, no 3, pp328–349.

Mount, D.C and Bielak, A.T. (2011) *Deep Words, Shallow Words: An Initial Analysis of Water Discourse in Four Decades of UN Declarations*, UNU-INWEH, Hamilton.

Nicol, A. and Cascao, A.E. (2011) 'Against the flow: new power dynamics and upstream mobilisation in the Nile Basin', *Review of African Political Economy*, vol 38, no 128, pp317–325.

Norwegian Water Resources and Energy Directorate (NVE) (2010) *Water and Energy: Sustainable Development of Hydropower Involving the Private Sector in Research Collaboration in the Lower Mekong Region*, NVE, Oslo.

Petersen-Perlman, J.D., Veilleux, J.C., Zentner, M. and Wolf, A.T. (2012) 'Case studies on water security: analysis of system complexity and the role of institutions', *Journal of Contemporary Water Research & Education*, vol 149, pp4–12.

Phillips, D. J. H. (2012) 'The Jordan River basin: at the crossroads between conflict and cooperation', *International Journal of Sustainable Society*, vol 4, no 1/2, pp88–102.

Phillips, D. J. H., Attili, S., McCaffrey, S. and Murray, J.S. (2007) 'The Jordan River basin: 2. Potential future allocations to the co-riparians', *Water International*, vol 32, no 1, pp39–62.

Phillips, D. J. H., Jägerskog, A. and Turton, A. (2009) 'The Jordan River basin: 3. Options for satisfying the current and future water demand of the five riparians', *Water International*, vol 34, no 2, pp170–188.

Raadgever, G. T., Mostert, E., Kranz, N., Interwies, E. and Timmerman, J.G. (2008) 'Assessing management regimes in transboundary river basins: do they support adaptive management?', *Ecology and Society*, vol 13, no 1.

Räsänen T.A., Koponen, J., Lauri, H. and Kummu, M. (2012) 'Downstream hydrological impacts of hydropower development in the upper Mekong Basin', *Water Resources Management*, vol 26, no 2, pp3495–3513.

The Royal Academy of Engineering (2010) *Global Water Security: An Engineering Perspective*, The Royal Academy of Engineering, London.

Salman, S. M. A. (2007) 'The United Nations Watercourses Convention ten years later: why has its entry into force proven difficult?', *Water International*, vol 32, no 1, pp1–15.

Swatuk, L.A. (2012) 'Water and security in Africa: state-centric narratives, human insecurities', in M.A. Schnurr and L.A. Swatuk (eds) *Natural Resources and Social Conflict: Towards Critical Environmental Security*, Palgrave Macmillan, Basingstoke.

Tarlock, D. and Wouters, P. (2010) 'Reframing the water security dialogue', *The Journal of Water Law*, vol 20, no 2/3, pp53–60.

Teasley, R. and McKinney, D. (2011) 'Calculating the benefits of transboundary river basin cooperation: Syr Darya basin', *Journal of Water Resources Planning and Management*, vol 137, no 6, pp481–490.

United Nations Development Programme (UNDP) (2006) *Human Development Report 2006: Beyond Scarcity—Power, Poverty and the Global Water Crisis,* Palgrave Macmillan, Basingstoke and New York.

Wegerich, K. (2008) 'Passing over the conflict: the Chu Talas Basin agreement as a model for Central Asia?', in M.M. Rahaman and O. Varis (eds) *Central Asian Waters: Social, Economic, Environmental and Governance Puzzle,* Water & Development Publications, Helsinki University of Technology, Espoo.

World Economic Forum (WEF) (2011) *Water Security: The Water-Energy-Food-Climate Nexus,* Island Press, Washington DC.

Zeitoun, M. (2011) 'The global web of national water security', *Global Policy,* vol 2, no 3, pp286–296.

Zeitoun, M. and Mirumachi, N. (2008) 'Transboundary water interaction I: reconsidering conflict and cooperation', *International Environmental Agreements: Politics, Law and Economics,* vol 8, no 4, pp297–316.

Part III

Water Security as Practice Debates

12 Easy as 1, 2, 3?

Political and Technical Considerations for Designing Water Security Indicators

Nathaniel Mason

A Numbers Game: Understanding What Indicators Do, and Do Not, Tell Us

Numbers, as well as words, can be used as markers around which to debate the meanings of water security. Where agreement can be reached, numbers can be used as guides by decision makers and to assess progress towards common goals. This is especially the case for *indicators*—in effect, the headline numbers into which larger amounts of data are distilled. But as Molle and Mollinga (2003) have pointed out in relation to indicators of water poverty and scarcity, while the distilling and simplifying power of indicators is essential to their utility, it also carries risks. In the messy business of shaping public policy or business strategy, indicators can appeal as shortcuts to empiricism, compared with costly context-based research. Too often, this means that indicators can obscure as well as disclose meaning, and value judgements through which one variable is selected over another are glossed over. Even where those judgements are revealed, there commonly lurks an 'information iceberg' of data and statistics below an indicator (Jesinghaus, 1999) with which audiences are rarely able to engage.

This chapter takes as a premise the idea that indicators, used in the right way, can contribute to improved outcomes—including those associated with the concept of water security. But a second key premise is that, just as different verbal framings of water security tend to reflect the interests of the constituencies that advocate them, so, too, do quantitative indicators.

The chapter does not attempt to propose an all-encompassing indicator, or set of indicators, for water security. Indeed, a central contention is that such an ambition can only be fulfilled contextually and collaboratively, incorporating the widest possible range of perspectives. Instead the chapter explores technical and political lessons from development of water-related indicators to date, in anticipation of such a collaborative effort.

The argument is structured around five 'indicator' themes that have previously proven contentious, and are therefore likely to become the fault lines when it comes to developing indicators of water security—namely water availability, human water needs, water risks, environmental sustainability, and institutional capacity. The chapter proposes that each of these five themes is an important consideration for water security, and, taken together, they provide a working definition: Water security requires a sufficiency of the physical resource, the ability to access and utilise it, resilience to water risks, safeguarding of environmental needs and ecosystem services, and institutions to balance competing claims. Under each theme, a number of relevant indicators are discussed to tease apart the interests and values imported by different constituencies in the public and private sectors, and in civil society. In each case, more technical considerations are also touched on, associated with data availability and quality, and with computing indicators from the data.

The penultimate section discusses the limited examples of indicators already developed, which expressly claim to measure water security—though whether it is better to incorporate all these themes, or data pertaining to them, under a single water security indicator is a question returned to. The chapter closes by identifying a number of key considerations, based on the experience with existing water-related indicators discussed throughout, to help guide development of water security indicators.

To set the scene, the chapter first considers the water-related indicators, and their associated targets, in the Millennium Development Goal (MDG) framework: the percentage of the population with access to 'improved' drinking water supply and sanitation.

Interested Parties: What's at Stake in an Indicator?

The rise to prominence of the drinking water and sanitation access indicators followed a political decision about which issues would be prioritised as MDG targets (in the case of both water supply and sanitation, that target is to reduce the population without access by half, compared to a 1990 baseline). Yet, over time, the indicators themselves have become widely accepted and utilised, subject to the interests of a number of constituencies—notably donors, developing country governments, and international NGOs.

Given the implicit priority of the targets is to increase access, a simplistic expectation might be that the MDG indicators would be used by public agencies to direct the greatest funding and associated capacity support to areas with the lowest access rates. But this is not necessarily evident from historic patterns of progress and resource allocation. De Waal et al. (2012) found that, from 1990 to 2008, one group of low-income countries had, on average, made lower levels of progress in increasing access to drinking water and sanitation. But this same group received a third of the aid per head of population without water supply and sanitation services, compared to that which flowed to a second group of countries. So what could explain the apparent mismatch between the signal presented by the indicators, and the patterns of resource allocation? While both groups of countries were categorised as low income, the second group was also classified as 'fragile', a contentious label which nonetheless appears to have at some level diverted effort from these countries, despite their need.

This does not, however, mean that the indicators have been ignored entirely: An alternative reading is that funds have flowed to those countries, areas, and sectors that demonstrate the greatest progress according to the MDG indicators of access. Yet even in this case, it is difficult to imagine that resource allocation has been decided by donor governments systematically examining access rates and agreeing which countries to prioritise. More likely, resources have been directed on the basis of existing political relationships and a general sense that it's easier to 'get things done' in stable countries where there is existing local capacity in government and the private sector to develop and maintain services. Similar patterns are observed at the subnational level, where hard-to-reach populations have been neglected despite the evidence from indicators of access that this is where need is greatest, while the same has been argued of sanitation in general compared to water supply (WaterAid, 2011).

It is impossible to prove either of these explanations: Considerable time has elapsed since original funding decisions were made, and it is in any case rare to get transparent answers about why large volumes of finance are spent in certain ways. Nonetheless, the example suggests that even if an indicator is widely ascribed and referred to by those holding the purse strings, it can be used to support very different courses of action. Other information signals, including the perennial attraction of 'low-hanging fruit', may trump an objective assessment of needs based on indicators.

It is not only those constituencies with significant resources to at their disposal whose interests can coalesce around indicators. Civil society campaigns such as End Water Poverty have used the water supply and sanitation indicators as points of leverage to hold donors and governments accountable. The integration of sanitation within the MDG targets and indicators is seen as a major advocacy victory by NGOs such as Tearfund and WaterAid (O'Connell, 2007). This hints at an 'ecology' of indicators, whereby those with resources seek metrics to help achieve impact or value for money and demonstrate that achievement; while those with interests in particular agendas, from NGOs to industry lobbyists, seek to shape the metrics and, in so doing, the direction of resources. The research community may meanwhile serve its own interests by acting as a broker, developing ambitious, well-funded programmes to provide new (and often complex) indicators.

In practice, many stakeholders would acknowledge the imperfections and perverse incentives inherent in such a market. But this caution can be forgotten when a new opportunity arises. For those seeking indicators for relatively novel concepts like water security, awareness that 'we measure what we value, and we also come to value what we measure' (Hoon et al. 1997, in Molle and Mollinga, 2003) is important, but can be further caveated in two respects. First, different constituencies will embed different values in what they measure. Second, while the greater value placed in what is measured may materialise in terms of greater resourcing, there is less evidence that this additional resourcing translates into greater impact. In the case of the MDGs, while an increase in aid is apparently correlated with buy-in to the goals and targets, 'the impact of that aid on outcomes is difficult to assess and plausibly muted' (Kenny and Sumner, 2011, p24).

At a more technical level, too, the MDG indicators of access have not been free of contention and differences of interest. Disputes have arisen over the relative validity of international estimates (compiled from household surveys by UN agencies) and those of national governments (often extrapolated from infrastructure built) (de Waal et al., 2011). The Joint Monitoring Programme (JMP) of UNICEF and the World Health Organization argues that household survey responses provide a better guide to the level of service people actually have access to. Meanwhile, some national governments argue that infrastructure built is a fairer (and potentially more frequently updated) metric on which to hold them to account.

Both national government and JMP approaches to measuring access take certain technology types—such as wells, boreholes, and piped water supply—as proxies for the safety of the water actually received and general adequacy of the service (WHO and UNICEF, 2000). The choice of proxy is sensible given the cost of assessing the quality (as well as the affordability, quantity, proximity, and consistency) of water at point of use. But within a given technology bracket there is potential for huge variation—for example, poorly designed and constructed boreholes, especially in developing countries. Moving from indicators of technology type to more direct measurement of the user experience itself has been a recurrent concern for the JMP.

For development of water security indicators, then, a second lesson to be drawn from the MDG indicators of access is that interests can diverge not only around 'what' to measure, but also 'how' to measure it (i.e., how to define indicators at a relevant scale in time and space and ensure they are underpinned by available data of sufficient quality). Paradoxically, moreover, once interests do converge around what to measure, the resulting increase in political significance of the chosen indicator can increase contention over how it is measured.[1]

Investing now in working through competing interests and potential unintended consequences in a transparent manner could avoid costly mistakes later. For example, Simon Kuznets, an originator of the Gross Domestic Product (GDP) concept within the U.S. national accounting framework, cautioned in 1934 that measures of national income should not be equated with welfare (Costanza et al., 2009). But GDP is still used today, by governments around the world, as a key indicator for the health of the economy, and, by extension, society as a whole.

Bigger Is Better? Data, Indicators, and Meta-Indicators

Before turning to the five component themes that this chapter proposes are integral to water security, however, it's worth pausing to consider in more detail a fundamental tension: Incorporating more variables into an indicator does not necessarily make it more useful.

A first issue is data. As will be observed for many of the variables discussed under the five themes, data on the economic, social, and environmental aspects of water is frequently piecemeal. It may be compiled at a certain scale, giving an impression of robustness, but from unreliable underlying sources. Wherever there is aggregation or averaging of data, from finer to courser granularity, it is important to understand the processes involved. The same is especially true as different variables are combined together into indicators.

All indicators require a mathematical function to translate component variables into a single score. Where this function is straightforward, it is relatively easy to discern how constituent variables contribute to the overall score, and to interrogate how each has been estimated together with the quality of underlying data. But where more complex functions (notably weightings) and a greater number of variables are involved, it becomes harder for the end user to understand what an indicator score requires in terms of response.

This has not, however, limited the appeal of complex multidimensional indicators, and proponents have argued that reducing many variables to a single number can help to engage audiences, even if underlying data components and functions need then to be interrogated to gain a more rounded interpretation. This is an argument advanced for the water poverty index (WPI) developed by Sullivan et al. (2003): 'however imperfect a particular index, especially one which reduces a measure of development to a single number, the purpose is political rather than statistical' (p193). The WPI incorporates numerous variables into subindicators relating to the resource (availability, variability, and quality), access, capacity for management (using income and access to education and health as proxies), use (domestic, agricultural, and industrial), and environment. The subindicators are amalgamated using expert weightings.

The WPI experience shows the challenges of deriving a multidimensional indicator—the work was funded by a major donor and involved a large number of researchers from northern and southern countries, but has yet to be taken up as part of a regular monitoring framework or to inform policy decisions in the manner its authors envisaged.[2]

Since the publication of the original article, various authors have critiqued, applied, and, in some cases, suggested improvements to the WPI (reviewed in Fenwick, 2010). Highlighted challenges include the lack of universal agreement on component variables (making cross-comparison difficult); the subjectivity involved in weighting the indicators (in most cases, undertaken by researchers rather than broader stakeholders); problems of correlation between component variables (between themselves and with the overall score, undermining the additional informative contribution of each variable); and the normalisation technique that leads to distortions where a small number of situations are compared (Fenwick, 2010). Another important aim of the index, to provide an indicator which can be computed at a range of scales (with a preference for upscaling from more locally derived scores) has also proved challenging, due to relative availability of data and inconsistencies between the boundaries on which hydrological and socioeconomic data are assembled.

This is not to say that indicators incorporating only a few variables are without difficulty. However, if the main aim is, in Sullivan et al.'s terms, to convey a political message, then they must be judged on these terms. Ohlsson (1998) proposed a composite water and social resource scarcity index, using total renewable water resources as a measure of 'first order' scarcity (pertaining to water resources themselves), set against countries' Human Development Index (HDI) score as a measure of 'second order' scarcity or 'adaptive capacity' (i.e., as a proxy for social capacity to access the resource). Similarly, the measure of 'economic water scarcity' proposed in the Comprehensive Assessment of Water Management in Agriculture uses malnutrition rates as a proxy: Economic water scarcity is observed to occur where 'water resources are abundant relative to water use, with less than 25% of water from rivers withdrawn for human purposes, but malnutrition exists' (Molden, 2007, p63). If the purpose is to compare progress between countries and prioritise resources, the HDI or malnutrition rates may not be the most appropriate proxies for explaining societal capacity to manage and access water (for example, malnutrition may signify inadequacy of food distribution and storage systems). But Molden's index, in particular, has nonetheless had considerable communications influence, aided by a global map showing areas of 'economic water scarcity', which illustrates the need to look beyond physical insufficiency of supply relative to demand. In this sense, its political impact has plausibly been greater than the WPI.

Comparing the WPI with these simpler alternatives, the lesson for those seeking to develop water security indicators may be that, if they are to convey the maximum intelligible information, the priority should be to select the *minimum number* of the *most appropriate* variables, as well as to find ways to aggregate them so that underlying changes in component variables, and how these relate to overall scores, are clear. These are both exercises that statistical techniques can usefully support (Fenwick, 2010), but that require open dialogue about what the priorities and needs are at a political level.

Potential Fault Lines: Five Themes for Water Security Indicators

As noted, this chapter proposes that water security requires sufficient water, the capacity to access and use it, ability to manage water risks, provision of water to support ecosystem services, and an overarching and underlying institutional capacity to realise these requirements. Other considerations have their proponents, of course, and this typology is proposed more to structure the discussion than as an attempt to conclusively define water security.

Indicators of Water Availability

According to a simple logic, increasing shortage or scarcity of water might be expected to correlate with decreasing water security. But while simple indicators of availability can be used to deliver compelling and emotive messages about scarcity, they are less suited to articulating the complex spatial and temporal distribution of water. Moreover, where the physical resource features centre stage, there is a risk that attention is drawn away from localised 'scarcities' confronting poor and vulnerable people and communities (Mehta, 2011). For example, on measures of physical water availability, the Democratic Republic of the Congo (DRC) has abundant water resources compared with other countries, but scarcity of water of sufficient quality for the most critical purpose—to sustain human health—is a daily reality for more than half the population (WHO and UNICEF, 2012). Lack of infrastructure and institutional capacity further constrains the benefit that poor communities, and society as a whole, derive from water via industry and agriculture. At the limit, emphasising physical availability may, therefore, draw attention away from society's responsibility to address these 'anthropogenic' scarcities through improved water management and more equitable investment.

One of the most widely used indicators of water availability, 'total renewable water resources per capita per year' (m^3/person/year), is compiled by the UN Food and Agriculture Organization (FAO) from data submitted by national governments. Broadly, it computes the annual volume of renewable water theoretically available through runoff and groundwater recharge arising from precipitation within a country, and water flowing in from, or shared with, other countries, set against population (Margat et al., 2005). The fact that even this summary definition is cumbersome hints at the difficulty in communicating the subtlety of what an indicator actually describes. Since a landmark paper by Falkenmark et al. (1989), total renewable water resources (TRWR) per capita has been used as a simple marker of whether a country faces water scarcity or stress; according to the Falkenmark water stress index, a country faces water scarcity at the TRWR per capita threshold of 1,000m^3/person/year.

The FAO's estimates of TRWR change over time, but for most countries this tells us more about population change ('crowding' of the available volume) than about any change in renewable water availability in an absolute sense. This is because while national estimates of renewable water resources are updated by the FAO only when new information becomes available from a country, population estimates are modified yearly. That said, trends showing increased 'water crowding' (i.e., declining TRWR per capita) can still serve to direct attention towards the anthropogenic nature of water scarcity or insecurity in many situations.

Raskin et al. (1997) proposed an alternative indicator, the ratio of total annual withdrawals to total renewable water availability (m^3 withdrawn/m^3 available; also often expressed as a percentage) to more explicitly emphasise that scarcity is a function of demand as well as supply. On this relative measure, scarcity occurs where withdrawals exceed 20% of available resources. Nonetheless, the standardised per capita thresholds of both Falkenmark's and Raskin's indicators may, if read simplistically, be taken to imply that people have identical water needs regardless of their socioeconomic and cultural circumstance—for example, the extent of their dependence on irrigated farming. In both cases, too, the thresholds emphasise the problem of overexploitation, and thus do not fully reveal the water insecurity that arises with underexploitation in countries such as DRC.

Turning to technical aspects, calculating the physical availability of water for a given territorial unit is not straightforward. Variations in surface waters from season to season, and year to year, are readily apparent. But even so-called 'fossil' groundwater is characterised by flow, with water lost to, or gained from, surrounding geology, though residence times must be measured in millennia (Foster, Tuinhof, et al., 2003). The challenge of measuring a fugitive resource is compounded by infrequent updating of hydro-meteorological estimates (for precipitation and surface runoff, for example) by national agencies, following historic underinvestment in monitoring capacity globally (WWAP, 2012a).

Most estimates of available water compiled at country level overlook water quality issues, notably salinity, which in reality would further lower the available water if part of the latter had to be set aside for dilution purposes. The prevalent water availability indicators also omit soil moisture ('green' water) which remains the key form of the resource for rain-fed agriculture—the norm in sub-Saharan Africa.

Groundwater, which is for many countries far more significant in volumetric terms than surface water, and globally accounts for the vast majority (96%) of freshwater not bound up in ice, represents the most serious challenge to estimating available water resources. Insofar as groundwater is included in country-level assessments of renewable water resources, the portion that interacts more rapidly with surface water (recharge from precipitation and river flow, as well as contribution from groundwater to surface water baseflow) tends to be emphasised over that part of the storage volume with longer residence times (Margat et al., 2005). At root, here, is the challenge of designing indicators that can capture system dynamics—namely the dual characteristics of water as a stock (where it accumulates for a time, notably in aquifers but also in lakes and reservoirs) and as a flow. The characteristics of groundwater, notably the time lag between ground and surface water variability, make it a critical buffer or stabiliser against annual and interannual variation in rainfall (and, in the longer term, climate change). As such, groundwater sources provide reliable supply for rural communities in Africa in dry seasons and droughts, and supplementary or full irrigation for farmers in the Indo-Gangetic plain and semi-arid northern China (Shah, 2007; MacDonald et al., 2009; Calow et al., 2010).

At the same time, overexploitation is a real risk due to the difficulty in obtaining reliable data, and, hence, indicators, of groundwater availability, especially in a dynamic sense. And even if indicators are available to characterise the long-term viability (or 'security') of groundwater abstraction, interests may resist or seek to manipulate the message that they present. The sustained pumping of groundwater for irrigation by millions of small farmers in South Asia and the North China plains has driven rapid economic development, aided by generous subsidised energy. Levels of groundwater, meanwhile, are in some areas significantly depleted, threatening livelihoods.

Since this 'groundwater economy' emerged in the 1970s, political interests have accrued around it, and they are reluctant to jeopardise their rural support base by seeking to regulate through permits (in any case difficult with so many users drawing on a poorly mapped, subsurface resource) or to reduce subsidies (Shah et al., 2003).

Indicators of physical availability are thus conceptually and technically challenging, and are particularly open to contestation and shaping by different constituencies in ways that might not be expected. But this does not remove the need to better understand water availability, especially if in parts of the world we are indeed transitioning 'to a new era where finite water constraints are starting to limit future economic growth and development' (WWAP, 2012a, p124).

Indicators Relating to Access to Water

As noted, abundant physical water availability will have no bearing on individuals' experience of water security if they lack the wherewithal to access and manage the resource. In many contexts, the message given eloquent expression in the 2006 Human Development Report (HDR) still rings true: 'There is more than enough water in the world for domestic purposes, for agriculture and for industry'; the issue is rather that some, especially poor and marginalized people, are 'systematically excluded' from accessing water (UNDP, 2006, p3). Most of the indicators relating to access are concerned to incorporate measures of society's capacity to capture and distribute water. Indeed, the most obvious indicator relating to access, the MDG drinking water indicator, effectively excludes the resource base and focuses entirely on society's capacity to provide a service. Given the extensive discussion of this indicator above, this section is focused on various other indicators that attempt to relate societal capacity to the resource base in various ways.

The simplest example of this is arguably withdrawals as a proportion of available resources but, as noted, this indicator does not readily capture water insecurity that arises from insufficient development and exploitation of water resources. Partly in response to this challenge, Seckler et al. (1998) proposed to modify calculation of the indicator, attempting a forecast to 2025 using projections of 'reasonable future requirements' to incorporate the need for some countries to develop water resources. Projections of future demand were made using population trends and likely increased food production; the supply-side estimates projected on the basis of plausible infrastructure development and irrigation efficiency trends. As such, the indicator is also calculated as m^3 withdrawn/m^3 available, but projected for 2025. The authors, however, admit that poor data availability constrains use of the indicator for detailed planning.

The WPI, discussed above, as well as the simpler indicators proposed by Ohlsson (1998) and Molden (2007) using the HDI or malnutrition rates as proxies, revealed other approaches to incorporating measures of societal capacity to manage and use the resource. A somewhat different approach is to focus on the benefits derived from that capacity—for example, indicators of water productivity. The logic here is that if a society is increasing the value it obtains from each unit of water, measured in terms of economic output (or in the case of agricultural uses, yield or kilocalories), then it is more likely to possess adaptive capacity, and thus is less likely to be increasing water insecurity. The difficulty, as ever, is that top-line numbers can conceal more-nuanced realities. The simpler indicators of water productivity, which for example compare value added from industry or irrigated agriculture to water withdrawals (US$/m^3 withdrawn) (UN-Water, 2009), may not convey important considerations. These include the risk of degradation in the quality of the resource that often accompanies water use intensification, or the equity with which any benefits are distributed between different members of society (and the related consideration of how to properly value benefits from water use in the 'informal' economy, such as smallholder irrigation).

In the case of indicators of agricultural water productivity, irrespective of whether the output is measured in terms of money, kgs, or kcals, a further consideration is what unit to use for the water input. Volume of water withdrawn is less appropriate for areas where agriculture is predominantly rain fed, as in much of sub-Saharan Africa. And even for areas in which irrigation predominates, what is often more relevant is consumptive use (i.e., water evaporated and transpired which is not then available to other users downstream). Reviewing these considerations around water productivity, the issue, as ever, is not so much the indicator itself, but careful choice of variables, and ensuring the resulting indicator is interpreted alongside important contextual information.

Indicators of Water-Related Risk

The word *security* implies resilience to instability and extremes. Prevalent definitions of water security usually capture this with reference to risk (e.g., Grey and Sadoff, 2007). Such risks can arise from both insufficiency and overabundance—and in terms of quality as well as quantity. Indicators are, naturally, more readily available to characterise variation where it falls within boundaries defined by historic experience, for example, the coefficient of variation of precipitation or runoff (de Stefano et al. 2010) (the coefficient of variation is the ratio of the standard deviation to the mean and may be expressed as a percentage). According to one school of thought, climatic variability, particularly within years, is a more important determinant of economic growth for agricultural, low-income economies than average availability (Brown and Lall, 2006).

As well as the likely bounds of variation for a water-related phenomenon, the probability of an event of certain magnitude occurring can be estimated. Such probabilistic metrics have been applied to disaster risks for some time, including hydrological extremes, such as floods. The concept of security implies freedom from harm, and an individual or collective sense of water security is likely to be intimately bound up with the risk of an extreme hydrological event occurring. Stochastic datasets of extreme events, which aim to simulate the likely range of possible events based on probability distributions, are an important development, though it is still difficult to model the effects of future climate change (Michel-Kerjan et al., 2012). And despite the increasing sophistication of global climate models, linking and downsizing them to the single river basin or subcatchment scale of hydrological models remains a challenge.

Even the most accurate modelling of the magnitude and likelihood of hazards leaves an incomplete picture of risk. As the metrics on disaster risk acknowledge, risk is also shaped by the exposure and vulnerability of people and assets. Given the extent of uncertainty about future hydrological extremes, understanding these societal dimensions of risk is especially important. The multidimensional Mortality Risk Index, developed for the Global Assessment Report on Disaster Risk Reduction, attempts to build a composite indicator around hazard, vulnerability, and exposure components for floods and cyclones. Drought risks are, however, harder to model in this way, due to the lack of historic data and difficulty distinguishing between socioeconomic and meterologic causes (Peduzzi et al., 2010).

Indicators of water storage and conveyance capacity (e.g., m^3 capacity per capita) have been proposed as simple proxy indicators for society's ability to cope with hydrological impacts of climate change, drawing on databases of impounding dams, groundwater as a share of total renewable water resources, and extent of irrigated land (UN-Water, 2009). It can, of course, be argued that this is more simplistic than simple, and the global datasets have serious shortcomings: for example, overlooking groundwater that has long residence times and smaller built water storage infrastructure.

Both the complex multidimensional indicators and simpler proxies mentioned above, however, arguably disguise the fact that 'risk analysis is a political enterprise as well as a scientific one', deeply intertwined with issues of public perception, power, and trust (Slovic and Weber, 2002, p2). This is to some extent acknowledged in recent framings of water risk by and for the private sector, within which regulation and reputation are important dimensions. The World Resources Institute's Aqueduct Water Risk Framework, for example, recommends integrating measures of media coverage of water issues, drinking water supply coverage, and threatened amphibious species as proxies to give companies an idea of the extent to which their water-related operations may prove contentious in a given location (Reig et al., 2013).

'Water footprinting', although not always viewed as a risk-oriented approach, has been used by business actors as a means to understand how water-related risks can propagate through value chains, exposing them to water insecurity in remote locations. Chapagain and Tickner (2012) review the evolution of water footprinting methodologies and reflect on the utility for communicating risks to high-level stakeholders, such as business leaders. The genesis of water footprinting shows, once again, the potential for a market to arise around indicators. Several NGOs recognise the advocacy and communications value and have supported various footprinting efforts.

With a number of major businesses undertaking water footprinting, the approach has acquired political capital and is now used by several companies. Meanwhile, the research community continues to respond enthusiastically, with standards developed by The Water Footprint Network and the International Organization for Standardization, in an effort to consolidate the analytical space (and potentially to secure future business). There are, however, as yet few demonstrated operational impacts arising from business water footprint estimates (Chapagain and Tickner, 2012). At a technical level, too, application and interpretation of water footprinting is challenging and requires a large number of assumptions. The impact of each component of a given water footprint needs to be seen in the light of the specific temporal and spatial context in which it occurs, and the value of the alternative uses foregone (the opportunity cost) (Chapagain and Tickner, 2012).

While these initiatives from both public and private actors provide important direction for attempts to conceptualise and measure water security, the full implications of water-related risk need to be more fully explored. Improved understanding of the political and public perception issues around water-related risk (and how indicators, e.g., of flood risk, can reflect and influence those perceptions) will be critical for tackling key collective action dilemmas, including climate change. Yet, at the same time, such understanding has the potential to be misused: Measures of exposure to flood risk, for one, could be used to justify very different ends, from costly flood defences, to evicting informal settlement residents from flood-prone land, to restoring natural infrastructure such as wetlands. What remains critical, if water security is to have universal relevance, is to ensure that all stakeholders' risks count equally—a concern that arises especially with the entry of new powerful actors, such as multinational corporations, into the water management space (Mason, 2013; Hepworth and Orr, Chapter 14).

Indicators of Environmental Sustainability

Despite the focus on human needs and water security described in the sections above, an understanding of water security that ignored environmental water requirements would be short-sighted. The concept of ecosystem services has helped articulate the inseparability of societal welfare and ecosystems, and underlying natural capital (Hassan et al., 2005), but environmental water requirements remain both difficult to calculate and politically contentious. The challenge is exemplified by South Africa's 'ecological reserve', the volume of water of certain quality, which the Department of Water Affairs must progressively assign to sustain ecosystems in each basin. The ecological reserve is intended to be calculated taking account of seasonally fluctuating requirements of different water bodies, as well as other needs in the given catchment or basin, with the actual figures to be agreed on a consensual basis by a range of stakeholders (Brown et al., 2010). Where the ecological reserve has been successfully calculated, it has often nonetheless proved contentious, where reductions in current allocations are required. The quality dimension is, meanwhile, proving difficult to implement because of limited technical and financial capacity for wastewater management in municipalities, further stretched by growing societal demand for flush toilets (Muller, 2012). Some progress has been made, for example in wastewater reuse, suggesting that definition of the environmental reserve has contributed, or at least reinforced, policy change towards a version of water security that recognises the dependence of society on ecosystems.

Given its relatively evolved institutions and macroeconomic trajectory, South Africa represents a key test case for resolving the political and normative tensions that arise between long- and short-term developmental objectives. These wicked problems require pragmatic negotiation, in which appeals to ecosystem services and environmental water indicators are important bargaining tools, but not the only tools available.

So what are the prospects for integrating an ecological reserve component into indicators of water security in other contexts, potentially even on a global basis? Borrowing from the methodology developed in South Africa, Smakhtin et al. (2004) made an attempt to calculate environmental water requirements in river basins worldwide. The premise is to modify the conventional indicator of the ratio of withdrawals to renewable available water resources, to take account of environmental needs (m^3 withdrawn divided by m^3 available, less m^3 environmental water requirement). Smakhtin et al. make an important distinction between low-flow requirements to sustain fish and other aquatic species throughout the year, and high-flow requirements (essentially wet-season flows) that have important roles in ecosystem processes, including fish spawning and wetland flooding. The calculation is undertaken using generic rules and global datasets, which do not permit differentiation of management classes to which different aquatic ecosystems might be (politically) assigned. The authors also acknowledge their inability to account for quality aspects of environmental water requirements, due to limitations in available data.

The data availability issue impacts other environmental water considerations that might be integrated into water security indicators, such as river fragmentation (number and location of dams, especially smaller dams), pollutants (point-source and diffuse), and even indices of key species, such as the Freshwater Living Planet index (ZSL and WWF, 2012).

Despite significant limitations in data, methodological advances continue. The System of Environmental-Economic Accounts for Water (SEEA-W) developed by the UN Statistical Division offers a standardised methodology for accounting for stocks and flows (abstraction, supply, use, reuse) in economic as well as hydrological terms. While it does not provide any single indicator or index, the utility of SEEA-W is to provide a coherent framework for collating, organising, and interpreting data, 'enabling a consistent analysis of the contribution of water to the economy and of the impact of the economy on water resources' (UN DESA, 2012, p12). Systems like SEEA-W may yet help bring about the increased data quality, availability, and harmonisation needed to properly represent the environmental dimension of water security.

Indicators of Institutional Capacity

Institutions 'are the prescriptions that humans use to organize all forms of repetitive and structured interactions' (Ostrom, 2005, p5). Put more crudely, institutions for the purpose of this chapter are the rules of the game—formal and informal, explicit and implicit. Understanding these rules, and how they condition outcomes, is an important part of understanding if and how we are making progress on any societal goal, including water security. Indicators of institutional capacity are commonly oriented towards process, or the building blocks required to get us towards an end state, rather than the end-state itself. In the case of water security, examples might relate to the extent to which mechanisms for assigning, monitoring, and safeguarding secure water rights exist; the number of disputes that arise and are resolved over water rights; or the number of catchments with water user associations in place.

Some argue that indicators of institutional performance and readiness distract us from what really matters, emphasising process over outcomes. The desire to have defined items to enumerate also means that institutional indicators often focus on formal organisations sanctioned by law or state, rather than informal equivalents, which may operate below the radar of the bureaucracies that often compile and use the indicators.

Monitoring of progress on embedding Integrated Water Resources Management (IWRM), still the dominant paradigm in terms of process, demonstrates both these tendencies. To date, broad assessments of progress on IWRM have focused on the presence of generic, formal organisational components—policies, laws, plans, strategies, assessments, programs and budgets—with limited consideration of whether these are leading to better outcomes. A 2012 global assessment invites governments to self-report on numerous, predominantly qualitative indicators regarding the status of certain institutional components, which as a sum are proposed to constitute IWRM. These range from 'cost recovery mechanisms/progressive tariff structures for all water uses' to the 'involvement of the private sector in water resources management and development at the national level' (UNEP, 2012). Respondents are invited to state progress on each component, from 'underdevelopment' to 'fully implemented'.

Inviting consideration of whether an institution is functioning, rather than simply whether it exists on paper, is an important step—even if there are question marks about self-reported responses. A further challenge is to understand whether these components, even if implemented, do contribute to enhanced water management outcomes, or indeed to water security. The core problem is that we have comparatively little empirical evidence for the impact of different institutional reforms in water management. To take the above example indicators, while the case for the importance of sustainable cost recovery is reasonably clear, both intuitively and empirically, this is less so for private sector involvement.

The complexity of institutions means that properly diagnosing functionality, let alone the extent to which they underpin outcomes, is expensive and time consuming. The approach adopted in the above-mentioned assessment of IWRM, using simple self-reported questionnaires, is therefore natural. Relatively simple, high-level indicators of water resources management institutions may also be used to target more in-depth research. In a 2010 World Bank study, de Stefano et al. (2010) mapped indicators of present and projected hydrological variability against indicators of relevant institutions, such as treaty mechanisms for water allocation or adapting to variable flow, and river basin organisations. The resulting hotspots were proposed as targets for more in-depth research on water-related climate change impacts.

Nonetheless, given the complexity of institutions in the water domain and the common difficulty in translating generic blueprints across contexts, the ultimate ambition should ultimately be two-fold. First, indicators of the performance and functioning of selected institutional components must be evolved, rather than just assessing whether they are present on paper. For example, if transparency of information is considered to be key, the aim would be to incorporate information on the rate of response to access to information requests for water-related data, or the proportion of that data made available, rather than whether there is an access to information policy. Second, and simultaneously, more work is urgently needed to establish what, if any, generic institutional functions for water resources management *do* contribute to better social, economic, and environmental outcomes.

Indicators of Water Security: Have Lessons Been Learnt?

None of the above-mentioned indicators actually claims to measure water security—a sign that it is a relatively new concept, and one for which verbal and numerical definitions are still in contention. A few attempts have been made, however, that encounter many of the same challenges discussed in relation to the five thematic areas discussed. A brief review nonetheless provides some important further pointers for future efforts.

Vörösmarty et al. (2010) developed a human water security threat index encompassing a wide range of variables ('drivers of stress'), including nitrogen-loading, dam density, and the ratio of population to river discharge (a 'localised' version of total renewable water resources per capita). As such, the index is open to many of the concerns levelled around other water-related multidimensional indicators, such as the WPI. Vörösmarty et al.'s study is nonetheless worth considering for several reasons, particularly the underlying techniques for data acquisition and the presentation of the index.

Data is primarily obtained from geospatial databases and models, permitting the authors to compute subindicators on a more closely disaggregated basis compared to many previous studies, using hydrological rather than administrative boundaries. River corridors are also mapped to account for the downstream transmission of hydrological impacts. Their use of modelling and remote sensing to overcome gaps in data acquired at ground level is not new, but points to the possibility for technology to help address some of the major data gaps that hamstring the application of water-related indicators. This does not remove the need to verify data from ground-level sources. It can, however, provide a relatively inexpensive means to obtain more dynamic and comprehensive estimates of certain variables, including evapotranspiration and land use (with water use by association). Water quality is more challenging, though spectral analysis can be used to assess certain types of water quality problems, for example, eutrophication of standing water bodies. Remote sensing can also be applied to estimating groundwater volume, though currently the gravimetric technology is useful only for measuring average change in volume across large areas (WWAP, 2012a).

In terms of presentation, Vörösmarty et al.'s human water security threat index has not been regularly updated to provide a 'live' decision support tool, despite being to some degree proposed as assisting in prioritising resources. But the greater utility of these indices, and arguably the primary purpose, is as a communications tool, with the human water security and biodiversity threat indices presented using visually compelling global maps. The maps portray a situation in which many transitioning and developing regions are struggling to mitigate human water security challenges. Equivalent maps illustrating a biodiversity threat index (using the same underlying indicators weighted differently) show that developed countries are succeeding in mitigating human water security threat at the expense of ecosystem integrity.

Another water security index, developed by Lautze and Manthrithilake (2012), selects component data items under the categories of basic household needs, food production, environmental flows, risk management, and independence (internally generated water resources). Compared to Vörösmarty et al.'s indices, the index is more 'user-friendly', with a much smaller number of underlying variables that are more readily apparent to the user, and with subscores that are first normalised then aggregated on a linear basis using equal weightings, rather than expert-judged weightings. As such, Lautze and Manthrithilake's index captures far less of the nuance in Vörösmarty et al.'s index, but represents a credible attempt to focus down on a number of readily available data items and compile these in an easily intelligible manner.

Comparing the two, the lesson is perhaps to take careful stock of the end purpose to which an indicator is directed. If the aim is to provide a snapshot of the issue that is as comprehensive and up-to-date as possible, and to communicate this in engaging visual terms, then some of the approaches used by Vörösmarty et al. will be relevant. Meanwhile, while Lautze and Manthrithilake's index and subindicators would benefit from further testing and refinement, their basic approach is perhaps more promising for the development of indicators primarily concerned to support decision making in real time.

Recommendations for Developing Water Security Indicators

As long as water security continues to gain traction as a normative concept, the desire to develop indicators to help communicate it, monitor progress, and support planning and prioritisation will remain. Consequently, this chapter closes with a few key considerations, drawn from the above discussion, to help navigate the political and technical challenges that have been encountered with the many water-related indicators already in existence.

The first is, simply, that water security is an extremely challenging issue for which to construct indicators. Even the physical resource itself presents numerous conceptual and technical difficulties, characterised as it is by spatial and temporal variability, but also possessing simultaneous stock and flow characteristics. Water security indicators must also characterise how hydrological systems interact with social and ecosystems, which are similarly innately unpredictable and resistant to generalisation. For those contemplating development of water security indicators, complexity needs to be acknowledged without letting it overwhelm the endeavour.

The second consideration is that decisions about both what to measure within a concept of water security, and how to measure it, need to be informed by a recognition of different interests. These interests can be at work at both the technical and political levels; assigning numbers to a concept does not mean that contention and subjectivity are somehow stripped away. While empiricism should be the watchword, those developing indicators need to be alert to the potential for different interests to be in play, and for even the most seemingly neutral numbers to be understood and used in different ways.

The importance of interests points to a third key consideration for the design of indicators for a concept such as water security: the need for open dialogue. This chapter proposed certain key themes as the most important components of water security. But the appropriateness of these themes, beyond providing a structure for the chapter itself, can and should be debated in a manner that gives voice to the broadest possible range of stakeholders. Open dialogue on the development and discussion of indicators for water security may in turn help to reach some consensus about what the concept really means in different contexts.

Whether multiple sets of interests, let alone all interests, can be captured in any single indicator is a different question. The fourth consideration, arising from this chapter's discussion of complex, multidimensional indicators, suggests that a starting point should be to identify the indicator's primary purpose, and then to design the indicator in a manner fit for that purpose.

If the primary purpose is communication, then more complex multidimensional indicators, derived in a data-intensive, one-off exercise, may have a role, as long as the results are presented in a compelling manner (for example, through maps) and as long as underlying methods can be interrogated if necessary.

If the primary purpose is to monitor progress and support forward planning on a recurrent basis, then the emphasis should rather be on reducing the variables of concern, and the complexity in how they are amalgamated, to a minimum. Each additional data item and computational step further removes the end user from underlying trends, and makes it harder to judge what response is appropriate. As such, a collection of simple indicators for different dimensions of water security, each comprising only two or three variables, may be more appropriate than an all-singing, all-dancing multidimensional indicator that combines everything but means nothing to most people.

Inadequacy of water-related data will continue to hamstring the utility of indicators. Consequently, a fifth and final consideration is the need to iteratively make the case for improved monitoring, in parallel to the process of agreeing on and utilising indicators. Technology, ranging from the high end of satellite imaging and cell phones, to the low end of rainfall and well-depth gauges, will play its part. But monitoring and data management are as much institutional challenges as technical ones. Support and soft investments are needed to arrest the long decline in institutions and capacity for monitoring that has afflicted many parts of the water sector over the past decades, especially in developing countries.

'Water security' is not a conceptual panacea: It does not, in and of itself, boil down all the ways that water interacts with livelihoods, economies, and environments into two words, which can attract resources to the sector and aid integration with others. Similarly, developing indicators for water security will not suddenly operationalise an otherwise abstract concept. It is not usually indicators themselves that galvanise action, but rather targets and objectives, which are often agreed at a level removed from the indicators. This said, we stand little prospect of leveraging greater effort and resources to tackle the most serious water challenges, now and in the future, if we do not have broadly agreed on tools with which to assess and communicate our water security priorities, and progress against these. As such thoughtfully designed and used indicators will play their part.

Notes

1. At the time of writing, interest in a further set of goals and targets to succeed the MDGs after 2015 is growing. These 'post-2015' goals and targets, which may attempt to unify socioeconomic and environmental concerns, will necessarily be accompanied by indicators. As such, a properly framed concept of water security may yet be used to articulate the social, economic, and environmental aspects of water's centrality to human well-being, and the language of water security included in a post-2015 water goal. Readers of this chapter in the post-2015 world will be better placed to assess how far the positions advanced are already an anachronism.
2. It will not escape attention that the five WPI subindicators are not dissimilar from the five themes discussed in this chapter. The chapter does not question whether water security is a complex issue comprising multiple concerns, many of which can be measured, but rather whether we are better off amalgamating those considerations and measurements or keeping some degree of separation.

References

Brown, C. and Lall, U. (2006) Water and economic development: The role of variability and a framework for resilience. *Natural Resources Forum*, vol 30, no 4, pp306–317.

Brown, C.A., van der Berg, E., Sparks, A. and Magoba, R.N. (2010) 'Options for meeting the ecological reserve for a raised Clanwilliam Dam', *Water SA*, vol 36, no 4, pp387.

Calow, R.C., MacDonald, A.M., Nicol, A.L. and Robins, N.S. (2010) 'Ground water security and drought in Africa: linking availability, access and demand', *Ground Water*, vol 48, no 2, pp246–256.

Chapagain, A.K. and Tickner, D. (2012) 'Water footprint: help or hindrance?', *Water Alternatives*, vol 5, no 3, pp563–581.

Costanza, R., Hart, M., Posner, S. and Talberth, J. (2009) 'Beyond GDP: the need for new measures of progress', *The Pardee Papers* No.4, University of Boston, MA.

Falkenmark, M., Lundqvist, J. and Widstrand, C. (1989) 'Macro-scale water scarcity requires micro-scale approaches: aspects of vulnerability in semi-arid development', *Natural Resource Forum*, vol 13, no 4, pp258–267.

Fenwick, C. (2010) *Identifying the Water Poor: An Indicator Approach to Assessing Water Poverty in Rural Mexico*, PhD thesis, University College London.

Foster, S., Tuinhof, A., Kemper, K., Garduno, H. and Nanni, M. (2003) 'Characterization of groundwater systems: key concepts and frequent misconceptions', *GW-MATE Briefing Note*, World Bank, Washington, DC.

Grey, D. and Sadoff, C.W. (2007) 'Sink or swim? Water security for growth and development', *Water Policy*, no 9, pp545–571.

Hassan, R., Scholes, R. and Ash, N. (2005) *Ecosystems and Human Well-Being: Current State and Trends, Volume 1*, Island Press, Washington, DC.

Jesinghaus, J. (1999) 'Case study: the European environmental pressure indices project', Paper prepared for the workshop *Beyond Delusion: Science and Policy Dialogue on Designing Effective Indicators of Sustainable Development*, May, International Institute for Sustainable Development, Costa Rica.

Kenny, C. and Sumner, A. (2011) *More Money or More Development: What Have the MDGs Achieved?*, Center for Global Development, Washington, DC.

Lautze, J. and Manthrithilake, H. (2012) 'Water security: old concepts, new package, what value?', *Natural Resources Forum*, vol 36, pp76–87.

MacDonald, A. M., Calow, R. C., Macdonald, D. M., Darling, W. G., and Dochartaigh, B. E. (2009) 'What impact will climate change have on rural water supplies in Africa?', *Hydrological Sciences Journal*, vol 54, pp690–703.

Margat, J., Frenken, K. and Faures, J.-M. (2005) 'Key water resources statistics in Aquastat', Paper presented at *IWG-Env, International Work Session on Water Statistics*, 20–22 June, Food and Agriculture Organisation, Vienna.

Mason, N. (2013) 'Uncertain frontiers: mapping new corporate engagement in water security', *ODI Working Paper*, Overseas Development Institute, London.

Mehta, L. (ed) (2011) *The Limits to Scarcity: Contesting the Politics of Allocation*, Earthscan, London.

Michel-Kerjan, E., Hochrainer-Stigler, S., Kunreuther, H., Linnerooth-Bayer, J., Mechler, R., Muir-Wood, R., Ranger, N., Vaziri, P. and Young M. (2012) *Catastrophe Risk Models for Evaluating Disaster Risk Reduction Investments in Developing Countries, Risk Management and Decision Processes Center*, The Wharton School, University of Pennsylvania, Philadelphia.

Molden, D. (ed) (2007) *Water for Food, Water for Life: A Comprehensive Assessment of Water Management in Agriculture*, Earthscan, London.

Molle, F. and Mollinga, P. (2003) 'Water poverty indicators: conceptual problems and policy issues', *Water Policy*, vol 5, pp529–544.

Muller, M. (2012) 'Lessons from South Africa on the management and development of water resources for inclusive and sustainable growth', Background Report to the European Report on Development, ODI, DIE, ECDPM.

O'Connell, M. (2007) *The Advocacy Sourcebook*, WaterAid, London.

Ohlsson, L. (1998) *Water and Social Resource Scarcity*, Food and Agriculture Organization, Rome.

Ostrom, E. (2005) *Understanding Institutional Diversity*, Princeton University Press, Princeton, NJ.

Peduzzi, P., Chatenoux, B., de Bono, H., de Bono, A., Deichmann, U., Guiliani, G., Herold, C., Kalsnes, B., Kluser, S., Løvholt, F., Lyon, B., Maskrey, A., Mouton, F., Nadim, F. and Smebye, H. (2010) 'The Global Risk Analysis for the 2009 Global Assessment Report on Disaster Risk Reduction', International Disaster and Risk Conference (IDRC), Davos 2010, 30 May, 3 June 2010, on-line conference proceedings, IDRC, Davos.

Raskin, P., Gleick, P., Kirshen, P., Pontius, G. and Strzepek, K. (1997) *Water Futures: Assessment of Long-Range Patterns and Problems*, Stockholm Environment Institute, Stockholm.

Reig, P., Shiao, T. and Gassert, F. (2013) 'Aqueduct water risk framework', Working paper, World Resources Institute, Washington, DC.

Seckler, D., Amarasinghe, U., Molden, D., de Silva, R. and Baker, R. (1998) *World Water Demand and Supply, 1990 to 2025: Scenarios and Issues*, International Water Management Institute, Colombo.

Shah, T. (2007) 'The groundwater economy of south Asia: an assessment of size, significance and socio-ecological Impacts', in M. Giordano and K.G. Villholth (eds) *The Agricultural Groundwater Revolution: Opportunities and Threats for Development*, CAB International, Wallingford, UK.

Shah, T., Deb Roy, A., Qureshi, A.S. and Wang, J. (2003) 'Sustaining Asia's groundwater boom: an overview of issues and evidence', *Natural Resources Forum*, vol 27, pp130–140.

Slovic, P. and Weber, E.U. (2002) 'Perception of risk posed by extreme events', Paper prepared for discussion at the conference *Risk Management Strategies in an Uncertain World*, 12–13 April, Palisades, NY.

Smakhtin, V., Revenga, C. and Doll, P. (2004) *Taking into Account Environmental Water Requirements in Global-Scale Water Resources Assessments*, Comprehensive Assessment Secretariat, Colombo.

de Stefano, L., Duncan, J., Dinar, S., Stahl, K., Strzepek, K. and Wolf, A. (2010) 'Mapping the resilience of international river basins to future climate change-induced water variability', *Water Sector Board Discussion Paper Series*, No. 15, World Bank, Washington DC.

Sullivan, C., Meigh, J., Giacomello, A., Fediw, T., Lawrence, P., Samad, M., Mlote, S., Hutton, C., Allan, J.A., Schulze, R.E., Dlamini, D.J.M., Cosgrove, W., Delli Priscoli, J., Gleick, P., Smout, I., Cobbing, J., Calow, R., Hunt, C., Hussain, A., Acreman, M.C., King, J., Malomo, S., Tate, E.L., O'Regan, D., Milner, S. and Stely, I. (2003) 'The water poverty index: development and application at the community scale', *Natural Resources Forum*, no 27, pp189–199.

UN DESA (2012) *System of Environmental-Economic Accounting for Water*, Statistics Division, United Nations Department of Economic and Social Affairs, New York.

UNDP (2006) *Human Development Report 2006. Beyond Scarcity: Power Poverty and the Global Water Crisis*, United Nations Development Programme, New York.

UNEP (2012) *Status Report on the Application of Integrated Approaches to Water Resources Management*, United Nations Environment Program, Nairobi, Kenya.

UN-Water (2009) *Final Report of the UN-Water Task Force on Indicators, Monitoring and Reporting. Monitoring progress in the water sector: A selected set of indicators*, UNESCO, Perugia, Italy.

Vörösmarty, C., McIntyre, P., Gessner, M., Dudgeon, D., Prusevich, A., Green, P.G., Glidden, S., Bunn, S.E., Sullivan, C.A., Reidy Liermann, C. and Davies, P.M. (2010) 'Global threats to human water security and river biodiversity', *Nature*, no 467, pp555–561.

de Waal, D., Hirn, M. and Mason, N. (2011) *Pathways to Progress: Transitioning to Country-Led Service Delivery Pathways to Meet Africa's Water Supply and Sanitation targets*, Water and Sanitation Program—Africa, Nairobi, Kenya.

WaterAid (2011) *Off-Track, Off-Target. Why Investment in Water, Sanitation and Hygiene Is Not Reaching Those Who Need it Most*, WaterAid, London.

WHO and UNICEF (2000) *Global Water Supply and Sanitation Assessment 2000 Report*, World Health Organization and UNICEF Joint Monitoring Programme for Water Supply and Sanitation, Geneva.

WHO and UNICEF (2012) *Progress on Drinking Water and Sanitation: 2012 Update*, World Health Organization and UNICEF Joint Monitoring Programme for Water Supply and Sanitation, Geneva.

WWAP (2012a) *Managing Water under Uncertainty and Risk. The United Nations World Water Development Report 4*, World Water Assessment Programme, UNESCO, Paris.

ZSL and WWF (2012) *The Living Planet Index—An indicator of the State of Global Biodiversity*, World Wildlife Fund and Zoological Society of London, London.

13 Water Security Risk and Response

The Logic and Limits of Economic Instruments

Dustin Garrick and Robert Hope

Introduction: Water Security as a 21st-Century Challenge

Water security addresses multiple water-related risks for people, economies, and the environment and has multiple meanings and disciplinary traditions (Cook and Bakker, 2012; Grey and Garrick, 2012). Acknowledging the extensive political ecology literature on market environmentalism and neoliberal environmental policy (e.g., Bakker, 2002, 2005), this chapter takes a pragmatic look at economic instruments through the lens of risk science and institutional economics. This chapter focuses on (1) a risk-based framework to elaborate key concepts, categories, and dimensions of risk; and (2) the logic (design, sequencing, and performance) and limits (transaction costs and governance failures) of economic instruments to reduce and manage water-related risks to society. This chapter will consider both established and emerging economic instruments and consider how the instruments address specific water-related risks. Examples from the United States, Australia, India, Latin American, and Africa illustrate the risk implications, necessary conditions, and lessons learned from economic instruments used to manage multiple water security risks. This analysis will be used to argue that managing multiple risks and tradeoffs is integral to water security and requires a mixture of governance and economic instruments to enhance efficiency and cost effectiveness while ensuring equitable outcomes and accountability (Hope et al., 2012)

A framework is developed to examine the opportunities and limits of economic instruments to reduce risk and manage residual threats. A risk-based framework may inform tradeoffs by reconciling social, physical, and engineering perspectives on water security risk. In the next section, this framework categorises water security risks and identifies economic instruments often proposed in response. Specifically, this framework uses risk concepts to align risks and possible responses. A risk-response logic can guide the design, sequencing, and evaluation of economic instruments and related policy tools, as illustrated by a series of case studies later in the chapter. Following the case studies, it is argued that economic instruments are not a panacea and have involved high transaction costs, extensive institutional reform, and collective action at multiple levels. A transaction costs perspective is presented to identify the institutional and governance arrangements needed for economic instruments to manage risks effectively. This analysis demonstrates that economic instruments require investment and political will for institutional reform, information, and infrastructure; the opportunities and limits to apply economic instruments will vary across geographic and political economic differences.

A Risk-Based Framework for Water Security

This section elaborates a risk framework in order to:

- Introduce risk concepts applied to water security
- Categorise the principal types of water-related risks, including their driving forces, impacts (including scale), and affected actors
- Examine the logic and limits of economic instruments to manage risks and tradeoffs

Risk concepts are increasingly used to understand and inform behavioural change and public and private sector investments to enhance water security (Loucks, 2011). The prevalence of multiple, interacting water-related risks and uncertainties increasingly requires tradeoffs to share risks. *Risk* is defined as a threat to social, economic, and environmental values associated with multiple water-related challenges, for example, water stress, pollution, climate variability, and unreliable water supply and sanitation services. Quantitative risk assessment defines risk as a function of probability and harm to calculate an expected cost. Probability, uncertainty, and social perceptions are important influences on water security risk, however. Importantly, water-related events include (1) recurrent or chronic threats, such as competing uses, pollution, or drought; and (2) episodic shocks from low-probability, high-impact, and unpredictable events, such as extreme floods. Risk management principles have been particularly advanced in the case of flood management, including the development of insurance schemes to manage extreme events (Hall, 2011). Some design insights from flood management can inform the application of economic instruments to manage other water security threats.

The multiple types of water security risk can be categorised in terms of their driving forces, impacts (including scale), and affected actors. Risks combine the sources of risk and their driving forces.

Water Security Risk

The water security risks are framed in terms of quantity and quality, as well as climate risks and reliability of water services. These risks include (1) water stress (natural aridity, scarcity, and competition, e.g., supply–demand imbalances due to natural aridity and growing demand, and over-extracted groundwater aquifers); (2) water pollution; (3) climate risks, including impacts of climate variability, change, and extremes such as floods and droughts; (4) threats to the reliability of water supply and sanitation services; as well as (5) the cumulative and synergistic effects and feedbacks of multiple, interacting risks. A final category of risks relate to erosion of existing infrastructure and institutions due to neglect, insufficient financing, and lack of management and maintenance. It should be carefully noted that these categories of risk are not mutually exclusive; they interact and overlap. For example, water stress and climate risks both capture drought; unreliable water services may amplify and exacerbate risks tied to water stress, pollution, and so forth. However, the selected risks identify dominant ways of classifying risks to manage water resources and water services and to design appropriate governance and economic responses. These risks are driven by multiple, interacting forces including demographic and economic change, technology and infrastructure, social values and behaviour, disasters and climate, and hydrology.

Water security risks interact and need to be understood in terms of both their independent and synergistic impacts at multiple scales. For example, the United Nations World Water Assessment IV released in 2012 acknowledges 'coordinated and synergistic management in related domains improves overall outcomes' (UNESCO, 2012). Risks rarely act in isolation. For example, water stress and pollution are interrelated and exacerbated by vulnerability to climate risks and unreliable water services.

A Risk-Based Framework

A risk-based framework is useful to classify water-related risks in terms of their likelihood and impacts, including scale. The framework identifies the driving forces, risks, likelihood, and affected actors. This framework informs decision makers by matching management tools and economic instruments to different types of water-related risks. Decision makers cannot know the exact outcome of their choices, which increases the importance of information, risk, and uncertainty, as well as potential for economic instruments to manage residual threats when coupled with effective multilevel governance.

Two broad types of risks are distinguished by their likelihood and impact: (1) predictable, high probability events with high cumulative impacts (e.g., chronic water scarcity or water pollution); and (2) unpredictable, low probability events with very high impact shocks (e.g., flood, prolonged drought, industrial accident causing water contamination). Predictability requires an examination of the role of uncertainty, social perceptions, and behaviour in risk mitigation and response. In theory, economic instruments can coordinate information and incentives matched to these two broad types of risk. For example, insurance schemes are best suited for the least predictable water security risks. Economic instruments could also be used to encourage and incentivize actors to build their adaptive capacity.

Quantitative assessment of risk relies on predictable variation with a known distribution, and, therefore, highlights the importance of uncertainty (Quiggin, 2011). Uncertainty refers to variation with low predictability and an unknown probability distribution. A bell curve (an example of a probability density

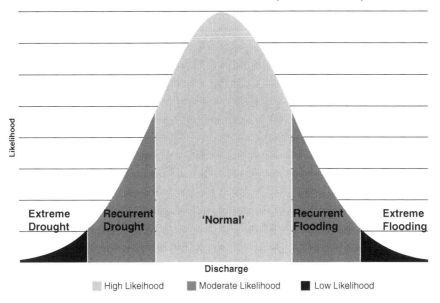

Extreme
Drought

Recurrent
Drought

'Normal'

Recurrent
Flooding

Extreme
Flooding

Discharge

High Likelihood   Moderate Likelihood   Low Likelihood

Figure 13.1 Climate extremes and the likelihood of water security risks

function, see Figure 13.1) is a tool used to understand water-related risk. It is important to note that many risks do not follow a normal distribution (e.g., a bell curve) (Pindyck, 2006; IPCC, 2012). Finally, climate change, ecosystem degradation, and other natural and human processes require a focus on 'tipping points' to manage complex water-related risks. Climate change adaptation strategies therefore emphasise no-regrets investments that are attractive across a range of future scenarios and avoid irreversible impacts.

The information on likelihood and impacts informs the development of economic instruments to internalise risk for extreme events (e.g., flood insurance), hedge risk for water stress and pollution (cap-and-trade), and mitigate risk of unreliable public water supply and sanitation through marginal costs pricing and infrastructure financing schemes. Risks can have high impacts, due to extreme events caused by low extremes (prolonged drought) and high extremes (extreme flooding). These 'tail-of-distribution' events have a low probability and are difficult to predict. Recurrent droughts (low extremes) and floods (high extremes) are more probable. Predictive capacity has improved for moderate likelihood events—for example, forecasting conditioned by El Nino-Southern Oscillation data, as well as other hydro-meteorological advances. High likelihood risks combine high probability and predictability, where the impacts may be relatively low compared to extremes but lead to significant cumulative impacts (for example, they are tied to the lack of water supply and sanitation and associated impacts on human development).

Affected actors and ecosystems are fundamental parts of a risk-based framework for designing and implementing economic instruments. The affected actors can be considered in terms of exposure to water-related risk, vulnerability to impacts, and adaptive capacity to limit exposure and reduce vulnerability. The IPCC's Special Report on Managing the Risks of Extreme Events and Disasters to Advance Climate Change Adaptation (SREX) was released in 2012. It characterises these dimensions of a risk-based framework in the context of disasters and extreme events. The water security risks are broadly analogous to these extreme events.

- *Affected actors* include households and economic sectors at multiple levels, from the local to the global. For example, the 2011 flooding in Thailand displaced people, disrupted livelihoods, and had spillover global impacts on electronics and car supply chains.
- *Exposure* refers to the presence of people, livelihoods, infrastructure, and ecosystems in areas susceptible to water security risks, including extreme events (IPCC, 2012; Gasper, 2010). Exposure of affected actors is determined by physical processes, human behaviour, information, infrastructure, and policy measures.
- *Vulnerability* to impacts depends on the ability of affected actors to 'anticipate, cope with, resist, recover from' water security risks (IPCC, 2012). Vulnerability, like exposure, is linked to biophysical, social, and policy conditions.
- *Adaptive capacity* refers to the capacity of the system and actors to adapt to changing environmental conditions, including extreme events.
- *Individual and social perceptions of risk* are fundamental to decision making to manage water security risks and tradeoffs. Key attributes include the likelihood of negative impacts and vulnerability, which raises the importance of high quality and trusted information to support good governance and effective economic instruments (UNESCO, 2012). The perceptions of risk often vary according to different actors (e.g., rich, poor; young, old; level of education; etc).

If coupled with multilevel governance, economic instruments may enhance adaptive capacity while ensuring efficiency and equitable outcomes. *Efficiency* is defined as maximizing total welfare, while *equity* refers to the fairness of the allocation and resource consumption, including exposure to residual risks tied to shock events (see later in this chapter). Effective instruments must establish incentives matched to the characteristics of the risks (impact, likelihood) and of the affected actors (exposure, vulnerability). This discussion underscores the influence of development on adaptive capacity, which can lower vulnerability to water-related risks.

A Risk-Response Logic

A risk-response logic (see Figure 13.2) maps the connection between economic instruments and water security risks. This logic starts from an understanding of the water security risks as the basis for effective responses. It should be

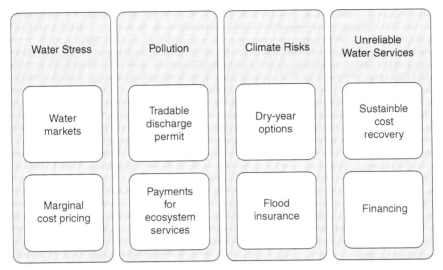

Figure 13.2 Risks and selected economic instrument responses

carefully noted that the list of responsive economic instruments is not meant to be exhaustive of the range of available tools. In this framework, economic instruments respond to water security risks in concert with collective action and governance, as such economic instruments are a complement—not a substitute—for effective governance, regulation, and political engagement at multiple levels, as outlined below.

Risk of Water Stress. Intensified competition for scarce and variable water supplies elevates the risk of supply–demand deficit and misallocation. There is a potential role of marginal social cost pricing to increase water-use efficiency, reduce demand, and stimulate private investment in supply infrastructure. Misallocation results in a loss in economic welfare when water use is locked into historic patterns, despite the emergence of new demands that may be considered higher in (economic) value than existing uses. Water markets are an economic instrument to allow buying and selling of tradable water rights. Water trading is an economic response to risks posed by scarcity/drought and competition within and across cultural, domestic, urban, industrial, and ecosystem uses (Chong and Sunding, 2006; Grafton et al., 2011). Water markets depend on adequately defining property rights, licensing, and accounting systems, and ensuring the appropriate institutions and adequate financial and technical resources to monitor the inflow and diversion patterns. Innovative insurance schemes have been tailored to manage drought and floods (reservoir storage or water rights options contracts). Water markets must overcome several market failures associated with social and environmental externalities caused by changes in water use patterns, infrastructure needs, and return flows. There are further public goods provision failures (e.g., in freshwater conservation) and market power that cause unintended consequences, uncertainty, and institutional risks. Regulatory safeguards and governance capacity are preconditions for water trading (Bauer, 2004) and be updated regularly (Garrick and Aylward, 2012) to address these social, economic, and environmental weaknesses, as outlined below.

Risk of Water Pollution. Point and nonpoint pollution and associated with biological, chemical, and physical impacts on water systems create risks for ecological and human health. This risk results in a loss of welfare and equity, particularly for vulnerable groups, where the lack of safe drinking water and sanitation contribute to public health risks. Tradable permit systems are economic instruments to manage both point and nonpoint pollution systems by establishing incentives for polluters to adopt clean technologies or purchase offsets from more efficient producers. Approaches to deal with (low-likelihood, high-impact) risks of accidental pollution are mostly regulatory (e.g., EU's Seveso II Directive). The risks of water pollution are sensitive to context. For example, semi-arid regions address chronic risks of salinity impacts, while other regions may be more exposed to risks tied to pesticide use. There are significant governance challenges associated with implementing these systems and achieving their long-term impacts, particularly given the complex biophysical and socioeconomic risks and uncertainties in such schemes.

Risk of Climate Variability, Change, and Extremes. Recurrent and extreme drought and flooding events are inherent features of climate systems. Impacts are shaped by social and economic choices that interact with natural fluctuations, such as choices about land use and infrastructure in floodplains. Climate change is projected to exacerbate extremes and make impacts more difficult and unpredictable to manage. Insurance schemes are economic instruments able to manage extremes. The most extreme events have required significant reinsurance or government intervention and backing, such as Hurricane Katrina and the Thailand Floods. Innovative insurance schemes have been tailored to manage floods (zoning-based insurance), but require careful consideration. Poorly designed schemes can encourage development in high risk areas and undermine incentives to adapt to long-term climate change and economic conditions, for example, the U.S. Federal Flood Insurance programme in some instances.

Risk of Unreliable Water Supply and Sanitation Services. These risks result in time, income, health, education, and welfare costs. Globally, 783 million people do not have reliable access to an improved water supply and 2.5 billion people do not have basic sanitation services (UNICEF/WHO, 2012). The enormous morbidity, mortality, and development burden on insecure water and sanitation services is well-documented for developing regions (UNDP, 2006). The full scale of this threat is underestimated by these metrics; billions have access to improved water supply, but still do not have the standard of water services considered acceptable by many. The economic and social consequences are significant (Payen, 2012; Cairncross et al., 2010). Further, neglect and underfinancing of existing infrastructure pose risks for many who currently have reliable water supply.

The spatial, temporal, and political economic dimensions of risk are also important and may require responses to share risks between individuals and society at multiple scales, including at the local and global levels (Marshall, 2005; Ostrom, 2012). Risks (stress, pollution, climate risks, and water services) can have significant effects on global trade. This has already been documented through recent flooding in Thailand and drought impacts in Russia. In contrast, little is known about the possible effects of trade liberalisation on risks. This includes, for example, the extent to which domestic food security objectives distort production and trade of agricultural commodities and the resulting global distribution of water demand.

Economic Instruments and Risk Implications

Economic instruments have important risk implications, including the ability to reduce and share water risk. Such instruments also depend on multilevel governance arrangements to assign roles and responsibilities for risk management between public and private sectors. The following review considers these implications systematically across illustrative economic instruments. A suite of economic tools is examined by applying this risk-response framework to define each instrument in terms of its necessary conditions, sequencing and scaling, and risk implications. It is important to underscore that this set of case studies is not exhaustive but nonetheless illustrates the promise and pitfalls of economic instruments in managing water security risks.

Case Studies: Risk Implications, Necessary Conditions, and Lessons Learned

Water Stress Risks and Allocation Response: Water Markets

In theory, water markets establish price signals to allocate water to its highest valued use (Grafton et al., 2011; Olmstead 2010). Competition for scarce water supplies is a necessary condition. In the context of scarcity and competition, a set of institutional, technological, socioeconomic, and environmental preconditions must develop. Institutional requirements include tradable water rights and associated licensing, measurement, and enforcement systems. Technological conditions include supply and distribution infrastructure to deliver and transport water. Water distribution infrastructure is often a precondition for tradable water entitlements and associated administrative capacity to measure and enforce them. Socioeconomic and environmental preconditions include mechanisms to manage negative impacts of water markets on other users, communities, or the environment. The Murray-Darling Basin is an example where initial cap-and-trade reforms enabled several hundred million dollars of welfare gains from trade but had some unintended consequences (Box 13.1). Initial reforms created tradable water entitlements but did not fully specify water accounting rules to close groundwater basins or retire unused water rights. The rising economic value of well-defined entitlements established incentives to activate previously unused water rights, leading to downstream environmental and economic impacts.

Rarely are the necessary conditions fully met at the outset, which requires sequencing of reforms to allow for scaling up. Pilot water trading in local catchments allows for adaptive learning until regional or national frameworks are developed. Piloting has been prevalent in countries where water markets have been introduced, such as in Australia, the western United States and, more recently, China. In the latter case, China has experimented with pilot water entitlement trading systems in the Jiao River Basin to coordinate irrigation and environmental water use, including a permitting system linked to caps on water extraction for large irrigation districts comprised of thousands of water users (Speed, 2009).

Box 13.1 Water Markets and Risk Assignment: An Example from the Murray-Darling Basin of Australia

The Australian water reform experience in the Murray-Darling Basin has been widely cited as an example of a mature water market. Less well understood are the efforts to manage and assign risk. In 1994, a cap-and-trade system of water markets allowed for the management of supply risk and financial risk for irrigators dependent on variable water supplies. In 2004, the Council of Australian Governments (COAG) adopted the National Water Initiative in part to strengthen market enabling reforms by establishing water rights registers, accounting standards, and pricing for water storage and delivery.

Quiggin (2011) outlines two risks addressed through these market-enabling reforms. First, water allocations were defined as proportional shares rather than as fixed volumes. Users share risk proportionally during periods of reduced inflows. The second class of risks arises from long-term changes in the total consumptive pool of water available. The risks of changes in this total consumptive pool are assigned based on the factors causing the change. Risks due to improved knowledge about long-range renewable supplies were assigned to the users. Risks due to regulatory or policy changes were assigned to the public, which may require the government to compensate users. Recent policy and planning efforts under the 2007 Water Act have introduced uncertainty about the risk assignment provisions due to changes in the sustainable diversion limits (e.g., cap). The National Water Initiative illustrates that risks of short-term reductions in water availability have been managed by defining tradable water rights as proportional shares that allow farmers to manage their production and financial risks through trading. However, long-range reductions in availability have required multilevel governance reform to adjust the cap on sustainable diversions and share the risks between the public and private sector.

The risk implications of water markets include efforts to reduce risks to buyers and sellers, particularly during drought conditions when temporary trades allow the reallocation of water from relatively low value, annual crops to permanent, high-value plantings. Water markets also depend on clear regulatory frameworks to limit and reduce risks to the environment and other users and communities, for example, by defining water entitlements as shares that are adjusted for water inflow conditions and environmental impacts.

Water Pollution Risk and Financing Response:
Payments for Ecosystem Services

Payments for ecosystem services (PES) are an economic instrument that compensates ecosystem service providers conditional on the external ecosystem benefits supplied (Chomitz et al., 1999). By internalising upstream, land-use decisions to downstream externalities, PES reduces risks of suboptimal outcomes for society associated with deforestation (sedimentation, biodiversity loss), poorly managed agriculture (nutrient transport leading to poor water quality), and industrial pollution (river health), *inter alia*.

Necessary conditions for PES are defined by five principles: (1) at least one willing ecosystem service seller, (2) at least one ecosystem service buyer, (3) a clearly defined ecosystem service, (4) voluntary transactions, and (5) payments conditional on ecosystem service supply (Wunder, 2005). Global reviews indicate that all five principles are rarely, if ever, met (Porras et al., 2008). PES schemes operate at different scales. National PES schemes operate in Costa Rica, China, and Mexico, with local initiatives identified more widely, including in the United States, France, the UK, Colombia, India, and South Africa. Institutional arrangements vary widely in terms of financial structures, with global funding bodies (World Bank, Global Environmental Facility) as key players in piloting major projects linked to carbon offsetting projects in the late 1990s and early 2000s. Initiatives to ensure financial sustainability beyond a project cycle have emphasised the need for local sources of finance, which have included private sector engagement. Examples include Nestlé Water's Vittel scheme in France (Perrot-Maître, 2006), linking farmers' incentives with wetland benefits in India (Hope et al., 2008) and farmer income and reduced water supply treatment in Brazil (Bassi, 2002).

The risk implications of PES for water security manifest in terms of financial sustainability, attribution, and behavioural change. Without a guaranteed flow of finance, many schemes have failed or been abandoned (Porras et al., 2008). Where finance is guaranteed through charges to domestic water users protecting source catchments (Ecuador) or through a small surcharge on petrol prices (Costa Rica), it remains unclear if the 'conditionality' condition is met. Attribution of ecosystem services reflects that the majority (>90%) of PES schemes are based on land-use contracts rather than on the measurement of downstream ecosystem services, which are both complex and expensive to measure, given the mosaic of land uses in any large catchment and the associated problem of attribution (Porras et al., 2008). For example, water quality has proven to be a more realistic ecosystem service to measure and value than is water quantity (Calder, 2005; Bassi, 2002). The moral hazard of paying land users to 'improve' their behaviour has led to perverse outcomes, where farmers have demanded compensation for ecosystem services outside the project boundary (Pagiola et al., 2004). Ex-ante analysis of user preferences to differing scenarios of PES arrangements has modelled heterogeneity in user choices, which provides an evidential platform to avoid costly mistakes that are rarely borne by external proponents of PES schemes (Hope et al., 2008).

Unreliable Water Services Risks and Financing Response: Sustainable Cost Recovery

Sustainable cost recovery reflects an appropriate mix of tariffs, taxes, and transfers (OECD, 2010). An emerging instrument to increase tariff efficiency and flexibility, including targeting affordable tariffs for low-income water users, is mobile money (Pickens 2010). Mobile money is an electronic payment system that allows real-time, secure, and reliable financial transfers for anyone with a mobile phone. Since 2009, water service providers in Kenya, Uganda, Tanzania, and Zambia (serving over 12.5 million urban, piped water consumers) have piloted mobile water payments, which allow water users to pay their bill using their mobile phone (Foster et al., 2012). With low collection-efficiency rates, the potential financial benefits to African utilities with a US$9 billion annual deficit are significant (Banerjee and Morella, 2011). For example, in Kenya, collection efficiency averaged 83% in 2008–2009, resulting in at least US$15 million per year remaining uncollected. With one in six water bills unpaid, fewer than one-third of Kenyan water service providers managed to recover operation and maintenance costs. This contributes to a deadweight loss for society in higher water prices, but lower water consumption with welfare impacts, operational inefficiencies, and revenue transfer from consumers to water vendors (Hope et al., 2012).

Necessary prerequisites for mobile water payments include high mobile network coverage, extensive mobile phone penetration, electronic billing systems, user acceptability, utility adoption/ promotion, metered connections, and financial regulation. With three-quarters of the estimated 5.5 billion mobile subscriptions in developing countries and 90% of Africa estimated to have network coverage by 2014, technological conditions are largely met. However, social acceptance has varied, with average adoption rates around 10% of the customer base in Africa. Key barriers to adoption include delayed reconciliation of billing systems, limited customer awareness, lack of physical proof of payment, high transaction tariffs, and convenience of alternative pay points. All of these barriers can be overcome, as illustrated by a small and privately-run scheme in Nairobi, where 85% of customers have adopted the mobile bill payments (Foster et al., 2012).

Water security risks in urban settlements in developing regions cut across financial, operational, sociocultural, and governance issues (Banerjee and

Morella, 2011; Bakker, 2010). Mobile payments offer no singular solution, but can lower transactions costs for water users and utilities. Where markets are more competitive, consumers and utilities have more choice, which can lower transaction costs and contribute to the significant operational and financial risks for water security for Africa's rapidly expanding urban population. Emerging evidence illustrates significant savings for water users in time and financial costs, and improved collection efficiency for water utilities (Foster et al., 2012). Institutional responses are emerging across urban Africa as mobile payment platforms are being rapidly adopted in differing ways in differing contexts. Understanding Africa's urban water governance challenge through a transaction costs lens provides a promising avenue to lower financial risks to promote water security. Importantly, mobile network operators are sharing the risks and transaction costs of implementation and promotion to gain or maintain market share. What is less clear is how the benefits are then distributed between water users, utilities, and mobile network operators.

Institutions Are Not Free: A Transaction Costs Perspective on Economic Instruments

The previous discussion demonstrates that economic instruments to manage water security risks depend on effective governance reforms at multiple scales (OECD, 2011, 2012). However, establishment of new institutions is costly because the political economy of water reform is often contentious. This requires attention to cultural, social, environmental, and economic barriers to reform. A transaction costs perspective is helpful to identify, measure, and reduce these barriers (McCann et al., 2005). To what extent and why do transaction costs impede the establishment and scaling up of economic instruments? Transaction costs occur because tradeoffs involve multiple goals other than efficiency and stem from a range of technological, social, cultural, and geographic factors.

Transaction costs affect water reform in at least four key areas:

- Technological, socioeconomic, institutional, and biophysical attributes create lock-in, particularly associated with historic water use patterns and infrastructure. Reversing or shifting paths is costly.
- Water rights are difficult to define, and information is costly. Water is different from other economic commodities for several reasons. Property rights in water are difficult to define with pervasive externalities and multiple uses and values, including public and private goods (Hanemann, 2006). Hydrological, social, and economic information is costly to acquire and analyse and it requires careful stakeholder engagement and public participation.
- Transaction costs reflect costs of making tradeoffs among multiple goals when resolving externalities, for example, those tied to downstream users and return flows.

- The costs of exchange must be considered along with the costs of enacting policy alternatives and economic instruments, as well as the opportunity costs of doing nothing. The costs of water reform can be significant and require decades without strategic design and sequencing. Reform happens when the opportunity costs of doing nothing outweigh the transaction costs of reform (Saleth and Dinar, 1999).

In the water context, it is helpful to consider transaction costs in terms of a nested set of costs, including: (1) costs of implementation, (2) costs of institutional development, and (3) the costs of reversing past water-use patterns and infrastructure. This approach is helpful to establish a platform of enabling conditions and sequence reforms that overcome barriers. Many proposals for water reform and economic incentives focus on the comparative costs and benefits of implementation without addressing the fundamental institutional capacity and path dependencies that can inhibit reform. Therefore, a multilevel understanding of transaction costs demonstrates the importance of piloting economic instruments, limiting potential for maladaptive lock-in, sequencing reforms, and, when necessary, harnessing crises to overcome lock-in (Garrick et al., 2013).

Concluding Remarks

There has been a nearly 40-year history of policy interest in economic instruments to manage environmental and natural resource challenges, with particular application to water security issues tied to water resource allocation and investments in water supply and sanitation. There has been increasing adoption and implementation experience with such tools in diverse settings from advanced economies to developing contexts, albeit with varying levels of success, failure, and political acceptance. However, evidence demonstrates that effective incentive-based instruments depend on a strong regulatory framework, property rights, and institutional capacity for measurement, monitoring, and information systems. Economic instruments and governance are, therefore, mutually supporting and require attention to institutional development and context.

A risk-based framework identified multiple, interacting water security risks (stress, pollution, climate risks, and unreliable water services), their likelihood, impacts, and affected actors. This framework was used to (1) map the connection between economic instruments and different risks of varying likelihood and affected actors, and (2) examine the logic and limits of economic instruments. The analysis demonstrated that the use of economic instruments should go hand in hand with effective governance. However, institutional development is costly. The political economy of water reform is often contentious and requires attention to cultural, social, environmental, and economic barriers to reform. A transaction costs perspective diagnosed governance failures impeding economic instruments and demonstrated the wider reforms needed to address these barriers. A risk-response typology of economic instruments identified a portfolio of options to reduce and manage threats posed by water stress, water pollution, climate risks, and the reliability of water services. If paired with multilevel governance reforms, economic instruments can respond to both high and low likelihood risks, as exemplified by the development of water markets to allocate scarce and variable water supplies (Australia), payments for ecosystem services to address water pollution (Latin America and India), and sustainable cost recovery to enhance the reliability of water services (Africa). Case studies identified a common platform of necessary conditions to build appropriate regulatory and governance capacity, information and monitoring systems, and appropriate sequencing to move from pilots to scale.

References

Bakker, K. (2002) 'From state to market: Water mercantilización in Spain', *Environment and Planning A*, vol 34, pp767–790.

Bakker, K. (2005) 'Neoliberalizing nature? Market environmentalism in water supply in England and Wales', *Annals of the Association of American Geographers*, vol 95, no 3, pp542–565.

Bakker, K. (2010) *Privatizing Water. Governance Failure and the World's Urban Water Crisis*. Cornell University Press.

Banerjee, S.G. and Morella, E. (2011) *Africa's Water and Sanitation Infrastructure: Access, Affordability and Alternatives*, World Bank, Washington, DC.

Bassi, L. (2002) 'Valuation of land use and land management impacts on water resources in Lajeado Sao Jose Micro-watersheds, Chapeco, Santa Caterina State, Brazil', *Land-Water Linkages in Rural Watersheds Case Study Series*, FAO, Rome.

Bauer, C.J. (2004) *Siren Song: Chilean Water Law as a Model for International Reform*, RFF Press, Washington, DC.

Cairncross, S., Bartram, J., Cumming, O. and Brocklehurst, C. (2010) Hygiene, Sanitation, and

Water: What Needs to be Done. PLoS Medicine 7 (11): 1000365.doi:10.1371/journal.pmed.1000365.

Calder, I.R. (2005) *Blue Revolution: Integrated Land and Water Resources Management* (Second edition), Earthscan, London.

Chomitz, K., Brenes, E. and Constantantino, L. (1999) 'Financing environmental services: the Costa Rican experience and its implications', *The Science of the Total Environment*, vol 240, pp157–169.

Chong, H. and Sunding, D. (2006) 'Water markets and trading', *Annual Review of Environment and Resources*, vol 31, pp239–264.

Cook, C. and Bakker, K. (2012) 'Water security: debating an emerging paradigm', *Global Environmental Change*, vol 22, no 1, pp94–102.

Foster, T., Hope, R.A., Thomas, M., Cohen, I., Krolikowski, A. and Nyaga, C. (2012) 'Impacts and implications of mobile water payments in East Africa', *Water International*, pp1–17.

Garrick, D. and Aylward, B. (2012) 'Transaction costs and institutional performance in market-based environmental water allocation', *Land Economics*, vol 88, no 3, pp535–560.

Garrick, D., Whitten, S. and Coggan, A. (2013) 'Understanding the Evolution and Performance of Water Markets and Allocation Policy: A Transaction Costs Analysis Framework', *Ecological Economics*. 88(3): 195–205.

Gasper, D. (2010) 'The idea of human security', in K. O'Brien, A.L. St. Clair, and B. Kristoffersen (eds) *Climate Change, Ethics and Human Security,* Cambridge University Press, Cambridge, UK.

Grafton, R.Q., Libecap, G., McGlennon, S., Landry, C. and O'Brien, B. (2011) 'An integrated assessment of water markets: a cross-country comparison', *Review of Environmental Economics and Policy,* pp1–22.

Grey, D. and Garrick, D. (2012) 'Water security as a 21st century challenge', *Oxford Water Security Briefing Series,* www.eci.ox.ac.uk/watersecurity/downloads/briefs/1-grey-garrick-2012.pdf.

Grey, D. and Sadoff, C.W. (2007) 'Sink or swim? Water security for growth and development', *Water Policy,* vol 9, no 6, pp545–571.

Hall, J. (2011) 'Policy: a changing climate for insurance', *Nature Climate Change,* vol 1, pp248–250.

Hanemann, W.H. (2006) 'The economic conception of water', in P.P. Rogers, M.R. Llamas, and L. Martinez-Cortina (eds) *Water Crisis: Myth or Reality?,* Taylor & Francis plc, London.

Hope, R.A., Borgoyary, M. and Agarwal, C. (2008) 'Smallholder farmer preferences for agri-environmental change at the Bhoj Wetland, India', *Development Policy Review,* vol 26, no 5, pp585–602.

Hope, R.A., Foster, T., Money, A. and Rouse, M. (2012) Harnessing mobile communications innovations for water security. *Global Policy.* 3(4): 433-442.

IPCC (Intergovernmental Panel on Climate Change) (2012) *Managing the Risks of Extreme Events and Disasters to Advance Climate Change Adaptation,* A Special Report of Working Groups I and II of the Intergovernmental Panel on Climate Change, Cambridge University Press, Cambridge and New York.

Loucks, D.P. (2011) 'Risk and uncertainty in water planning and management: a basic introduction', in R.Q. Grafton and K. Hussey (eds) *Water Resources Planning and Management,* Cambridge University Press, Cambridge.

Marshall, G. R. (2005) *Economics for Collaborative Environmental Management: Regenerating the Commons,* Earthscan, London.

McCann, L., Colby, B., Easter, K.W., Kasterine, A. and Kuperan, K.A. (2005) 'Transaction cost measurement for evaluating environmental policies, *Ecological Economics,* vol 52, pp527–542.

OECD (2010) *Pricing Water Resources and Water and Sanitation Services,* OECD, Paris

OECD (2011) *Water Governance in OECD Countries: A Multi-Level Approach,* OECD, Paris.

OECD (2012) *OECD Environmental Outlook to 2050,* OECD, Paris.

Olmstead, S.M. (2010) 'The economics of managing scarce water resources', *Review of Environmental Economics and Policy,* vol 4, no 2, pp179–198.

Ostrom, E. (2012) 'Nested externalities and polycentric institutions: must we wait for global solutions to climate change before taking actions at other scales?', *Economic Theory,* vol 49, no 2, pp353–369.

Pagiola, S., Agostini, P., Gobbi, J., de Haan, C., Ibrahim, M., Murgueitio, E., Ramirez, E., Rosales, M. and Ruiz, J.P. (2004) 'Paying for Biodiversity Conservation Services in Agricultural Landscapes', Environment Development Paper No. 96, World Bank, Washington, DC.

Perrot-Maître, D. (2006) *The Vittel Payments for Ecosystem Services: A "Perfect" PES Case?,* International Institute for Environment and Development, London

Pickens, M. (2010) Mobile money 101. Washington, DC: CGAP.

Pindyck, R. (2006) 'Uncertainty in Environmental Economics', NBER Working Papers 12752, National Bureau of Economic Research, Inc., Cambridge, MA

Porras, I., Grieg-Gran, M. and Neves, N. (2008) 'All that glitters: a review of payments for watershed services in developing countries', *Natural Resource Issues,* no 11, International Institute for Environment and Development, London.

Quiggin, J. (2011) 'Managing risk in the Murray-Darling Basin', in D. Connell and Q. Grafton (eds) *Basin Futures: Water Reform in the Murray-Darling Basin*, Australia National University E-Press: Canberra.

Saleth, R.M. and Dinar, A. (1999) 'Water Challenge and Institutional Response (A Cross-Country Perspective)', World Bank Policy Working Paper No. 2045, Washington, DC.

Speed, R. (2009) Water Entitlements and Trading Project, China. In *The Nuts & Bolts of Flow Reallocation*. Proceedings of a workshop held February 22nd, 2009 in Port Elizabeth as part of the International Conference on Implementing Environmental Water Allocations. Garrick, D., R. Wigington, B. Aylward, and G. Hubert, editors. Boulder, CO: The Nature Conservancy.

Tietenberg, T. (2006) *Environmental and Natural Resource Economics*, Prentice Hall, New Jersey.

UNDP (United Nations Development Programme) (2006) *Beyond Scarcity: Power, Poverty and the Global Water Crisis*, World Development Report, United Nations Development Programme, New York.

UNESCO (United Nations Educational, Scientific and Cultural Organisation) (2012) 'UN World Water Development Report IV: Managing Water under Risk and Uncertainty', Volume 1, UNESCO, Paris.

UNICEF/WHO (United Nations Children's Fund/ World Health Organisation) (2012) 'Progress on Drinking Water and Sanitation', 2012 Update, UNICEF/WHO, New York.

Vorosmarty, C. et al. (2010) 'Global threats to human water security and river biodiversity', *Nature*, vol 467, no 7315, pp555–561.

Wunder, S. (2005) 'Payments for Environmental Services: Some Nuts and Bolts', CIFOR Occasional Paper No. 42, CIFOR, Bogor.

14 Corporate Water Stewardship

Exploring Private Sector Engagement in Water Security

Nick Hepworth and Stuart Orr

Dank and foul, dank and foul, by the smoky town with its murky cowl,
Foul and dank, foul and dank, by wharf and river and slimy bank
Darker and darker the further I go, baser and baser the richer I grow.

Charles Kingsley, 1862

Introduction

The last decade has seen a rapid upsurge of interest in water by the private sector. The Water Footprint Network (WFN), Alliance for Water Stewardship (AWS), CEO Water Mandate, Water Futures Partnership (WFP), and the 2030 Water Resources Group (WRG) are banners for what is presented as a new paradigm for private sector engagement in water security—corporate water stewardship (CWS). Water stewardship and the debates around it are distinct from the discourse and controversies around privatisation of water utilities and public–private partnerships for water infrastructure (see Hall and Lobina, 2004; Marin, 2009). Although adopted by some water utilities, water stewardship is primarily driven by large commercial, multinational entities that rely on fresh water to produce and trade goods and services. They are supported by an expanding cadre of international NGOs, small businesses, consultants, financial service providers, international finance institutions, and bilateral and multilateral aid agencies within a growing multimillion pound-per-year water stewardship enterprise (Gasson, 2011).

As will be explored in this chapter, the set of concerns driving CWS require companies to grapple with both public decision-making processes, and decision making about public, collective goods of which water is a prime example. A central concern is whether this new engagement represents an opportunity for greater water security for all/the many (through large water users facilitating improved resource management), or a manifestation of water securitization for some/the few (via the mobilisation of multiple forms of power to achieve or maintain control of the resource [Buzan, 2001; Turton, 2003]).

The reach, influence and resources available to the private sector mean that the potential for enlightened CWS to contribute to greater water security is significant. As evidence of this reach, the 58 companies reporting to the Carbon Disclosure Project's (CDP) Water Initiative represent a market capitalisation value of US$2.49 trillion, equivalent to the GDP of a G5 country. According to Money (2012), they collectively abstract more than 1,598 billion litres of water per annum, equal to 0.6 litres per day for every person on the planet. Some of these companies operate in almost every country in the world, spending comparable amounts to, and exerting political and financial influence in excess of, nation-states.[1] Given the world's spiralling water challenges (UN-Water, 2009; Gleick et al., 2011), the underwhelming performance of integrated water resource management (IWRM) reforms (Biswas, 2004, 2008; GWP 2009), and a stark lack of evidence for where progress may lie for water security (Hepworth et al., in press; Lele, 2012), water security practitioners 'cannot afford not to engage' with the corporate sector (Hoekstra and Mekonnen, 2012).

But corporate engagement in water management and policy provokes significant concerns, including polemical critiques, rooted in fears about the private accumulation of the means of production and corporate takeover of global resource governance and its institutions (Bruno and Karliner, 2000; Barlow and Clarke, 2002; Kay and Franco, 2012; Mehta et al., 2012). Other critiques are more pragmatic, driven by anxiety that corporate actions might in fact be based on inappropriate processes and assumptions, or misdirected effort and the unintended consequences of these for social equity and the functioning of water management institutions, particularly in developing countries (Hepworth, 2012; Morrison et al., 2010). At the core of these concerns are two issues. First, as a common pool resource with multiple, changing, and socially defined functions and values, water is a highly complex public good. Its effective management requires the continual reconciliation of tradeoffs between private interests towards a collective well-being. Second, although they must 'have regard to' wider social and environmental interests, companies, by their nature, are legally obliged to prioritise the needs of a set of narrow shareholder or other private interests (Newborne and Mason, 2012). The role of the corporate sector in shaping public water policy is therefore problematic, because the neutrality of their engagement is far from assured.

In resolving these interests, CWS necessarily engages with the sometimes murky political economy of water, contested interpretations of water security, and inevitable tradeoffs. Most companies have little experience with common-pool resource management and the challenges of collective responsibility and co-management. The risk is that without good guidance, specialist capabilities, and clear boundaries and principles for engagement, well-meaning corporate initiatives on water may have unintended negative impacts both internally for the company and externally for water users and the environment.

Although the policy literature on water stewardship is well established, rigorous academic reflection on this new corporate water agenda—characterised by Sojamo and Larson (2012) as an 'emerging transnational water governance regime'—is still needed (though see Hepworth, 2012; Mason, 2013). Likewise there has been limited effort to evaluate the outcomes of water stewardship activity (though see AWS, 2011).

In this chapter, we introduce the terminology and conceptual foundations of CWS, consider the driving forces behind it, and, through summarising some contemporary examples of CWS initiatives, the current landscape of water stewardship activity. To conclude, we consider the implications of CWS for water security and equity and flag emerging priorities for action and research to assist the navigation between corporate securitisation and stewardship of the water commons towards a more genuinely shared, wider water security.

Defining Water Stewardship

Companies face incoming risks to operations (their own water security), and generate outgoing risks for society (influencing the water security of others), through changes in freshwater quality, quantity, flow, and ecology. In order to minimise risk, water-using companies are taking action internally and externally in the wider community, basin, state and their governance institutions. Water stewardship can be considered as this response to the mounting legal, financial, and political duty of care obligations faced by water users to ensure the sustainable use and equitable management of water both within and beyond the 'fence line' of their operations.

Although wider debate and consensus over a formal definition of water stewardship is still required, the Alliance for Water Stewardship (AWS, 2011) proposes that:

> Water Stewardship entails the internal and external actions that ensure water is used in ways that are socially equitable, environmentally sustainable and economically beneficial.

Commentators may ask how this definition sets water stewardship apart as a new paradigm, distinct from IWRM and its foundational principles of equity, sustainability, and efficiency. The answer is aided by the notion that water stewardship embodies 'taking care of something which one doesn't own' (A. Sym, personal communication) or 'of looking after an asset or resource on behalf of others' (SWSC, 2011). At its core, water stewardship is differentiated because of whom it infers is contributing to water resource management and taking action on behalf of other users. If IWRM is considered as actions by an authority mandated by the state (within which ownership of the resource is vested by law) to manage water resources on behalf of all water users, then water stewardship can be considered as actions by water users themselves to contribute to the management of the shared resource towards public-good outcomes. Water stewardship is, therefore, about nontraditional, private actors increasingly involving themselves in the management of the common pool—public good regarding water. As a progression from IWRM, with its emphasis on participation, this shift in parlance and the loci of action can arguably be considered a success. But, embedded in this success are the incumbent hazards imposed by large power asymmetries among disparate and variably engaged water users. In particular, a strengthened voice for the private sector has not necessarily seen a parallel strengthening of less powerful voices in resource management and decision making. Nevertheless, water stewardship is now the primary vehicle through which corporate entities are channelling responses to their own and wider public water challenges (AWS, 2011).

Although the role of CWS in water security and securitisation is increasingly aired (Mason, 2013), the interpretations of water security reviewed by Cook and Bakker (2012) are yet to explicitly refer to the dependencies between business and economic water use and water security. In the next sections, we explore the evolution of CWS, the concepts set out in the policy literature as rationale for engagement, and introduce three contemporary CWS initiatives before returning to consider their implications vis-à-vis water securitisation for corporations or water security for the many.

Conceptual Foundations and Driving Forces of Corporate Water Stewardship

Early Beginnings

Corporate engagement on water is by no means new. The formation of water corporations in 19th-century Britain to oversee investment in water supply and sanitation and protect catchments was driven in part by private owners of textile mills and foundries (Cornish and Clark, 1989). These technology entrepreneurs were motivated as much by a need to keep local communities and their labour force in good health as by a need to ensure uninterrupted clean water for manufacturing. More recently, the Mersey Basin Campaign initiated in 1985, in response to the 'dirtiest river in Europe' epitaph, delivered urban renewal, waterside regeneration, and helped return pollution-sensitive species like otters and salmon to the basin. It was driven financially and politically via business partnerships, notably with United Utilities, ICI, Shell, and Unilever (EKOS Consulting, 2006). In a further example, Water-Aid, now UK's 'most admired charity', providing almost 16 million people with safe water in 27 countries, began life in 1981 as a water industry response to the UN Decade of Drinking Water and Sanitation (WaterAid, 2012). WaterAid continues to derive the majority of its untied funding from its links to the privatised UK water industry.

These legacies are testament to the power of private industry to drive innovation and demand and support improved performance from the public sector. Private industry has always shared water risks with communities, government, politicians, and the environment, but in the past, pressures and responses were different from those emerging in today's highly branded, globalised, and increasingly water stressed world.

Evolution and Role of Water Footprint Accounting, Risk Metrics, and Stewardship Tools

With the advent of water footprint (WF) accounting, companies and their critics have been able to ask new and far-reaching questions about corporate water use and its impacts. Although water accounting and productivity analysis have long been used by water scientists (see Dinius, 1972; Molden, 1997), work by Hoekstra and Chapagain (2008) to develop Tony Allan's virtual water concept into an accounting framework for the 'embedded' water in products has helped place water security onto the global business agenda. In 2011, the Water Footprint Network (WFN), created by a collection of businesses and interested organisations, published a water footprint (WF) manual that laid out, for the first time, a standardised approach to WF accounting (Hoekstra et al., 2011). Other accounting methods have been proposed, most notably for water use within Life Cycle Assessments (LCA) (Mila i Canals et al., 2009).

As soon as the footprints of cola drinks, hamburgers, and T-shirts were calculated, both consumers and companies began to react to these metrics. WF consumption in the UK (Chapagain and Orr, 2008) made front page news by showing how 64% of the UK's agricultural and industrial WF is embedded in virtual water used outside the UK. The application of WF helped companies and consumers to understand and visualise the scale, geography, and vulnerability of water use across global supply chains. At the same time, investigations into the impacts of the UK's WF-linked virtual water consumption to catastrophic impacts for the poor in developing countries, for example, through the supermarket retail of horticultural produce irrigated through unsustainable water use (Hepworth et al., 2010). Applied at the product level, WF has allowed comparison of resource requirements across production sites and across business portfolios, and this has been used to identify priority locations for investment to minimise water risks. A compelling case study is work by SABMiller and WWF (2009), which established that 98.3% of the company's WF in South Africa was related to crop production. Overlaying the distribution of this WF on maps of water stress has since driven it into new stakeholder partnerships to address water scarcity in those locations (Chapagain and Tickner, 2012). Loftier goals of using WF analysis to drive international trade, governance agreements, and hydrologically sustainable patterns of consumption (Hoekstra, 2006) have yet to gain traction, though notably WF is being used to explore economic development scenarios in water-stressed basins (Pegram, 2010; Hepworth, 2012).

As a tool for communicating and stimulating action on water by new audiences, WF has proved formidable and has directly motivated business leaders to address water risk (Chapagain and Tickner, 2012). But WF accounting is not without its critiques, which include: the decoupling of volumetric water use and quality issues from the local contexts that determine sustainability and equity; data availability and reliability challenges; and misinterpretation of hazards brought by its reductive approach (see Wichelns, 2010, 2011).

Debates around the social and economic relevance of WF have raised a new, more refined set of questions about corporate water use and its impacts. In response, at least 20 water stewardship tools—defined as 'numeric and narrative techniques for characterising water consumption, impacts and risks'—have emerged (Larson et al., 2012). Whilst their use may not directly influence public policy, their application, limitations, and utility within corporate engagement warrant critical review (Daniel and Sojamo, 2012). Key questions include whether these tools guide appropriate actions by consumers and companies, and whether their influence on water governance leads to equitable, sustainable, and sensible outcomes.

Corporate Water Risks

Discussion has also focussed on how companies actually incur 'risk' and impose 'impacts'[2] through their water use, and their incentives for addressing these (Orr et al., 2009; Barton, 2010). Water risks differ among sectors and companies, affect direct operations as well as supply chains, and ultimately affect costs, profits, and future growth (JPMorgan, 2008). Studies have explored water risks facing the public (Orr et al., 2009), the insurance sector (Lloyd's, 2010), business (Morrison et al., 2010), and financial institutions (JP Morgan, 2008; Barton, 2009; WWF and DEG, 2011; UNEP-FI, 2012). Below we consider three distinct but related domains of corporate water risk: physical, reputational, and regulatory.

Physical Water Risks: Disruption to Production

Physical, or operational, water risk concerns the direct risks facing any operation because of changes in the flow, quality, or availability of water. Examples include the output reductions brought on by drought and water shortages in Texas, India, Pakistan, and Brazil in 2011 when cotton prices reached an all-time high, prompting the Gap to cut its annual profit forecast by 22% (Larson et al., 2012). Other examples include the salinization of groundwater and its impact on the beer brewing infrastructure in Dar es Salaam, which is driving engagement in Tanzania by SABMiller and its local partners (Hepworth, 2008). Corporates are engaging with public water policy because of such physical risks, including food and beverage manufacturers concerned about production and agricultural water requirements, household chemical manufacturers concerned about negative water impacts through their products' use, and financial institutions concerned about investment risk because of unreliable supplies (Orr and Cartwright, 2010).

Reputational Water Risks: Damage to Brand or Company Image

Reputational risks affect brand value and market share and are associated with increased visibility of negative impacts on communities and ecosystems because of water use by business (Orr and Cartwright, 2010). The

revolution in global communications and growth of social media activism—
where images can move from field to front page within minutes—has the
potential to support much greater public scrutiny of corporate water use.
Whether it concerns water use by drink manufacturers and bottled water
companies, the impact of East Africa's cut flower industry, or the supply of
Peruvian asparagus to UK supermarkets, greater media coverage of water
problems has given rise to business concern over reputations and reactions
in the market. Reputational impacts have significant, long-term financial
implications for a company (WBCSD, 2012) and do not always need to be
accompanied by legal proceedings or material environmental impacts. As a
response, a great deal of water-related activity by companies, particularly
those with international brands to protect, targets public relations exercises
and philanthropy to protect reputations. They seek to ensure their 'social
licence' to operate through maintaining legitimacy in the eyes of water institutions, local communities, and consumers.

Regulatory Water Risks: Concerning Legal Action

Regulatory risk drives business to protect its legal licence to operate through
compliance with relevant legislation, and to understand and influence policies and regulations that apply to their operations. On the one hand, companies voice the concern that unless they 'get their act together' on water
at operational, strategic, and advocacy levels they may face fines, prohibitive laws, loss of water access, and increasingly stringent water regulation.
On the other hand, they see the failure of public entities to regulate fairly,
enforce laws, and create level playing fields as obstacles to economic growth
(Porter and Kramer, 2011). There is a wide variation in maturity in how
companies engage with government over these issues, highlighting not
only sectoral differences in water stewardship but also the idea that many
industries are favoured by government because of their contribution to the
economy (Pegram et al., 2009). A further issue is the apparent confusion
about the nature and direction of regulatory risk with companies and their
advisors interpreting the lack of effective, vociferous regulatory activity as
either a boon or a bane.

Secondary Drivers: Perceptions of Communities,
Shareholders, Consumers, and Financiers

Physical, reputational, and regulatory water risks affect the financial bottom line, market share, and attractiveness to investors. An understanding
of the secondary drivers behind these risks is important for water security
outcomes because the ways in which stakeholders (such as purchasers,
investors, or consumers) express their needs and wants of a company's performance will tend to determine the nature of the response. For example,
the Financial Services Industry (FSI), interested in investment, equity, and
bond performance, is an important secondary driver, increasingly requiring

business to demonstrate due diligence on water (Larson et al., 2012). But if the FSI is asking the wrong questions or applying clumsy metrics to understand water-related investment risks, such as whether a business is located in a hydrologically water-stressed area, then perverse outcomes may result. For example, in many circumstances, hydrological water stress may be less important than scarcities caused by poor water governance and allocation. Consumer choice for ethically produced goods plays another increasingly important role in corporate behaviour. Ethical labelling schemes adopting volumetric indicators such as WF metrics or efficiency savings are unlikely to capture the complexity of water management in a way that would drive desirable results. Third-party certification standards also currently fail to reflect whether a company is using water in ways that are genuinely sustainable. It remains to be seen whether production standards, certification, and labelling can incentivise enlightened stewardship behaviours through the power of consumer preference.

The role of NGOs in driving water stewardship and as brokers of corporate action should not be underplayed. Larger NGOs such as IUCN, WWF, and TNC have established partnerships with corporate actors on stewardship and play a major role in shaping the debate and response. These relationships are based on opportunities to leverage corporate interest and risk towards greater societal and environmental gains. Objective third-party evaluation of any such gains would support the legitimacy of these relationships and opportunities for adaptive learning, but this has received limited attention to date.

Shareholders appear to be silent on water issues and it might, therefore, be assumed that water risks facing dividend payments are rarely considered or even acknowledged. Through the interest of the FSIs, this may change, as tools to assess water risk become more mature and relevant to financial decision making around due diligence and portfolio risk assessment (WWF-DEG, 2011; UNEP-FI, 2012; CDP-Water, 2013).

Shared Water Risk

The idea of shared water risk aims to reflect that degraded and depleted water resources and inadequate supply and sanitation impose shared impacts across society. Therefore, action to address these problems stems from these shared interests and requires shared, collaborative action. This emphasises the need for stability and cooperation, as opposed to merely competing over a resource that is becoming scarcer and, therefore, more socially, ecologically, and economically valuable (Orr and Cartwright, 2010). The conceptualization of 'shared water risk' (Pegram et al., 2009) represents a giant leap in shifting attention away from only internal efficiencies, to explore and act on water issues 'outside the fence line'. As a corollary of the concept of Creating Shared Value (CSV) (Porter and Kramer, 2011), addressing shared risks through CWS is a compelling idea and a unifying concept. The NGO and company literature emphasises that corporate actors are motivated towards delivering public-good outcomes in response to shared risk (for example, Morrison et al., 2010; SABMiller and WWF, 2009; WFP 2011; CEO Water Mandate, 2012a, 2012b, 2012c; The Coca-Cola Company, 2012). However, if we apply a traditional understanding of risk, neither the magnitude nor the probabilities of impact are equally distributed among water users. Shared water risks are unlikely to be shared equitably due to varying levels of hazard exposure and resilience among water users, and reducing one user's water risk may result in increased vulnerability for others. The presence of these conjoined inequalities and interdependences is echoed by Warner and Johnson (2007) in their reflections on water security, where water security for some means vulnerability for others. Although the concepts of shared risk and CSV can therefore be questioned in terms of how risks and values play out for different stakeholders, their central tenet—that companies share a need with the public for reliable water services and sustainable water resource management—is sound (Hepworth, 2012). This shared need goes beyond the provision of adequate water for production, consumption, and ecosystem services; it also extends to a need for public water managers to regulate water use by businesses and corporations in ways that are considered fair and equitable in order to maintain their legitimacy and social licence to operate (although there are many contexts where these may be overlooked, or where incentives for these do not exist). The implication is that if the public sector is doing its job in overseeing the sustainable management of water and effectively managing shared risk, there is little or no justification for business engagement in water policy.

Examples of Corporate Stewardship and Engagement Initiatives

Here we explore three contemporary CWS initiatives to contextualise the discussions in this chapter and then elaborate some priorities for research and action.

The CEO Water Mandate

Launched by the UN Secretary General in July 2007, the UN Global Compact's CEO Water Mandate is a public–private initiative designed to assist companies in the development, implementation, and disclosure of water sustainability policies and practices. Comprised around 90 of the world's largest corporations, 'endorsers' meet in working conferences with the support of a secretariat (the Pacific Institute) to progress thinking and action across six work areas: Direct Operations; Supply Chain and Watershed Management; Collective Action; Public Policy; Community Engagement; and Transparency.

Alongside guidelines for collective action and a web-based 'Action Hub' to broker partnerships in response to demand from endorsers, the CEO Mandate has drawn up 'Guidelines for Responsible Business Engagement on Water Policy' (Morrison et al., 2010) that outline five broad principles for corporate conduct:

1.   Advance sustainable water management
2.   Respect public and private roles
3.   Strive for inclusiveness and partnerships
4.   Be pragmatic and consider integrated engagement
5.   Be accountable and transparent

These principles are neither codified, nor have they been exposed to multi-stakeholder deliberation, review, or monitoring of uptake. Nonetheless, they remain the primary guidance in an arena in which most companies have little experience. Given that policy engagement has traditionally entailed companies 'fighting their corner', to the potential detriment of other resource users, these guidelines support a sea-change in how businesses advocate for improved policy and regulation, in the public interest.

Despite the commitment to hyper-transparency, there remains scepticism on the part of some NGO's as to the real intent of big business and collaboration between the UN and transnational corporations in the water policy realm (see Sierra Club, 2008; Polaris Institute, 2010). The advent of binding standards or codes, rather than mere guidelines, and objective evaluation of the outcomes of corporate engagement in public policy processes, could do much to rebut this scepticism.

Alliance for Water Stewardship

The Alliance for Water Stewardship (AWS) is an international collaborative effort to develop a certifiable standard for water stewardship that will guide, differentiate, and reward responsible water users through preferential treatment in the market place (AWS, 2011). Analogous to stewardship standard programmes in forestry and fisheries, the AWS has been driven initially by environmental, social, and water sector NGOs. Together with a growing number of business supporters, they seek to realise the potential for incentivising transparent and independently certified attainment against water stewardship criteria, through a universal benchmark of good and continually improving corporate performance on water. Early pilots illustrate the potential offered by water stewardship standards. For example, in Kenya, they have driven investment in improved institutional functioning, legal compliance, and municipal waste treatment (Hepworth et al., 2011). Difficult questions remain and include how a standard should operate in basins with low levels of data, infrastructure, or governance without penalising producers based there; be accessible and relevant to smaller producers; agree and interpret a universal definition of equitable and sustainable water use; and interact with local statutory frameworks where they are inadequate. The AWS has made a commitment to procedural justice through establishing a 14-member committee of geographically and sectorally balanced stakeholders to deliberate these questions and the wider standard content and requirements. It remains to be seen whether this will translate to distributive justice once the standard is implemented, or whether demand will exist for a standard that transparently holds corporates to account against an international benchmark of good practice. Companies, consumers, and investors are likely to judge the standard on whether it demonstrably reduces risk at a cost that is not prohibitive, and a programme of international field testing is now underway.

The 2030 Water Resources Group

The 2030 Water Resources Group (WRG) is a consortium of large companies[3] hosted by the International Finance Corporation (IFC) that aims to 'engage in fact-based, analytical approaches and coalition building initiatives that help governments to catalyze sustainable water sector transformations in support of their economic growth plans' (2030WRG, 2013). The main thrust of the group's work so far has been via the report 'Charting Our Water Future' (Addams et al, 2009). The report analyses the nature and scale of the global water challenge and proposes solutions to 'close the demand-supply gap' through a series of basin-by-basin analyses of future water availability against projected demand. The report presents a series of Marginal Abatement Cost (MAC) curves to illustrate the costs and benefits of options available to maximise economic productivity of water.

The WRG differs from many company engagements in water in that it doesn't specifically address the specific water risks of the supporting companies, but rather seeks to inform and raise awareness of water challenges at a national level. The WRG seeks to instigate water supply, demand, and governance debates with high levels of government, but, in doing so, raises some important questions relating to technical rigor, privileging of knowledge, and assurance of sustainability outcomes. First, the use of MAC curves has been criticised for a lack of transparency (the methodology applied is often proprietary), its poor handling of uncertainty, intertemporal dynamics, interactions between sectors and ancillary benefits, and because the options considered would take decades to implement, during which conditions are likely to change (Vogt-Schilb and Hallegatte, 2011). Second, the report and methodologies developed are now being promoted internationally and to developing country governments to inspire policy, investment, and actions that release water for 'allocation to higher value uses'. The value of such analysis and lobbying is double-edged. By initiating a debate about 'water for the economy', the initiative may well incentivise new investment in water management, but by advocating for prioritisation of water allocation to highest value uses, it risks disenfranchising local water users in multiple basins who may have prior use rights, or favour noneconomic priorities for water use. Important questions also arise about the ultimate beneficiaries of 'high value' water use, particularly given the ubiquity of very generous tax regimes as a strategy to attract foreign investment in developing countries.

The WRG has had particular traction in South Africa, where the South Africa Strategic Water Partners Network (SWPN) has been established. This is a partnership between the Department of Water Affairs (DWA) and the 2030 WRG to help address three priorities: (1) water use efficiency/leakage reduction, (2) supply chain in agriculture, and (3) effluent partnerships (WRG, 2013). A challenge for the 2030 WRG is to bridge the legitimacy gap by expanding its analysis beyond a narrow financial treatment to include multiple values, functions, and uses of water, and perhaps more importantly, by being much more transparent and inclusive in its methodologies and engagement with government.

Discussion: Seeking Progress for Shared Water Security

The CEO Mandate, AWS, WRG, and other corporate initiatives invoke questions on whether the water risks faced by companies can really drive them towards shared public water security outcomes, particularly when conditions traditionally considered to deliver for a company's financial bottom line (e.g., preferential, sustained water access; permissive water quality objectives; and laissez-faire regulation) undermine wider public good and benefits at village, basin, and global scales. Without adequate transparency, scrutiny, and challenge, these corporate initiatives risk privileging the perspectives of the already powerful by utilising persuasive data, analysis, and knowledge that may be beyond the reach or comprehension of people whom decisions about water will affect. Within all these initiatives, a failure to prioritise procedural and distributive water justice, ecosystem integrity, and social well-being in the basins where corporates operate will severely limit their contribution to water security.

By reviewing these initiatives, the opportunities they present, and their implicit risks, we aim to emphasise why water scientists, governments, development practitioners, activists, and other water users who share resources with business should view the recent surge of interest in CWS with great interest. For those who have strived for greater user engagement in managing water, the rousing of the private sector represents a breakthrough: releasing potential to influence society and the global economy towards more sustainable means of production and resource use.

For others, it represents capital's unique ability to appropriate and sublimate critiques against it and a threat to future water equity and justice (Water Alternatives, 2012). Critics fear policy and regulatory capture that will prioritise water allocation for highest value economic value use over environmental and social well-being, livelihood, and cultural values and functions, enabling the already powerful to buy out or capture the resource. Vulnerability to capture is greatest in basins with weak and dysfunctional institutional arrangements, which also tend to be those in poor countries where the most is at stake in terms of human welfare and biodiversity conservation (Morrison et al., 2010). Ironically, then, the places where additional support via corporate engagement is most needed are the same places where this external support could most easily lead to unforeseen or undesirable consequences.

Appropriate strategies within water research, policy, and practice are required in order to avoid undesirable consequences and to assist and harness the progressive contribution offered by corporates. An important starting point for divining these is to revisit an important thread running through this chapter and to understand corporate motivations for engagement, summarised in the next paragraph.

We believe corporates are likely to engage in water policy and stewardship because of actual or perceived future risk of impacts to their operations, reputations, or legal status, and the secondary drivers discussed. Enlightened engagement will recognise that reputational risks cannot be mitigated unless shared risks and shared water security are delivered in ways and with outcomes considered to be socially legitimate, and will, therefore, be to promote and support improved governance, management, and resourcing in the sector to reduce water risks through clear, consistent, and proportionately applied policy, laws, and regulations. In order to gain legitimacy, companies will be required to show that they are part of a progressive and inclusive water agenda that delivers public-good outcomes through procedural and distributive justice in relation to water.

As large and powerful actors with often-privileged access to the ears of government, corporates have the potential to make important contributions to equitable and sustainable resource management. CWS is a rapidly evolving and complex field of activity, with few clear norms or guidance, and the numerous attendant risks of negative outcomes. These include those already mentioned in this chapter: diverging corporate and public interests where water is contested; policy and regulatory capture; privileging of economic over social perspectives; process inequities such as unrepresentative participation in rule or standard setting; and others such as confusion and displacement of existing water management priorities, and the risks of misguided interventions that undermine hydrological sustainability (Hepworth, 2012). There is also the risk that engagement may actually heighten reputational risks given the growing distrust of corporate motivations by the public—vulnerability that is underappreciated by some critics. Such hazards mean that even well-intentioned corporate engagement in public water management may lead to deleterious and harmful outcomes.

The alternative to publicly minded water engagement by corporates is that which explicitly or tacitly attempts the securitisation of water resources for their private benefit. There are multiple routes and forms of power available to corporate actors in such an endeavour, ranging from ideational power to brute force. The application of subtle or overt private sector influence towards inequitable and harmful ends plays to the worst fears of liberal observers. Moreover, despite this emerging optimism regarding incentives for action on shared risk, private sector securitisation remains the dominant paradigm in many contexts of private sector engagement on water (see Hepworth, 2009; Hepworth et al., 2010).

This two-way tension between private sector securitisation and public water security presents an interesting case of the 'prisoner's dilemma', where entities could gain important benefits from cooperating or suffer from the failure to do so, but find it difficult or expensive, but not necessarily impossible, to coordinate their activities to achieve cooperation. Further, corporate water use can have multiple negative externalities on wider society, but these might only become priorities for action by the company once they begin to inflict reputational harm. The additional problem is that this happens in a stochastic and complex water environment.

In response to this opportunity for corporate contributions to water security for all, and risks of resource securitisation for some, a constructive response from water security practitioners, policy makers, and the research community will include:

- Development of clear definitions, standards, methodologies, and benchmarks for sustainable and equitable water use and shared water security against which corporate actions can be gauged and actors differentiated
- Development and promotion of principles of engagement that promote procedural justice, balanced representation, and hyper-transparency in decision making
- Provision of sound guidance, support, and evaluation of actions and advocacy activity that are aligned in intent, in process, and that avoid perverse outcomes and policy capture
- Testing of claims and tracking actions and outcomes on the ground and in policy fora to hold to account and challenge the application of power within CWS

Delivering this response will require that three further priorities are met. First, the mainstream water sector needs to upscale its engagement with CWS. The research community, in particular, has been slow to provide the theoretical and empirical analysis required to navigate this new paradigm. The practitioner community has also yet to fully grasp the threats, opportunities, and appropriate modes of engagement with CWS, and this in turn risks unhelpful fragmentation in an already fragmented field.

Second, the important roles of civil society need emphasis and strengthening. The validity of the shared risk concept relies on the fear of reputational repercussions to keep abuses of corporate power in check. This in turn relies on tenacious social accountability monitoring: auditing performance, exposing impacts, and incidences of capture. Yet, in many countries, the individuals, groups, and institutions required to deliver these tasks are either absent, or are constrained by a lack of resources, capability, or political freedom. The validity of shared risk and the role of civil society also assumes that the media has the ability (and willingness) to report on water issues to consumers and stakeholders. It also depends on the consumers' willingness (and ability) to reward or punish water performance through the markets. There is sparse evidence to date that the degree of transparency required to substantiate this theory of change is in place.

Third, bilateral and multilateral development agencies need to reorient policy and funding frameworks on water based on cognisance of the threats and opportunities posed by CWS. In particular, there is a need to facilitate the research, practitioner, and civil society action suggested here, which currently lacks investment—a dangerous imbalance given the levels of financing for CWS from corporates themselves. With the exception of one or two donors, notably Germany and UNEP, the international development community has been slow to see and act on the implications of CWS.

Conclusion

In conclusion, given the very significant challenges facing global water security, dogmatic stances for or against corporate engagement are likely to be counterproductive. At the same time, capture of policy and regulatory activity by corporate interests are very real hazards, requiring that opportunities and threats posed by corporate engagement in the management of public goods be aired, understood, and navigated transparently. Addressing the priorities, answering the outstanding questions, and crafting the tools and institutions required to deliver on the opportunities of corporate water stewardship are exciting but formidable challenges for water security scholars and practitioners. Their resolution probes the question of whether the linkages between profit, growth, and resource overexploitation—alluded to in Kingsley's rhyme at the outset of this chapter—can be decoupled. Engagement with the corporate sector on water in the coming years will be an important 'water security crucible' within which to test whether lofty ideals, such as creating shared value or water stewardship, can guide our development trajectory towards a future of mutual rather than exclusive social, economic, and environmental well-being.

Notes

1. For example, the Coca-Cola company has invested almost US$2 billion on water initiatives in the past decade (Koch, 2012), compared to Department for International Development's estimated investment of US$3 billion over the same period (based on reported DFID investment on WASH between 2006 and 2011; DFID, 2012).
2. Water impacts are defined as 'human induced effects which bring about undesirable detriment to the functions, values and uses of a water resource' (Hepworth and Dalton, 2009).
3. The Barilla Group, The Coca-Cola Company, The International Finance Corporation, McKinsey & Company, Nestlé S.A., New Holland Agriculture, SABMiller plc, and Standard Chartered Bank.

References

Addams, L., Boccaletti, G., Kerlin, M. and Stuchtey, M. (2009) 'Charting our water future: economic frameworks to inform decision-making,' 2030 Water Resources Group: McKinsey & Company, www.2030waterresourcesgroup.com/water_full/Charting_Our_Water_Future_Final.pdf, accessed 20 August 2012.

AWS (2011) *Exploring the Value of Water Stewardship Standards in Africa: Results of the AWS Kenya Case Study*, Alliance for Water Stewardship, German Technical Cooperation (GIZ) and Marks and Spencer, Washington, DC.

Barlow, M. and Clarke, T. (2003) *Blue Gold: The Battle against Corporate Theft of the World's Water*, Earthscan, London.

Barton, B. (2010) *Murky Waters: Corporate Reporting on Water Risk*, CERES, Boston, USA.

Biswas, A.K. (2004) 'Integrated water resources management: a reassessment', *Water International*, vol 29, no 2, pp248–256.

Biswas, A.K. (2008) 'Current directions: integrated water resources management–a second look', *Water International*, vol 33, no 3 (September), pp274–278.

Bruno, K. and Karliner, J. (2000) 'Tangled up in blue: corporate partnerships at the United Nations,' TRAC (Transnational Resource and Action Center), www.corpwatch.org, accessed 26 September 2012.

Buzan, B. (2001) 'Losing control: global security in the twenty-first century', *International Affairs* vol 77, no 3, pp696–696.

CDP-Water (2013) *Collective Responses to Rising Water Challenges*, CDP, London, www.cdproject.net/CDPResults/CDP-Water-Disclosure-Global-Report-2012.pdf, accessed 12 March 2013.

CEO Water Mandate (2012a) *Corporate Water Disclosure Guidelines: Toward a Common Approach to Reporting Water Issues*, Pacific Institute, Oakland, CA.

CEO Water Mandate (2012b) *Guide to Water-Related Collective Action*, Pacific Institute/Ross Strategic/Pegasys Strategy and Development/Water Futures Partnership, Oakland, CA.

CEO Water Mandate (2012c) Water Action Hub, http://wateractionhub.org/organizations, accessed 3 October 2012.

Chapagain, A.K. and Hoekstra, A.Y. (2004) 'Water footprints of nations', *Value of Water Research Report Series* No.16, UNESCO-IHE, Delft, Netherlands.

Chapagain, A.K. and Orr, S. (2008) *UK Water Footprint: The Impact of the UK's Food and Fibre Consumption on Global Water Resources*, WWF-UK, Surrey, UK.

Chapagain, A.K. and Tickner, D. (2012) 'Water footprint: help or hindrance?', *Water Alternatives*, vol 5, no 3, pp563–581.

The Coca-Cola Company (2012) *Global Water Stewardship and Replenish Report*, Coca-Cola Company, Atlanta, USA.

Cook, C. and Bakker, K. (2012) 'Water security: debating an emerging paradigm', *Global Environmental Change*, vol 22, no 1, pp 94–102.

Cornish W.R. and Clark, G. de N. (1989) *Law and Society in England 1750-195*, Sweet & Maxwell, London, UK.

Daniel, M. and Sojamo, S. (2012) 'From risks to shared value? Corporate strategies in building a global water accounting and disclosure regime', *Water Alternatives*, vol 5, no 3, pp637–658.

DFID (Department for International Development) (2012) *Water, Sanitation and Hygiene Portfolio Review*, March, DFID, London, UK.

Dinius, S. H. (1972) 'Social accounting system for evaluating water resources', *Water Resources Research*, vol 8, no 5, pp1159–1177.

EKOS Consulting (2006) 'Evaluation of the Mersey Basin Campaign', Report to Government Office North West, http://merseybasin.org.uk/archive/items/MBC057.html, accessed 28 September 2012.

Gasson, C. (2011) 'The case for corporate water. Need to know and analysis', *Global Water Intel*, vol 12, no 9 (September), www.globalwaterintel.com, accessed 10 October 2011.

Gleick, P., Allen, L., Christian-Smith, J., Cohen, M.J., Cooley, H., Heberger, M., Morrison, J., Palaniappan, M. and Schulte, P. (2011) *The world's water: The biennial report on freshwater resources*, Pacific Institute, Island Press, Washington, DC.

Goubert, J.P. (1989) *The Conquest of Water: The Advent of Health in the Industrial Age*, translated by A. Wilson, Princeton University Press, NJ.

GWP (2009) 'A new vision for IWRM', Stockholm Water Week, from *IWRM in Practice—Lessons from Practical Experience*, GWP, Stockholm.

Hall, D. and Lobina, E. (2004) 'Private and public interests in water and energy', *Natural Resources Forum*, vol 28, no 4, pp268–277.

Hepworth N.D. (2008) 'Dar es Salaam Water Dialogue Technical Report', SAB Miller Ltd/Tanzania Breweries Ltd., Tanzania.

Hepworth, N.D. (2009) *A Progressive Critique of IWRM in Sub-Saharan Africa*, PhD thesis, University of East Anglia, UK.

Hepworth, N D, Agol, D, Von-Lehr, S, and O'Grady, K (2011) *Kenya Case Study*, Technical Report, GIZ/Marks and Spencer/Alliance for Water Stewardship, Water Witness International, Edinburgh, UK.

Hepworth, N. D. (2012) 'Open for business or opening Pandora's Box? A constructive critique of corporate engagement in water policy: an introduction', *Water Alternatives* vol 5, no 3, pp543–562.

Hepworth, N.D. and Dalton, J. (2009) *Understanding Impacts in Water Resource Management*, WWF UK/LTS International, Edinburgh, UK.

Hepworth, N.D., Hooper, V., Hellebrandt, D. and Lankford, B. (in press 'What factors determine the performance of institutional mechanisms for water resources management in developing countries in terms of delivering pro-poor outcomes, and supporting sustainable economic growth?', DFID/CEE review 11–006, Collaboration for Environmental Evidence, www.environmentalevidence.org/SR11006.html.

Hepworth, N.D., Postigo, J. and Guemes, B. (2010) *Drop by Drop: Understanding the Impacts of the UK's Water Footprint through the Case Study of Peruvian Asparagus*, Progressio/Water Witness International/CEPES, London.

Hoekstra, A.Y. (2006) 'The global dimension of water governance: Nine reasons for global arrangements in order to cope with local water problems', *Value of Water Research Report Series* No. 20, UNESCO-IHE, Delft, the Netherlands.

Hoekstra, A.Y. (2013) *The water footprint of modern consumer society*, Routledge, London, UK.

Hoekstra, A.Y. and Chapagain, A.K. (2008) *Globalization of water: Sharing the planet's freshwater resources*, Blackwell Publishing, Oxford, UK.

Hoekstra, A.Y., Chapagain, A.K., Aldaya, M.M. and Mekonnen, M.M. (2011) *The Water Footprint Assessment Manual: Setting the Global Standard*, Earthscan, London.

Hoekstra, A.Y. and Hung, P.Q. (2002) 'Virtual water trade: a quantification of virtual water flows between nations in relation to international crop trade', *Value of Water Research Report Series* No.11, UNESCO-IHE, Delft, the Netherlands.

Hoekstra, A.Y. and Mekonnen, M.M. (2012) 'The water footprint of humanity', *Proceedings of the National Academy of Sciences*, vol 109, no 9, pp3232–3237.

JPMorgan (2008) 'Watching water: A guide to evaluating corporate risks in a thirsty world', Global Equity Research.

Kay, S. and Franco, J. (2012) *The Global Water Grab: A Primer*, Transnational Institute (TNI), Amsterdam.

Koch, G. (2012) 'Risk and response: A business perspective on water security', *Water Security, Risk and Society Conference 2012*, 17 April, Oxford University, http://podcasts.ox.ac.uk/risk-and-response-business-perspective-water-security-video, accessed 22 September 2012.

Larson, W.M., Freedman, P.L., Passinsky, V., Grubb, E. and Adriaens, P. (2012) 'Mitigating corporate water risk: Financial market tools and supply management', *Water Alternatives*, vol 5, no 3, pp582–603.

Lele, U., (2012) 'Good governance for water and food security', Key note presentation, World Water Week, August 28 2012, Stockholm.

Lloyd's (2010) *Lloyd's 360 risk insight: Global water scarcity - Risks and challenges*, Lloyd's, London, http://www.lloyds.com/News-and-Insight/Risk-Insight/Reports/Climate-Change/Global-Water-Scarcity.

Marin, P. (2009) 'Public private partnerships for urban water utilities: a review of experiences in developing countries', *Trends and Policy Options Series*, #8, World Bank Publications.

Mason, N. (2013) 'Uncertain Frontiers: Mapping New Corporate Engagement in Water Security', ODI Working Paper 363.

Mehta, L., Veldwisch, G.J. and Franco, J. (2012) 'Introduction to the special issue: water grabbing? Focus on the (re)appropriation of finite water resources', *Water Alternatives*, vol 5 no 2, pp193–207.

Mila i Canals, L., Chenoweth, J., Chapagain, A., Orr, S., Anton, A. and Clift, R. (2009) 'Assessing freshwater use impacts in LCA: part i-inventory modelling and characterisation factors for the main impact pathways', *International Journal of Life Cycle Assessment*, vol 14, no 1, pp28–42.

Molden, D. (1997) 'Accounting for Water Use and Productivity', SWIM Paper 1, International Irrigation Management Institute (IWMI), Colombo, Sri Lanka.

Money, A.L.N. (2012) *Managing What You Measure: Corporate Governance, CSR and Water Risk*, April 19, http://ssrn.com/abstract = 2042564.

Morrison, J., Schulte, P., Orr, S., Hepworth, N., Pegram, G., Christian-Smith, J. (2010) *Guide to Responsible Business Engagement with Water Policy*, Pacific Institute/The CEO Water Mandate, UN Global Compact, Oakland, CA.

Newborne, P. and Mason, N. (2012) 'The private sector's contribution to water management: clarifying companies' roles and responsibilities', *Water Alternatives*, vol 5, no 3, pp604–619.

Orr, S. and Cartwright, A. (2010) 'Water scarcity risks: experience of the private sector', in L. Martinez-Cortina, A. Garrido, and E. Lopez-Gunn (eds) *Re-thinking Water and Food Security*, CRC Press, London.

Orr, S., Cartwright, A. and Tickner, D. (2009) *Understanding Water Risks: A Primer on the Consequences of Water Scarcity for Government and Business*, WWF-UK, London.

Pegram, G. (2010) *Shared Risk and Opportunity in Water Resources: Seeking a Sustainable Future for Lake Naivasha*, WWF and PEGASYS, Godalming, UK.

Pegram, G., Orr, S. and Williams, C. (2009) *Investigating Shared Risk in Water: Corporate Engagement with the Public Policy Process*, WWF-UK, Surrey, UK.

Polaris Institute (2010) *CEO Water Mandate Fact Sheet*, www.polarisinstitute.org/files/The%20CEO%20Water%20Mandate_0.pdf, accessed 2 May 2013.

Porter, M.E. and Kramer, M.R. (2011) 'Creating shared value: How to reinvent capitalism—and unleash a wave of innovation and growth', *Harvard Business Review*, January-February, pp62–77, www.riverfoundation.org.au/riverprize_international.php.

SABMiller and WWF (2009) *Water footprinting: Identifying and addressing water risks in the value chain*, Surrey, England: SABMiller and WWF-UK.

Sierra Club (2008) www.sierraclub.org/committees/cac/water, accessed 2 May 2013.

Sojamo, S. and Larson, E.A. (2012) 'Investigating food and agribusiness corporations as global water security, management and governance agents: the case of Nestlé, Bunge and Cargill', *Water Alternatives*, vol 5, no 3, pp620–636.

SWSC (Sustainable Water Stewardship Collaboratory) (2011) *Sustainable Water Stewardship: The Next Big Step Forward*, University of Cambridge Programme for Sustainability Leadership, Cambridge.

Turton, A. (2003) *The Political Aspects of Institutional Developments in the Water Sector: South Africa and Its International River Basins*, PhD thesis, University of Pretoria, South Africa.

UNEP-FI (2012) *Chief Liquidity Series 3—Extractives Sector*, Geneva.

UN-Water (2009) *The World Water Development Report, Water in a Changing World*, United Nations, New York.

Vogt-Schilb, A. and Hallegatte, S. (2011) 'When starting with the most expensive option makes sense: use and misuse of marginal abatement cost curves', *World Bank Policy Research*, 5803.

Warner, J. and Johnson, C. L. (2007) '"Virtual water"—real people: useful concept or prescriptive tool?', *Water International*, vol 32, no 1, pp63–77.

WaterAid (2012) www.wateraid.org/uk/about_us/financial_review/default.asp, accessed 25 January 2012.

Water Alternatives (2012) Special issue: Open for business or opening Pandora's Box? A constructive critique of corporate engagement in water policy, Eds, Hepworth N, Morrison J and Lall, U, http://www.water-alternatives.org/.

WBCSD (2012) *Water Valuation; Building the Business Case*, WBCSD Water, Geneva.

WFP (2011) *Water Futures - Addressing shared water challenges through collective action*, WWF, SABMiller, GIZ. Eischborn, Germany.

Wichelns, D. (2010) 'Virtual water and water footprints offer limited insight regarding important policy questions', *International Journal of Water Resources Development*, vol 26, pp639–651.

Wichelns, D. (2011) 'Assessing water footprints will not be helpful in improving water management or ensuring food security', *International Journal of Water Resources Development*, vol 27, pp607–619.

WWF (2012) *Shared Risk and Opportunity in Water Resources: Seeking a Sustainable Future for Lake Naivasha*, WWF International, Gland.

WWF and DEG (2011) 'The water risk filter', http://waterriskfilter.panda.org, accessed 26 September 2012.

15 The Shotgun Marriage

Water Security, Cultural Politics, and Forced Engagements between Official and Local Rights Frameworks

Rutgerd Boelens

Introduction

It is an increasingly accepted policy notion that water scarcity and water insecurity are not so much related to the precarious, absolute availability of sufficient fresh and clean water but rather to the ways in which water and water services are distributed, in contexts of unequal power. The United Nations Human Development Report, for example, emphasizes the need to 'debunk the myth that the crisis is the result of scarcity . . . poverty, power and inequality are at the heart of the problem' (UNDP, 2006). Similarly, the Asian Development Bank concludes that 'the key issue in almost all circumstances is not whether there is enough water: it is the factors that determine and limit whether the poor can gain access to the benefits that water resources provide' (Soussan, 2004, p20). Consequently, the notion of water security is necessarily politically contested and relates to power differentials. For instance, for millions of marginalized families around the world who face the powerful water interests of capitalist agribusiness, forest logging, mining, and hydropower companies, as well as nation-states' economic, political, and military objectives, defending control over water resources is a matter of life and death (Goldman, 1998; Swyngedouw, 2005; Bakker, 2010). Water is the liquid that feeds their livelihood systems and, often, the energizer that invigorates collective action. But despite the worldwide importance of smallholder production systems for livelihood security, the threats and water insecurities they face in a globalizing society are ever growing (Swyngedouw, 2000; Martínez-Alier, 2002).

Therefore, distribution and redistribution issues, as well as the recognition of smallholders' water rights, are core themes in water security debates. And it comes as no surprise that water rights and property relations have become central issues in global policy debates and rural development initiatives. But ironically, in most law and policy schools and intervention programs, this attention tends to be strongly biased toward modernistic, technical-economic water development prescriptions and universalistic, modernizing water laws, while insight into 'real' and 'living' in-the-field water rights and management forms is largely lacking. Commonly, production of water knowledge, ontologies, disciplines, and truths concentrates on the issue of how to align water services, use systems, and local users to imagined ('better') water governance hierarchies and supralocal water security discourses.

There seems to be an important shift, however, whereby official water laws and policies increasingly tend to 'recognize' smallholders' and customary rights to enhance their water security (CLEP, 2008; Meinzen-Dick and Nkonya, 2007). Incorporating local rights and right notions into the formal legal frameworks is a common way for this to happen—testified also by the worldwide support of international financing institutes to numerous water rights formalization programs. But, as I argue in this chapter, we need to challenge the widespread assumption that formally recognizing customary water rights simply leads to increased water security for local water user groups. The chapter explains how often the contrary may occur: Water rights formalization programs may reinforce local water insecurities.

I will explain the reasons by outlining various key elements. In the next section, I present a conceptualization of 'divergent water securities' as a plural notion that is given shape in and through unequal power contexts, one that by definition refers to a relational, political, and multiscale relationship of water access and control. The third section characterizes the dynamics of water rights as embedded in local livelihood security frameworks, and the ways they are challenged by dominant water players. The fourth section explains how, often, legalization of particular customary laws results not from benevolent policy strategies to support the poor but from the tactical installation of 'forced engagements' between official and local law systems in a mutual quest for societal legitimacy. It scrutinizes dominant groups' formalization interests, whereby local rights recognition may result simultaneously in containing and petrifying the dynamics of those customary rights that are recognized and in rendering the universe of nonrecognized local rights illegal. The fifth section examines why, therefore, apart from striving for official recognition of their rights, local collectives increasingly may turn toward other, multiscalar strategies to defend their water securities. The final section draws conclusions on the complex relationship among formal and alternative water securities, the cultural politics of recognizing water rights, and multiple forms of contestation.

Divergent Water Securities

In general terms, *water security* refers to the secure, adequate, and sustainable access that people and ecosystems have to water, including the equitable distribution of advantages/disadvantages related to water use and development opportunities, the safeguarding against water-based threats, and the ways of sharing decision-making power in water governance. Commonly, however, when such a notion is elaborated in, for example, national laws, policy documents, or development strategies, the issue is entirely depoliticized—as if water security mostly refers to win-win solutions in which all will benefit equally.

Simultaneously, the concept tends to be naturalized as if it were pertaining to the realm of naturally best, objective solutions and not to the one of human interests, choices, negotiation, intervention, material-economic distribution, and power plays.[1]

As Dimitrov (2002) observes, 'calls for holistic thinking on human security and for balancing various policy concerns . . . often rely on the assumption that values are compatible with each other. Yet when we compare different notions of water security, we see an inherent tension between the priorities established by these notions' (p688). Water security for ecosystems and urban or agricultural development are not always compatible; a national policy focus on water for hydropower security and large dams may challenge local villages' livelihoods; water security for mining companies in many countries endangers human consumption, health and subsistence, agriculture, and so forth (Harvey, 1996; Castro, 2008; Hoogesteger et al., 2012). Indeed, different dimensions of and interests in water security are often 'mutually incompatible and pursuing them simultaneously is impossible' (Dimitrov, 2002, p688).

Strongly linked to such depoliticized and naturalized water security notions and policies is the fact that most of them have a state- or market-centric bias. Depoliticization and naturalization tend to obscure policy measures' instrumentality for the controlling interests of state institutions, dominant market players, and local/national elites. Not surprisingly, therefore, the water security problems and solutions they frame are often entirely different from the water (in)security definitions by, for example, marginalized user groups (e.g., Hoogesteger, 2012). Consequently, around the globe, market-based water security strategies as the latest water rights privatization policies (Swyngedouw, 2005; Perreault, 2008; Bakker, 2010) or Payments for Environmental Services (PES) (Robertson, 2007; Sullivan, 2009) have encountered fierce resistance from local communities defending what they see as their water security and rights (Castro, 2008; Perreault, 2008; Bebbington et al., 2010). Dimitrov (2002) argues that security conceptions basically relate to four core elements: who or what is to be protected (the referential object of security), against what (the threats and sources of danger), how is should be protected (the means of pursuing security), and who needs to do this (the actors and institutions made responsible for securing water access/control). These core elements and their linkages are fundamentally different according to actors' interests and perspectives, often along lines of class, ethnic and gender, and other power differentials (Vos et al., 2006; Van der Ploeg, 2008; Hoogesteger et al., 2012).

Moreover, given its politically contested nature, the definition and pursuit of 'water security' by one entity or policy may often entail 'water insecurity' by the other. Therefore, I conceptualize the plural notion of 'divergent water securities' as an intrinsically relational, political, and multiscale relationship of water access and control that takes shape in contexts of unequal power relations. I also examine water securities as profoundly related to the pluralist, nonuniversal, and dynamic notion of 'water rights' following, similarly, a conceptualization that radically differs from mainstream notions that see clear, uniform, and enforceable water rights as tools and conditions for the rational exercise of state water control and/or exchange of water and services through market forces. Whereas such state- and market-centered conceptions preach uniformity and universality, this paper suggests an understanding that explicitly acknowledges the historical specificity and embeddedness of water securities and rights in particular cultural-ecological settings. Herein, locally existing water values, norms, meanings, and control practices, and the power relations that inform and surround them, are profoundly influential (Boelens, 2009; Zwarteveen and Boelens, 2011).

Despite discursive attention to pluralism and multiculturalism, peasant and indigenous federations and water user collectives have a hard time bringing their water views to the national and globalizing negotiation tables. The role of water expert communities, linking the local/national with the international, is influential here. They drive the construction of modern, advanced water rights and their implementation through water policy and technology intervention projects. Experts' knowledge embodies modernity, progress, and development. In many countries, legitimate water knowledge, rather than following from actual experiences with local water realities, importantly stems from economic foundations and cultural politics, banking on accreditation by institutional structures and officialdom. Water knowledge development, therefore, largely focuses on those water security and governance questions asked not by the water user collectives but by official water rulers and accreditation systems.

One enduring assumption of modernist water policy notions is that universalist rulemaking and standardization of agreements among all is for the benefit of all and produce efficient rights, mutually beneficial exchange, and rational organization. This universal ethical system of norms governing property and cultural rights is also key to the legal modernization thrust. For example, World Bank's influential policy-thinker, Hernando De Soto, explains that—contrary to the West—the lack of such universal norms in Southern, 'closed' countries is the main reason they cannot fully enter the capitalist world system.[2] The civilizing mission of advanced nations and academics would be 'to help governments in developing countries build formal property systems that embrace all their people' (2000, p180). A primary condition for prosperity and security, he argues, is to construct a world of uniform values and property relations precisely matching the imagined reality of civilized interaction and exchange. 'Common standards in one body of law are necessary to create a modern market economy' (De Soto, 2000, p164). Similarly, neglecting the ways in which collective and individual water rights actually function and are embedded in community water security rationalities, Hoekstra and Chapagain (2008, p141) argue for the 'need to arrive at a global agreement on water pricing structures that covers the full cost of water use. . . . Without an international treaty on proper water pricing it is unlikely that a globally efficient pattern of water will ever be achieved'. To concretize this uniform values and property framework, until recently, the liberal-positivist legal tradition largely ignored the construction and functioning of law in social action, denying that local and formal water law, rather than constituting an objective and rational system for designing societal life, is a cultural and political product developed by societal groups and governmental agents who strategize to foster their interests.

Thus, in fact, modernist water lawyers, economists, engineers, development planners, and others often saw, and still see, it as their mission to socially engineer rational, efficient water society by installing modern water rights and the effective rule of law. The latter are seen as both the tools for planning progress, the final objectives, and accordingly, the measuring stick to judge the chaos of existing water reality. However, as I show in the next section, numerous user groups do not restrict legitimate water authority and rights to those emanating from modern state law.

Everyday Water Rights and Water Security Struggles

Water rights exist in conditions of legal pluralism, where rules and principles of different origin and legitimization coexist and interact in the same water territory. Water rights systems comprise diverse, dynamic sets of hybrid rules, rights, and organizational forms, joining local, national, and global rulemaking sources and patterns (Benda-Beckmann et al., 1998; Roth et al., 2005). Understanding not just law- or project-driven rights concepts, but especially users' reasoning and rights expressions, is crucial for comprehending their water security claims and the ways in which local water control and livelihood defense interact with national and global water/power arenas.

In water use and control systems, diverse interest groups encounter and negotiate, reinventing and experimenting with rights definitions and normative codes that regulate day-to-day water practices. So, co-determined also by physical-ecological conditions, water rights development is interwoven with local user societies' cultural, political, economic, and technological histories. Particularly (but not exclusively) in user-constructed and managed systems, for instance of peasant and indigenous populations, this calls for, in each system and water territory again, a thorough comprehension of the particular nature of water rights, taking into account their multilayered bundles: rights to use and withdraw; rights to operate, supervise and manage; and rights to control (i.e., define, regulate, and represent water uses and users). Beyond reference rights as they are institutionalized in national and even local users' normative frameworks, examination of rights in action, as activated and materialized in actual social relationships and particular contexts, is central (Boelens & Zwarteveen, 2005; cf. Benda-Beckmann et al., 1998).

Water rights and distribution practices become manifest, simultaneously, in water infrastructure and technology, normative arrangements, and organizational frameworks to operate and maintain water control systems, all embedded in their political-economic and cultural-symbolic context. Beyond law and rights in a strict sense, technology, organizations, culture, economy, and ecology also fundamentally structure water access and control security (Roth et al., 2005; Bakker, 2010).

We have defined water rights, in the context of farmer-managed irrigation practice, as 'authorized demands to use (part of) a flow of water, including certain privileges, restrictions, obligations and penalties accompanying this authorization, among which a key element is the faculty to take part in collective decision-making about system management and direction' (Beccar et al., 2002, p3). The conditions, penalties, and decision-making privileges attached to water rights give shape to complex matrices.[3] The mutual bonds of obligations required to operate and sustain the system—together with common system ownership, in which each user's rights are embedded, and the shared water history, myths of origin and belonging, customs, rituals, and struggles—make users identify with their system, with each other, and engage in collective action. Most of these rights and rules are not written down or, as the notion of hydraulic property creation shows, they are 'written' in infrastructure and 'materialized' in bonds of mutual obligations. Though largely invisible to outsiders, such rights repertoires usually consist of clear, widely popularized patterns of norms that are part of collective reference frameworks. Water rights, and the struggles over their contents and authority, embody the ways in which water is closely connected to power, cultural meanings, and identities. Thereby, user-developed systems, far from romantic arrangements, result from ongoing internal struggles and harsh negotiations.

The construction of context-specific sociolegal repertoires and local water rights is an outcome of a historical process in which diverse actors strategically select and intercalate elements from a variety of normative sources. Therefore, local water rights, rather than referring to strict time- and place-related origins, relate to users' perceptions that water rights' access, definitions, and control are 'theirs'. They express that water rights belong to them and the locality, that they orient users' behavior, and that locally appointed authorities—rather than 'outside' rules and rulemakers—have legitimate enforcement power (Benda-Beckmann and Benda-Beckman, 2000; Roth et al., 2005). Therefore, local water rights systems constitute local-national-global hybrids whereby examination of the local requires focusing on the issue of localization/externalization of control on water rights' political use and convenience for either intervening agents and supralocal rulers, or for user groups struggling for livelihood defense and rulemaking autonomy.

Indeed, despite ongoing interaction and hybridization, official water rights and locally prevailing rights constructs differ fundamentally, in substance, range of application, sources of legitimacy, and modes of authority, but also in the ways they are constituted, reaffirmed, and frame water security. For example, investments made by ancestors—many years' blood, sweat, and tears that have flown through the canals before bringing home the first water drop, costs even in terms of human casualties all to create water rights—are not commensurable with water rights' payment or paperwork. Community irrigation rights, for example, in all their complexity and diversity, tend to have features[4] that directly clash with state law and market-oriented regulations (cf. Goldman, 1998; Perreault, 2008; Hendriks, 2010).[5] Clearly, local rights dynamics and control-localizing authority constitute a primordial obstacle for national and international rulemakers, planners, and interveners. Even though in practice state apparatuses and markets are far from monolithic entities but diverse and internally divergent, multiscale water regulators, both state and market institutions *require* a predictable, uniform playing field (e.g., Scott 1998; Assies, 2009, 2010). For this, context-based rights orders are not seen just as irrational systems that elude justice but, above all, as intangible, unruly disorder eluding control.

Because of their deviance from such models' equality imperative, local water collectives in many places of the world are confronted with both top-down and participatory strategies to adapt and transform. Notions such as equality and efficient and rational water use are prominent here. Modernist water policies fundamentally promote equalization to occidental, techno-centric, objectivist water management definitions and models, interspersed with implicit references to universalistic ideas about efficiency, effective organization, property rights, economic functionality, and social security (Lankford, 2012; Boelens and Vos, 2012).

But water user collectives do not always accept these dominant governance techniques. Since water rights intrinsically combine issues of material resource distribution with those of legitimate decision making and cultural-political organization, water governance policies are intimately linked with both the questions of socioeconomic justice and cultural justice. Here, the clash between efforts for control-externalization and control-localization may be subtle and hidden or fierce and violent, depending on the context. Sometimes, user collectives or federations strategize their resistance actions within the law, to achieve legal change. Other actions take place against the law, in the domains that officialdom labels 'illegality'. However, far more widespread and influential in many regions are the permanent, everyday water rights struggles outside (or on the margins of) the law: they occur in the deeper layers of local water societies, and relate to most rules and practices that user collectives apply when materializing their own rights repertoires. Often, these local water security norms and strategies are not accepted or denied by the law since they aim, precisely, to elude bureaucratic or market-based control.

So, in numerous water societies in the world, multilayered water rights and water authority struggles take place, profoundly colored by the cultural politics of particular places, nation-states, power geometries, and histories. Divergent parties' cultural political strategies aim to frame and impose their notions of morality versus immorality, inclusion and exclusion, superiority versus inferiority, identity and belonging, thereby shaping the water control arena in which rules, values, and social, economic, and political meaning are created and contested. As I will show in the next section, this does not imply or lead to a strict divide between officialdom and locality, or between the fields of positive and local justice.

Forced Engagements and Cultural Politics: The Shotgun Marriage between Positive and Local Justice

The above disputes over local water rights' access and legitimacy provide insight into how positive justice and local forms of equity (i.e., in their legal essence, respectively, official 'right-ness in general' vis-à-vis local 'fair-ness in particular cases')[6] are constructed and interconnected. For instance, the fact that local water rules and rights in many countries were actually official or colonial norms that have been locally appropriated, adapted, and internalized (cf. Benda-Beckmann et al., 1998; Hoekema, 2010; Sousa Santos, 1995) means that local law (whether indigenous, peasant, or other) cannot be conceived of as a set of normative repertoires that either preexisted or are autonomous vis-à-vis the state. Local water rights assume the presence of state law, and define themselves in contrast and relation to it (Benda-Beckmann et al., 1998; Boelens and Doornbos, 2001).

This works in both directions: State law also (though differently) grounds its existence and survival in the active functioning of multiple, locally particular sociolegal repertoires. Besides their interaction, state and local water rights systems have very different functions and characteristics: State law is formulated to regulate water control throughout national territories, whereas local law is context based. Therefore, in most (heterogeneous) cases, official justice is seen as inadequate and faces the problem of losing legitimacy by not 'doing social justice'. Totally ignoring these local fairness constructs and rights conflicts would endanger the state's legitimacy and possibly challenges the government or even the nation-states' broader political economic structure (cf. Foucault, 1975). Therefore, in many instances, customary law has been used, institutionalized, and codified by the legislators not to substitute positive law but to supplement and adapt it (Boelens, 2009; Hoekema, 2010). In fact, it is ironic that official justice has often been able to survive thanks to the 'equity' and 'acceptability' of customary laws that were incorporated (Schaffer and Lamb, 1981). This phenomenon is not typical of contemporary (e.g., Southern) countries. For instance, Roman Law (where *jus gentium* was added to *jus civile*) and English Law (where common law complemented statutory law) provide the same dualist rationality of protecting formal law systems. Hence both local and official orders base their existence on mutual interaction and strategic recognition and are partners to what I have called a 'shotgun marriage' (Boelens, 2009).[7]

But this state-institutionalized equity is a *contradictio in terminis*. It commonly leads to the decontextualization and depoliticization of local rules and rights, the latter becoming part of a general formalized system, which takes away their nature of suitability, acceptability, relevance, and being 'fair' in particular cases (cf. Schaffer and Lamb, 1981). Local sociolegal repertoires make sense only in their dynamic context, whereas national law demands order and stability. Indeed, there is a danger of freezing or even fossilizing customary rights systems by incorporating them into relatively static, universalistic state law, in which local principles lose their identity, functionality, and capacity for renewal (Benda-Beckmann et al., 1998; Roth et al., 2005). Moreover, as often occurs, local rights frameworks fall prey to expert-dominated redefinition or face assimilation and marginalization when legally recognized. Often, only those rights and principles that fit into official legislation and policies are recognized, thereby muzzling the complex variety of unruly rules. So, the shotgun marriage is above all a tactical one, forced from both sides, making it also unhappy and extremely complicated. 'Marriage' conflicts cast doubts on the effectiveness of official recognition policies in safeguarding customary water rights, and the recognition process that (at least discursively) aims to provide water security and legal backing to marginalized user groups often tends to reinforce the latter's water insecurities.

Indeed, as Hale observes, self-regulation comes with clearly articulated limits: It 'attempts to distinguish those rights that are acceptable from those that are not . . . defining the language of contention; stating which rights are legitimate; and what forms of political action are appropriate for achieving them' (2002, p490). Often, local water rights recognition policies are not in opposition to, but rather combine quite well with modernizing, disciplinary politics (Boelens, 2009). As once foreseen by Marx and Engels (1970/1848, p35) capitalism 'equalizes' and creates a world after its own image. In modern times, similarly, the neoliberal state does not simply recognize customary rights or local cultures but reconstructs them (Assies, 2010), reproducing its own relationships by differentiating between 'good' and 'bad' local rights. The former present demands compatible with the state's centralist or neoliberal project, whereas the latter call for redistribution of power and resources. As Hale argues, neoliberal multiculturalism, for instance, affirms cultural difference, 'while retaining the prerogative to discern between cultural rights consistent with the ideal of liberal, democratic pluralism, and cultural rights inimical to that ideal. In doing so they advance a universalistic ethic which constitutes a defence of the neoliberal capitalist order itself' (Hale, 2002, p491). Therefore, although such governance projects speak of decentralization and respect for pluralism and recognition of local norms, values and rights must not impinge on the model's foundations—that is, not interfere with state power and/or market rationality. Neoliberal multiculturalism opens up political space for local rights acknowledgment and, simultaneously, 'disciplines those who occupy them' (Hale, 2002, p490). The mechanisms to influence the consciousness of water users and redefine their frames of reference, reshaping local water security institutions to fit in with state administrative structures and/or market rules, are indeed increasingly subtle.

Legal formalization projects, as nationwide experiences from Peru, Ecuador, Chile, and Bolivia (among others) show, may be particularly dangerous for local water user communities when particular local rights are allowed, legalized, and institutionalized (often in their essentialized expression) at the expense of most others and at the cost of intensifying the repression of disobedient rules and rights. Legal recognition not only has repercussions for 'the recognized', but it particularly impacts peoples or management systems that do not have this new legal backing. As a direct consequence of non-recognition these suddenly have to suffer exclusion from basic services and rights and become explicitly illegal. The legalization of some is often accompanied by the illegalization and outright encroachment of others.

This relates to both 'cultural rights' (e.g., water management rules) and 'distributive justice' (e.g., water access rights). Chile's example of state-enacted Indigenous Law illustrates how legalization of particular indigenous water rights (naturalized as traditional) has direct implication for the nonprotection and encroachment of other local rights (e.g., Boelens, 2009). Del Castillo (2004) outlines how in Peru, similar to many other projects, the Pampas Verdes project uses the Caracha and Urubamba rivers' water to irrigate 218,000 hectares in the Nazca and Arequipa regions, for which some local rights were recognized. By building two dams, the territories of seven peasant communities are flooded, but since the latter have only local rights that have not been recognized, the territories are considered to be state property and are invaded as 'no-man's land'.

Aside from state water bureaucracy and neoliberal policymakers, codifying and freezing (or 'petrifying') local water rights is often a clear interest of powerful water interest groups who aim to intervene in local water control and contain existing water rights. The experiences in Chile, Peru, and Ecuador are telling, but can also be witnessed in, for instance, the United States. As Getches (2005) relates, after two centuries of limiting Indian rights geographically and politically in ever-smaller reservations, white settlers and investors are the prime defenders of 'clear Indian water rights' in order to know how far their encroachment practices can go without overstepping their legal backing. 'Because investments and property values are undermined by uncertainty, non-Indians and the western states that tend to support non-Indian interests have also urged that Indian water rights should be legally determined' (Getches, 2005, p48).

Lawmaking (in local, national, and supranational societies) aims to resolve historically and context-grounded contradictions and conflicts. Often, in societies with structural inequalities, dominant classes aim to naturalize their water law interests and rights framework as 'justice', putting the political-legal system to use to respond to particular rights conflicts that challenge their position rather than addressing the basic class, gender, and ethnic contradictions that underlie these conflicts (cf. Bourdieu, 1977). Thus, often, as part of the shotgun marriage, temporary (or ad-hoc) changes are introduced to safeguard the structure of domination and its rulemaking and enforcing legitimacy, but which do not address primary contradictions that involve the reproduction of the societal class system. Since the fundamental contradictions remain, this soon creates new dilemmas and triggers new conflicts (Chambliss & Zatz, 1993), as the ongoing water conflicts make manifest.

Water-user communities, in many places, do not remain silent or simply accept their unequal position in the shotgun marriage; they often challenge the structuring of 'their water society' by officialdom's powers. But, as I will elaborate briefly in the section below, notwithstanding their often contradictory objectives, it would be mistaken to suggest that local organizations, to defend their autonomy, try to avoid interaction with the state or development institutions.

Strategic Water Security Battles and Multiscalar Networks

Actual practice (i.e., the many communities requesting collaboration) shows that water-user communities actively look for interaction with the agents of formal justice and water governance. It is common to see, from Africa to Asia and in the South and the North, that they ask their leaders to negotiate with agency officials and 'get the engine running'.[8] How it should work, however, is quite another question. The interaction consciously sought by both parties is often only very partially based on a common interest. Realizing water security—through river basin management, irrigation, drinking water and sanitation, land conservation and water harvesting, flood-protection, or other programs—involves very high costs in terms of infrastructure, technology development, land acquisition, collective labor, and cash and managerial inputs; thus, both the state and the users (with different objectives) try to achieve for their purposes the most favorable ratio of investment versus control. Hereby, local groups try to gain more access to state resources and international funding without handing over local normative power. The question of who controls the activities and resources of whom and how is key; both try to align the other parties and their resources in the network and action program they desire, aiming to forge a chain of compliance that weakens the resistance of the other. In this interaction between entities with confronting interests and a mutual need to capture each other's resources (exemplified, again, in the shotgun marriage), both sides make use of each other's techniques, norms, rules, and even discourses—though under conditions of unequal clout.

Already in early history, the colonized or indigenous communities used the rulers' system of justice in which they selectively 'shopped around' to fight injustices or regain their water rights. Or, as Getches (2005, p44–65) strikingly described for North American cases, now and in history, they 'defended the indigenous water rights with the laws of a dominant culture' (cf. Getches, 2010). Commonly, such strategies of imitation and adaptation may both reinforce the legitimacy of the ruling system and, quite the opposite, serve counter-hegemonic objectives and hinder the water rights plundering practices of the invaders.

Therefore, it is important to take seriously that many local resource users, ethnic groups, and other minorities around the world actively strive for state legal recognition. Marginalized water users are often constrained by state law, but at the same time they approach it as a powerful resource for claiming or defending their interests and rights. At that moment, they recognize its legitimacy and power, but this does not mean that they accept its current manifestations or the power structures that sustain it. Their own strategies for claiming legal recognition, therefore, often go beyond 'recognizing' existing state legal hierarchies, and even may go against them or aim to conquer and transform them, putting recognition politics upside-down. As Bourdieu (1998) observed, 'the state in every country is the track in reality of social conquests'. The state is an ambiguous reality. 'It is not accurate to say that it is an instrument in the hands of the ruling class. The state is certainly not completely neutral, completely independent of dominant forces in society. . . . It is a battleground' (Bourdieu, 1998, p34). Current water security struggles, for instance in many Latin American countries, profoundly manifest how peasant and indigenous federations see state institutions, water laws, and policies as a battleground.

Next, water-user collectives know well that adopting official patterns of water control does not necessarily mean conforming to them. Recourse to formal law and using 'outside' water norms and official frames of organization may often also be a conscious strategy of local collectives. Under the disguise, sheltering and apparent adoption of outside rules, rights, and organizational constructs—what I referred to as *mimicry* (Boelens, 2009)—a diversity of local rights and hybrid norms are developed and exerted that—intangibly— act precisely against essentializing containment and universalizing take-over. In these covert territories, the rationality and styles of materially and culturally building water securities and water rights vitally diverge from official frameworks, and formal powers face huge problems penetrating and patrolling them. New, hybrid water rights and organizations strategically represent collectivities in their struggles against control externalization, harboring a world of difference below the outer appearances of uniformity and formality. Besides proliferating legal pluralism, such 'community undercurrents' or 'undertows' also enable action on broader political scales. They constitute flexible trans-local networks since local water conflicts increasingly involve global players. They dynamically ponder and strategize the manifestation of supralocal policies and markets in local water territories, embedding the local in the global and the global in the local.

Conclusions

This paper is not a plea against formalization of local rights in general. On the contrary, it challenges the mainstream, commonly ad-hoc incorporation of stereotyped customary law in positivist law and policies, whereby the fundamental power contradictions are left unchallenged. This has often led to subordinating local rights repertoires and illegalizing the huge variety of nonrecognized norms and rights, thus expanding local water insecurities. As a reaction, grassroots groups often do not demand for specific rules or rights to be legalized, but for the legal recognition of greater autonomy to develop their own management rules.[9]

Similar to other mainstream water governance discourses, the water security debate is often depoliticized by naturalizing shared problem definitions and common goals, while excluding the implicit values, ends, and material or power interests from the policy discussions, hiding in particular the cultural values of the framing experts or the control interests attached to state and market agents and institutional structures. Notwithstanding the fact that, in practice, they comprise heterogeneous entities themselves, these state and market institutions require the welding of a predictable, uniform playing field for water governance; thus, local water rights autonomy and diversity—including local water security conceptualizations—tend to be a primary obstacle for formal rulemakers and intervening agents. Commonly, the collective and territory-bound features of water rights, their hybrid origins, enormous diversity, and integration in community security structures makes them difficult to reconcile with the ideals of centralist control and modernist property regimes. Their multifaceted, rooted, and dynamic character makes them intangible and unrecognizable in positivist (bureaucratic and liberal) rights frameworks.

Not surprisingly, therefore, in many countries official legislation and policymaking have tended to deny or contain this water rights unruliness and water security disobedience. But ironically, as this chapter has shown, officialdom cannot afford to entirely neglect local water rights systems. There is a forced engagement: Overall state law—threatened to be unmasked as 'inappropriate'—bases its survival on the ability of local rights orders to adequately respond to local needs and contexts. To avoid losing legitimacy in the eyes of a heterogeneous society, and also in response to demands and uprisings by peasant, indigenous, and other marginalized water-user groups, there are many precedents of cases in which national governments have expanded their concept of unique, omnipresent national law and policy. But commonly, while incorporating local rights systems in its melting pot, the state legal system has done this in ways so as to not challenge the legal and power hierarchy. Such law and policy efforts to complement official law imply simultaneous recognition and negation of local water rights diversity; they are not simply responses to demands by subjugated groups for greater autonomy but also facilitate political control by the water bureaucracy and/ or help neoliberal sectors incorporate local water rights and organizations into the market system.

The dynamic and multiple manifestations of local water rights and security systems cannot be codified into blanket legal terms without jeopardizing their foundations. They refer to a broad range of diverse 'living' rights cultures and livelihood security systems that constantly reorganize their rules precisely to maintain their identity and capacity to negotiate and solve problems. Recognition policies often create essentialist constructs that do not represent this dynamic character; re-cognition points toward re-presentation (that is, 'identifying the object again' and 'portraying it again but in a modified way'), thus toward transformation of complex reality to discipline and make users' water behavior tangible. The cultural politics of these recognition policies imply that depoliticized water governance models and mainstream water security discourses differentiate those local rules that are 'good' and 'rational' from those that are 'inefficient' and 'not acceptable'. Hereby, by promoting best practices, rational water use. and good governance models, modernist water institutions aim—consciously or not—at self-reproduction in communities, creating a water world after their own image. Such all-embracing formalistic and expert framework often is the background for water user communities' participation and competition as presumed equals in a win-win exercise, vis-à-vis (trans)national water interest groups.

But these control-externalizing efforts to disembed water securities, render tangible local norms, and/or take over local water rights do not occur without resistance. Not all user collectives accept the individualistic, mercantile, or bureaucratic standards of being equal that file them in 'anomalous' or 'backward' categories (Boelens and Zwarteveen, 2005). They claim to respect different standards and demonstrate that not suiting the policy models is often a willful choice. They defend and re-embed their resources, rights, and decision-making faculties to keep water security from being dictated by outside institutions, power groups, and markets. The ongoing creation of own water rights and water security frames, connected to the subsurface proliferation of multilayered livelihood repertoires, broadens and deepens legal pluralism and inevitably questions the exclusiveness and self-evidence of uniform state- and market-based water governance institutions. In the end, there is an ongoing battle over the material control of water and water security systems, as well as over the right to culturally define, politically organize, and discursively shape their legitimate existence.

Notes

This chapter is based on the last decade's research in several Latin American countries, particularly within the framework of the international research and action alliance *Justicia Hídrica*/Water Justice (www.justiciahidrica.org).

1. It is also common to see here that attention to the (especially technical and economic) issues of water access gets priority over the ones related to water control, such as participation in decision making about water access, risks, and management.
2. The West would have overcome this presumed backwardness: 'Shifting the recognition of ownership from local arrangements into a larger order of economic and social relationships made life and business much easier. Formal property freed them from the time-consuming local arrangements inherent to closed societies' (De Soto, 2000, p174–175).
3. This is illustrated, for example, in the diverse ways in which a key mechanism for water rights' acquisition is given substance: In user-controlled systems around the world—unlike any official water law—the collective creation of infrastructure often generates collective and individual user rights and organizational forms ('hydraulic property'), resulting in the simultaneous, interrelated generation of infrastructure, organization, and rights. The re-creation and maintenance of user-managed systems re-creates and consolidates rights and organization (Boelens and Doornbos, 2001).

4. For example, individual rights embeddedness within collective property structures, territory-boundedness, community authority, nontransferability to outsiders, livelihood reproduction prioritization, hydraulic property creation, community-embedded obligations beyond the water domain, one vote per right-holder decision-making, and so forth.

5. Neoliberal policies, for instance, advocate private water rights separated from land and territory; state dominion and regulations that protect/foster 'free market' functioning; promotion of water transfer to 'higher values' (i.e., mostly outside community systems); rights auctioning to highest bidders; prioritizing market exchange value rather than obeying social priorities; rights acquisition and consolidation that are cut loose from system sustainability; individuals' voting weights that are proportional to their water shares and buying power, and so on.

6. *Equity* refers to location-, time-, and group-particular political constructs and concepts of fairness, and as such they cannot be reified or romanticized.

7. In the Andean countries, customary laws and water rights are sometimes incorporated into state law as 'special laws'—special norms that are enacted solely for certain societal (commonly essentialized) stakeholders and relationships, often in order to leave the official norm unmodified.

8. These 'negotiation missions' commonly receive much support from their communities, materialized in respect for community negotiators and in community gifts or bribes to persuade officials.

9. Thus, an important issue is how to give room to diverse local water rights and livelihood security systems, while not weakening their position in conflict with powerful exogenous interest groups. Another key issue related to providing local rule-enforcement autonomy is how to face existing gender, class, and ethnic injustices (which also may form part of customary sociolegal frameworks and practices). Answers point out directions where frameworks of collective rights and rulemaking autonomy for local collectives are combined with establishing democratic, pluralistic, supra-local institutions and rules that need to guarantee protection for individual and minority rights. These also need to offer opportunities for second-order conflict resolution and appeals in case local conflicts cannot be solved adequately.

References

Assies, W. (2009) 'Legal empowerment of the poor: with a little help from their friends?', *Journal of Peasant Studies*, vol 36, no 4, pp909–924.

Assies, W. (2010) 'The limits of state reform and multiculturalism in Latin America', in R. Boelens, D. Getches, and A. Guevara (eds) *Out of the Mainstream. Water Rights, Politics and Identity*, Earthscan, London and Washington, DC.

Bakker, K. (2010) *Beyond Privatization: Water, Governance, Citizenship*, Cornell University Press, Ithaca, NY.

Bebbington, A., Humphreys, D., and Bury, J. (2010) 'Federating and defending: water, territory and extraction in the Andes', in R. Boelens, D. Getches, and A. Guevara (eds) *Out of the Mainstream. Water Rights, Politics and Identity*, Earthscan, London and Washington, DC.

Beccar, L., Boelens, R., and Hoogendam, P. (2002) 'Water rights and collective action in community irrigation', in R. Boelens and P. Hoogendam (eds) *Water Rights and Empowerment*, Van Gorcum, Assen, Netherlands.

Benda-Beckmann, F. von, and Benda-Beckmann, K. von (2000) 'Gender and the multiple contingencies of water rights in Nepal', in R. Pradhan, F. Von Benda-Beckmann, and K. Von Benda-Beckmann (eds) *Water, Land and Law*, Wageningen University, Erasmus University, Wageningen and Rotterdam.

Benda-Beckmann, F. von, von Benda-Beckmann, K., and Spiertz, J. (1998) 'Equity and legal pluralism: taking customary law into account in natural resource policies', in R. Boelens and G. Dávila (eds) *Searching for Equity*, Van Gorcum, Assen, Netherlands.

Boelens, R. (2009) 'The politics of disciplining water rights', *Development and Change*, vol 40, no 2, pp307–331.

Boelens, R. and Doornbos, B. (2001) 'The battlefield of water rights. Rule making amidst conflicting normative frameworks in the Ecuadorian Highlands', *Human Organization*, vol 60, no 4, pp343–355.

Boelens, R. and Vos, J. (2012) 'The danger of naturalizing water policy concepts. Water productivity and efficiency discourses from field irrigation to virtual water trade', *Agricultural Water Management*, vol 108, pp16–26.

Boelens, R. and Zwarteveen, M. (2005) 'Prices and politics in Andean water reforms', *Development and Change*, vol 36, no 4, pp735–758.

Bourdieu, P. (1977) *Outline of a Theory of Practice*, Cambridge University Press, Cambridge and New York.

Bourdieu, P. (1998) *Acts of Resistance against the Tyranny of the Market*, The New Press, New York.

Castro, J.E. (2008) 'Water struggles, citizenship and governance in Latin America', *Development*, vol 51, no 1, pp72–76.

Chambliss, W.J. and Zatz M.S. (eds) (1993) *Making Law. The State, the Law and Structural Contradictions*, Indiana University Press, Bloomington and Indianapolis.

CLEP (2008) *Making the Law Work for Everyone, vol. I*, Report on the Commission on the Legal Empowerment of the Poor (CLEP), UNDP, New York.

Del Castillo, L. (2004) *Un Consenso Vital. Hacia un sistema de gestión compartida y descentralizada del agua*, Defensoria del Pueblo, Lima.

De Soto, H. (2000) *The Mystery of Capital. Why Capitalism Triumphs in the West and Fails Everywhere Else*, Basic Books, New York.

Dimitrov, R. (2002) Water, Conflict, and Security: A Conceptual Minefield. *Society & Natural Resources* vol 15, no 8, pp677–691.

Foucault, M. (1975) *Discipline and Punish. The Birth of the Prison*, Vintage Books, New York.

Getches, D. (2005) 'Defending indigenous water rights with the laws of a dominant culture: the case of the United States', in D. Roth, R. Boelens, and M. Zwarteveen (eds) *Liquid Relations*, Rutgers University Press, New Brunswick, NJ, and London.

Getches, D. (2010) 'Using international law to assert indigenous water rights', in R. Boelens, D. Getches, and A. Guevara (eds) *Out of the Mainstream. Water Rights, Politics and Identity*, Earthscan, London and Washington, DC.

Goldman, M. (ed) (1998) *Privatizing Nature. Political Struggles for the Global Commons*, Pluto Press, London.

Hale, C. R. (2002) 'Does multiculturalism menace? Governance, cultural rights and the politics of identity in Guatemala', *Journal of Latin America Studies*, vol 34, no 3, pp485–524.

Harvey, D. (1996) *Justice, Nature & the Geography of Difference*, Blackwell Publishers, Cambridge and Oxford.

Hendriks, J. (2010) 'Water laws, collective rights and system diversity in the Andean countries', in R. Boelens, D. Getches, and A. Guevara (eds) *Out of the Mainstream. Water Rights, Politics and Identity*, Earthscan, London and Washington, DC.

Hoekema, A. (2010) 'Community-controlled codification of local resource tenure: an effective tool for defending local rights?', in R. Boelens, D. Getches, and A. Guevara (eds) *Out of the Mainstream. Water Rights, Politics and Identity*, Earthscan, London and Washington, DC.

Hoekstra, A. and Chapagain, A. (2008) *Globalization of Water. Sharing the Planet's Freshwater Resources*, Blackwell, Malden, MA, and Oxford.

Hoogesteger, J. (2012) 'Democratizing water governance from the grassroots: the development of Interjuntas-Chimborazo in the Ecuadorian Andes', *Human Organization*, vol 71, no 1, pp76–86.

Hoogesteger, J., Manosalvas, R., Sosa, M. and Verzijl, A. (2012) 'Water security from a local perspective: understanding water struggles in the Ecuadorian and Peruvian Andes', SWAS, Unpublished document, Wageningen University, Netherlands.

Lankford, B. (2012) 'Towards a political ecology of irrigation efficiency and productivity', Preface to Special Issue, *Agricultural Water Management*.

Martínez-Alier, J. (2002) *The Environmentalism of the Poor. A Study of Ecological Conflicts and Valuation*, Edward Elgar, Cheltelham, UK, and Northampton, MA.

Marx, K. and Engels, F. (1970/1848) *Het Communistisch Manifest*, Progres Publishers, Moscow.

Meinzen-Dick, R. and Nkonya L. (2007) 'Understanding legal pluralism in water and land rights: lessons from Africa and Asia', in B. van Koppen. R. Giordano, J. Butterworth (eds) *Community-Based Water Law and Water Resource Management Reform in Developing Countries*, CABI, Cambridge, MA.

Perreault, T. (2008) 'Custom and contradiction: rural water governance and the politics of usos y costumbres in Bolivia's irrigator movement', *Annals of the Association of American Geographers*, vol 98, no 4, pp834–854.

Ploeg, J.D. Van der (2008) *The New Peasantries. Struggles for Autonomy and Sustainability in an Era of Empire and Globalisation*, Earthscan, London.

Robertson, M. (2007) 'Discovering price in all the wrong places: the work of commodity definition under neoliberal environmental policy', *Antipode*, vol 39, no 3, pp500–526.

Roth, D., Boelens, R. and Zwarteveen, M. (eds) (2005) *Liquid Relations. Contested Water Rights and Legal Complexity*, Rutgers University Press, New Brunswick, NJ, and London.

Schaffer, B. and Lamb, G. (1981) *Can Equity Be Organized? Equity, Development Analysis and Planning*, Institute of Development Studies, Sussex University, Brighton.

Scott, J. (1998) *Seeing Like a State. How Certain Schemes to Improve the Human Condition Have Failed*, Yale University Press, New Haven and London.

Sousa Santos, B. (1995) *Toward a New Common Sense. Law, Science and Politics in the Paradigmatic Transition*, Routledge, London and New York.

Soussan, J. (2004) *Poverty and water security: understanding how water affects the poor*. Water for All series, issue 2, vol. 2, Asian Development Bank.

Sullivan, S. (2009) 'Global enclosures. An ecosystem at your service', *The Land*, Winter 2008/9, pp21–23.

Swyngedouw, E. (2000) 'Authoritarian governance, power, and the politics of rescaling', *Environment and Planning D*, vol 18, no 1, pp63–76.

Swyngedouw, E. (2005) 'Dispossessing H2O: the contested terrain of water privatization', *Capitalism, Nature, Socialism*, vol 16, no 1, pp81–98.

UNDP/United Nations Development Program (2006) *Beyond Scarcity: Power, Poverty and the Global Water Crisis*, Human Development Report, Palgrave Macmillan, Houndmills, Basingstoke, New York.

Vos, H. de, Boelens R. and Bustamante, R. (2006) 'Formal law and local water control in the Andean Region: a fiercely contested field', *International Journal of Water Resources Development*, vol 22, no 1, pp37–48.

Zwarteveen, M. and Boelens, R. (2011) 'Thinking water justice: some inspiring concepts and theories', in R. Boelens, L. Cremers, and M. Zwarteveen (eds) *Justicia Hídrica. Acumulación, Conflictos y Acción Civil*, IEP, Lima.

16 Infrastructure Hydromentalities

Water Sharing, Water Control, and Water (In)security

Bruce Lankford

Introduction

This chapter addresses interrelationships between water security and water infrastructure. The premise of the chapter is that water security is defined by an ethos of, and attempts at, water sharing amongst users experiencing water variability and scarcity. As is explained below, sharing partially meets the Grey and Sadoff (2007) concept of security achieved through water sufficiency: 'the availability of an acceptable quantity and quality of water for health, livelihoods, ecosystems and production' (p545). The chapter proposes a water-management framework (termed *water meta-control*) that puts 'share management' alongside 'demand management' and 'supply management, and explores all three via an infrastructural lens. I argue that choices over structures for controlling water are greatly influenced by past and current infrastructural fashions and trends that I term *hydromentalities*. Cultural and sociopolitical influences arbitrate engineers' choices, while views held by water users are intangibly mediated by arrays of nearby and distant infrastructure. Without deeper reflection, water infrastructure will be unable to meet the growing challenges of climate change and water distribution (Pahl-Wostl, 2007; Giordano, 2013). By considering an infrastructural lens, I offer this definition: Water security seeks, and is consequent to, the sharing of water surpluses and deficits between different users mediated by the designed architecture of water infrastructure deployed to address the spatial, temporal, and scalar complexities of demand and supply. Although this definition ostensibly corresponds with a narrow volumetric deterministic concern for water (in)security (Zeitoun, 2011), this framework seeks to highlight the interplay between water security, the challenges and technologies of control, and cultures of water engineering.

In a context of growing water competition and variability, water insecurity may be characterised by volumetric shortfalls and poor timing and misallocation of water, resulting in uncertainty and inequity amongst users. These intertwined effects are mediated by the types of infrastructure selected for different sectors (e.g., irrigation). However, water insecurities (insufficient and poorly distributed water volumes) arise because these infrastructures are not considered coherently. Thus, water insecurity results not only from a lack of infrastructure to face water-related impacts of climate change (Jowitt, 2009), but also from how infrastructure is put together to serve increasingly interconnected users and sources. Because water is distributed by being divided (flows bifurcate), infrastructure ostensibly for one purpose jointly determines water supply for other sectors and thus affects the manageability of water apportionment. Moreover, present-day infrastructure as an expression of previous fashions is unlikely to have arisen as a single impartial coherent plan ready to face contemporary problems. Furthermore, infrastructure is costly, long lived, and not easy to alter (Stakhiv, 1998; Giordano, 2013).

Enhancing water supply and control across multiple users and scales is difficult, elusive, and socially mediated. A question arises: Can we design river basin, irrigation, environmental, and domestic/sanitation infrastructure in ways that fit together to promote the timely, transparent, and accurate placement and allocation of water for productive and protective human/ecological needs, while offering communities opportunities to reflect on their space, technology, and context-mediated water knowledge? To examine this question, a theoretical framework termed *water meta-control* is proposed. This framework addresses the control structures employed for apportioning water to multiple sectors, uses, and users from local to regional scales. I consider control in this comprehensive sense to be missing from engineering debates. Instead, water technologies and debates are diminished by reference to trends and fashions, such as 'small scaleness'. For example, small-scale irrigation received great interest in the 1980s and has continued to dominate donor and researcher interest.[1] This prompted Scott in 1996 to redefine 'appropriate technology' away from being simple, small-scale, low-cost and non-violent to being based upon technique, knowledge, organisation and product (Scott, 1996). This is less prescriptive and more accommodating of the complex interrelations between water, technology, and people.

Within the word limits of this chapter, the framework is far from comprehensive and cannot be used to reconfigure water infrastructure. Rather, the 'design biases' of the framework serve as a reminder for engineers and social scientists to think more cautiously about water engineering and water security outcomes. I move beyond storage as the means to solve water insecurities/scarcities because, without a parallel emphasis on sharing, water securitisation via supply augmentation may not remove water insecurities (and perceptions thereof) during periods of scarcity or when demand increases to take up the new supply.

The chapter is theory-based and informed by my PhD research conducted in the 1990s on irrigation design-management interactions. As such, this chapter is concerned with insecurities arising out of scarcity, not those associated with damage caused by floods (Stakhiv, 1998; Grey and Sadoff, 2007), for which different types and functions of infrastructure are required.[2] This analysis incorporates as only one factor the problem of individual or cumulative infrastructure sizing and capacity to address changing river regimes as a result of climate change (Rogers, 1997).

Water Control and Water Security

I use the term *water control* for describing the challenge of a water apportionment (rather than as meant in the political sense of 'taking control of
water', e.g., Boelens, 2008) that encapsulates an aspiration to improve the
sharing and placement of water for the many, not to securitise water supplies for the benefit of a few. This is water control in the manner explored
by the World Bank (Plusquellec, 1994) and FAO (Renault et al., 2007) in
their irrigation studies, but especially as defined by Bolding et al. (1995),
where the authors defined water control as central to the political economy
of water resources. This paper also suggests that without explicit reference
to infrastructure-induced control, different users gain at the expense of others, particularly benefitting during times of water contraction and scarcity.
Nevertheless, it would be naïve to suggest that the two senses of control are
unrelated; questions arise over whether poor apportionative control plays
into the hands of those wishing to 'take control' over water.[3]

Stepping back, I believe that four physical 'drivers' imprint themselves
on patterns of water distribution. This chapter seeks to highlight the role
of infrastructure as one of those. Three other drivers are: rainfall patterns
influenced by weather and climate (influencing the severity, location, and
unpredictability of droughts and floods), the topography and natural drainage patterns of the catchment (influencing the runoff characteristic and location of streams, floodplains, and wetlands), and the soils and geology of the
catchment (influencing runoff and water in soil and aquifers).[4] These four
suggest that water security emerges from a natural and human placement
of water in the volumes required in ways that communicate to water users
the manner, proportions, and volumes of water that can be stored, pumped,
abstracted, conveyed, and divided.[5]

I distinguish water control and meta-control. The former tends to see
apportionative control as sitting within sectors or in particular localities; it
draws on well-worn technological procedures and leads to discrete infrastructure packages such as canal systems. Meta-control encapsulates a broader
field of control covering all scales (from fields up to river basins) from local
to distant localities; it addresses all users and sectors; and it selects multiple
technologies from different schools of technology. It is the wider vision of
this water control that shapes river basin infrastructure architectures. Readers will note that I take the river basin as the unit of water management to
examine water meta-control.[6]

Supply and Demand Management—and Water Security

This chapter allies water security to water scarcity and takes the view that
water security comprises two main dimensions: improving the volumetric
sufficiency of the balance of supply over demand, and the distribution of
that adequacy (whether in surplus or deficit) more equitably to multiple
and disparate users. This dual approach argues that water security is not
solely predicated on creating positive surpluses. In parallel to this adequacy

puzzle, society also faces challenges related to the distribution of surpluses and shortages to many users and uses. Moreover, this distribution should be seen as fair and transparent, and, therefore, it has social 'equity' and informational dimensions beyond quantitative aspects.

One of the most common entry points for addressing the first part (adequacy) is to solve the equation of supply over demand, either through supply management or demand management (Tortajada, 2006; Lautze and Giordano, 2007). I shall explain their common understanding before explaining how they might also address water sharing. Supply management is the notion that in order to solve scarcity, more freshwater needs to be sourced, built, or otherwise obtained. Five examples of supply management include dam building (Lautze and Giordano, 2007), installing boreholes (or deeper boreholes), inter- or intrabasin transfers of water (Gupta and van der Zaag, 2008), catching rainwater in small storage bodies (Wisser et al., 2010), and desalinising salt water (Tortajada, 2006).

Demand management attempts to solve water scarcity/sufficiency by reducing demand and is usually understood by three parts. The first is the reduction of the net demand; in other words, the cultural, political, and economic decisions that affect the original source and amount of demand. Examples include decisions over the total number of houses built in a given area or publicity efforts to reduce garden watering during a drought. The second is the reduction of inefficiencies and losses in meeting net demand (in other words, the reduction of the gross demand for a given net demand). The third is pricing water so that the cost of using the resource pushes down net demand and nonbeneficial losses (exemplified by rising tariffs in Singapore described by Tortajada, 2006). Although infrastructure is more commonly associated with supply management, infrastructure is also required to deliver demand management—for example, equipment to reduce losses, to meter, and to precisely place water. However these two 'paradigms' are rarely analysed in terms of water sharing. For example, Pahl-Wostl (2007) refers to supply and demand infrastructure to reduce variation in supply and demand but omits the topic of water sharing elicited by the same infrastructure.

Infrastructure Design-Management Interactions

For the purposes of addressing catchment societal water security via a meta-control infrastructural lens, I adopt 'design-management interactions' from an irrigation perspective and apply them to the whole catchment. Design-management interactions were expressed adroitly by Bos in 1987: 'water management in future irrigation schemes could be improved if systems were designed in such a way that their proper management would be as easy as the mismanagement of existing systems' (Bos, 1987). This question asks if we can 'design in' water apportionment in ways that fit formal and informal property rights and supply and demand patterns over time and space (Lankford and Mwaruvanda, 2007). My rationale for applying this irrigation perspective to the catchment is threefold. First, canal systems are proxies of the problem of apportioning and scheduling water to different sectors within a catchment. They share features such as hierarchies of networks, limited (or absence of) within-system storage, the timing and times of flows, and social perceptions about water distribution in the face of lack of information or poorly designed networks. In other words, water sharing is not effected by giving storage to each irrigator but by the bifurcation of canal flows. This relationship between sufficiency and distribution was identified by Stakhiv (1998, p160):

> In fact, it can be said that much of water planning is inherently concerned with implementing anticipatory measures that are designed to either meet future demands or avoid future damage due to floods and droughts. It should be recognized that the origins of water resources management lie with the rise and evolution of civilizations, centering on the need to control and distribute water for agriculture.

Second, with increasing scarcity and basin closure, rivers take on canal-like properties of distributing water to competing sequential users with limited room for error and environmental flows. Third, irrigation is a major water consumer in semi-arid basins, influencing river hydrologies by placing abstractive and depletive infrastructure in the catchment.

In highlighting this topic, I reflect on an influential debate that originated with Lucas Horst in Wageningen University (Diemer and Slabbers, 1992), who argued that two solutions should be applied to the question of apportionment of water in canal systems: simplification and automation (Horst, 1983). Simplification argues that gates used to adjust flows on irrigation systems are built very simply so that humans can make adjustments 'easily'. Automation favours taking humans out of the picture by introducing the use of computer- and radio-controlled gates or self-actuating gates. Yet, other 'schools' of design can be considered, for example, structured systems (Lankford and Gowing, 1997; Albinson and Perry, 2002), which argue that water apportionment arises from more than just gate technology. Instead, matters such as canal density, the ratio of command areas to flows, stricter irrigation scheduling, and the fit between gate sizing at different canal levels also play a role. Choices between design schools provide the basis for a water meta-control framework.

A Proposed Framework for Water Meta-Control

A proposed framework for water control is given in Figure 16.1. This framework captures the concern phrased by Molle and Mamanpoush (2012): 'the actual apportionment and distribution of water and how management incorporates, and responds to, hydrologic variability and uncertainty. Water sharing may be more or less responsive to this variability, and diversely transparent/equitable and technically efficient'. This concern (and the framework of this chapter) distinguishes three dimensions to how society, working through engineers and artisans, approaches the sharing of varying

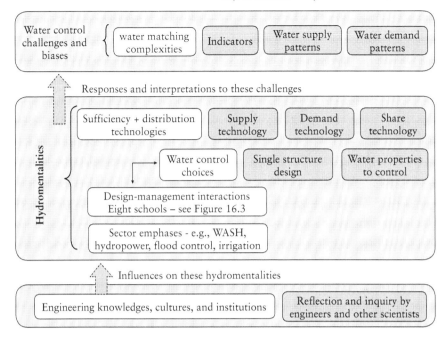

Figure 16.1 Water metacontrol framework leading to water architectures

amounts of river basin water. Each is covered by a grey cell in Figure 16.1 and by a subsection below.

Water Control Challenges and Biases

The challenge of water control arises from the complexities, uncertainties, and patterns of supply of and demand for water (Rogers, 1997). Individually, these are complex, but it is the overlay or matching of supply influenced by many factors (for example, slopes and relief, crops, agrometeorology) and demand influenced by many factors (for example, command areas, soil properties, population, households, and per capita demand; urban and industrial factors) that create complex patterns of excesses and deficits in turn to be managed by people using infrastructure. These patterns require indicators that assess how well management is provided; indicators such as equity, adequacy, and efficiency are commonly known. These indicators also establish a question over the design of the type and prevalence of measurement structures to allow performance to be assessed.

Hydromentalities

There are four types of hydromentalities that respond to the water control challenges discussed above: sufficiency and distribution technologies, water control design, design-management interactions, and sector emphases. I contentiously cast these hydromentalities as interpretable, subject to preferences, and, for the most part, lacking critical awareness. In other words, current water control architectures are mostly the product of accidental biases and fashions.

Hydromentality uses Agrawal's idea of environmentality: '[Environmentality] refers to the knowledges, politics, institutions and subjectivities that come to be linked together with the emergence of the environment as a domain that requires regulation and protection' (Agrawal, 2005, p226). *Hydromentality* refers to the water engineering knowledges, consultative references, politics, and institutions that emerge in response to water control challenges as a complex domain leading to water architectures that mediate the manageability of water apportionment in river basins. Hydromentality captures the linking and co-emergence of water control challenges, cultures of control design, overlapping infrastructure experimentations, and the extant control and distribution of water.

Sufficiency and Distribution Technologies: Towards Sharing

To review supply and demand management, introduced above, and to introduce share management (Lankford, 2011), I refer to Figure 16.2, which shows a typical unit of water management that contains structures that meet

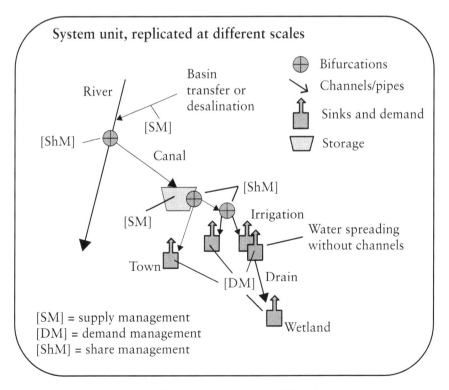

Figure 16.2 A basic model of water management and flow bifurcation

some of the functions mentioned in the previous section (water conveyance, bifurcation, storage, and depletion). Figure 16.2 shows that the distribution of water sufficiency comprises a complex mix of approaches. The meta-control framework asks: How do these different components fit together to influence the accuracy, transparency, flexibility, equity, and adequacy of water distribution in the face of variable supplies?[7]

This framework argues that structures closely associated with either supply or demand management also have 'share-type' functions. For example, a dam ostensibly acting to boost supplies also affects water sharing. It does this in two ways. First, it creates a body of water that is subject to claims over access, and, therefore, matters such as volume, function, and location shape those claims. (A small dam without hydropower will 'share' water between claimants differently than a large dam with hydropower installed.) Second, a dam disturbs an otherwise 'natural' river regime both by storing a flood volume and by creating an area of evaporation from the lake behind the dam wall. The implicit outcome is a subtraction from an otherwise pristine environmental flow and a distribution of water away from 'the environment'. Supply and demand structures are also structures with specific 'share' functions, making three types:

- Supply-share structures are designed to augment supplies. They work over time to even-out periods of deficit by storing or boosting water supplies and then sharing surpluses. Examples are storage dams, canals, or pipes that transfer water in from another basin and desalinisation plants or within basins. As explained above, the supply-side orientation also shapes the sharing of water between users and sectors. Thailand's putative water grid project (Molle and Floch, 2008), aiming to even-out deficits and surpluses at a regional scale, has both supply and sharing within its ambit.
- Demand-share structures intend to reduce net or gross demand via a variety of means. Conceptually, these are more difficult to characterise than supply structures. However, the aim of such structures or modifications is fourfold: to reduce leakages and spillage, to improve the precision placement of water where required, to consider the time and timing elements of water supply, and to assist with knowledge of demand such as metering. The latter recognises the additional pressures that can be brought to bear on demand by pricing and property rights. These structures, while encouraging the reduction of demand, are clearly involved in the sharing of limited water—one follows the other.
- Share management structures aim to distribute water rather than boost or reduce water supplies. The two main types are bifurcations/connections and conveyances, such as canals and pipes. They serve two main functions. The first is to divide varying flows between demands, introducing both a quantitative element but also a social issue of discerning this bifurcation and its transparency in the face of multiple competitive demands. The second is to place flows accurately for a given source and

volume of demand. The outcome of these two functions is to even out locations of deficits by dividing and moving supplies of water. Another way to think of these functions is to consider how water might be distributed in their absence; water would spread or spill across land or a landscape in a much more haphazard way.

Seeing structures in this way allows engineers and society to reconsider the various emphases and fashions that have influenced choices over structures—for example, whether to build storage or add metering. The aim of 'share' characterisation is to suggest that a focus on only demand or supply fails to deliver a coherent infrastructure that can accommodate ever-increasing complexities of distributing marginal water sufficiencies to multiple users.

Water Control and Single Structure Choices

Connected to design-management interactions (the next section) are a subset of water control complexities associated with single structures. An example is an irrigation control gate, part of a wider irrigation system. These control complexities comes from an interplay between water properties to be controlled and the structure's design parameters to control these properties. Water properties are: depth, velocity, volume, level/head/energy, flow rate, percentage division, location, timing, and duration. Physical structures control these properties by containing parameters and functions of convey, on–off, adjust/raise/lower, divide, join, maintain, clean, dispose/deplete, store, measure/be recorded, and be observed. These functions are interpreted often by 'schools of design management interactions' (below) to create user-facing and system-facing assemblages and arrays of structures throughout the river basin. However, hydromentalities also apply to single structures; irrigation managers experiment or stick with structures to control water, adopting (for example) modular gates, constant-head orifices, variable orifices, and wooden sluice gates. My PhD research in Swaziland found that sugarcane estates explored gate technology without considering the relationships between gates, flows, and the command areas they served.

'Schools' of Design-Management Interactions

Drawing on design-management interactions, Figure 16.3 captures eight 'schools' of engineering responses to water management. A school of design-management interaction is a particular type of hydromentality that packages a sufficiency-apportionment approach under a single label heralding a protagonist's expectation that 'their' school offers the optimum way to manage water.

- *Aggrandisation.* Here the emphasis is on bold large-scale projects and structures, such as very large dams and water transfer projects. As seen in Spain and a future Mekong, large dams boost supply while 'mega grids' ideas reveal supply-side thinking (Molle and Floch, 2008).

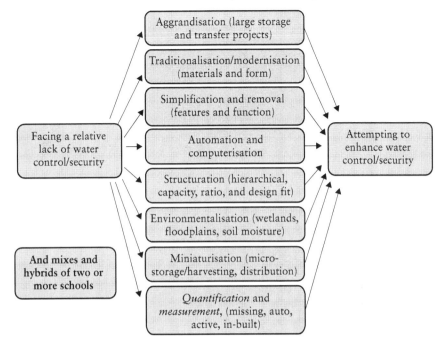

Figure 16.3 Water engineering schools of design-management interactions

- *Traditionalisation versus modernisation.* Engineers sometimes fail to distinguish that the form of, and materials used in, the structure are different to the functions contained within them. Irrigation 'modernisa-tion' assumes that existing local artisanal structures are without merit. Modernists might mistakenly improve traditional systems when not required (Pradhan and Pradhan, 2000; Lankford, 2004a).
- *Simplification and removal.* Horst (1983) argued that one of two solu-tions to the problem of water control was to simplify water architectures. Examples include moving to very simple division gates that required min-imal adjustment found on rotational 'warabandi' systems in Asia. While effective, the risk is such designs are unable cope with changes within those systems, such as cropping patterns, farm size, and expansion.
- *Automation and computerisation.* Horst's other water control solution was to move towards highly automated and computerised systems of control. With the advent of computers, radio control, telemetry, and remote activation in the early to mid-80s, this school was thought to represent the future of water control on large scale systems (e.g., Schuurmans et al., 1992). While this technology works well in some environments, it has not always lived up to its promise in remote semi-arid locations, also experiencing unreliable power supplies.
- *Structuration.* This mode of providing water control was missing from Horst's two-way solution and yet is seen on irrigation schemes worldwide.

Structuration offers water control by ensuring a high level of fit between upper levels of water supply (headworks, main and secondary canals) and the lower levels of supply and demand (tertiary canals, farms, fields, furrows, and earth canals). Albinson and Perry (2002) and Lankford (1992) explore a modular structured approach to water control. One criticism is that these are not flexible on-demand systems. In response, one can argue for the benefits of predictability of fixed schedules. Systems can be attuned to periods when evaporation and crop growth is at a maximum, making their operation during cooler or off-seasons feasible by on–off switching and rotating of flows.

- *Miniaturisation.* In parallel with other engineering is a trend towards micro-systems to promote local ownership and precision. Examples include micro-storage/harvesting of water on a farm, and the distribution of water to the crops using micro-systems, such as drip. Miniaturisation applies to all demand, supply, and share structures, but any optimism about the ability of this approach to deliver local solutions or highly efficient control must be set against the likelihood of an increase in resources required (human, construction, energy, plastics) and a highly atomised approach to water delivery and control that might undermine precision of control at the catchment scale.

- *Environmentalisation.* This examines the role that features of the environment (wetlands, floodplains, soil moisture) play in assisting water distribution. IUCN in their 2009 report 'Environment as Infrastructure' explored this topic but could have made more of the idea by inquiring of its ability to mediate water sharing and distribution under different conditions or of how additional built infrastructure extends and intensifies environmental benefits (Pittock and Lankford, 2010).

- *Quantification and measurement.* Alongside the storage and distribution of water are fashions about how to measure water supply and distribution. Designs may be missing, or require active human measurement, automatic measurement, or deliver 'built in' measurement (Lankford, 1992), or focus on water flows, depths, or proportions.

Other than the briefest of outlines given above, one other point can be made. At any one given place or time, hybrids (these are structures that merge ideas from the eight schools within one system), mixes (where separate but coexisting structures are utilised within one system) of the above are in action. I speculate that this hybridising and mixing needs to go further; water apportionment is best served by a selection of designs from all eight schools.

Sector Emphases

The final hydromentality arises from donor and government interventions steered towards particular water sectors in response to public and advocacy influence. Typical sectors include water and sanitation (WASH), irrigation, urban-industrial, environmental works, hydropower, and flood control. I have also included IWRM (integrated water resources management) and RBM (river basin management) for their place in shaping donor agendas, and recent interest in the water–energy nexus (SEI, 2011).

The inclusion of these sector emphases in the meta-control framework provides three insights on achieving water security by water distribution. The first is that one sector promoted at the cost of others creates lopsided water infrastructure. Grobicki (2009) argues for a more comprehensive approach to all water sectors. The second is that the professions active within each sector rarely reach across these sectoral boundaries. Thus, the implications of a sectoral focus on irrigation alone for meeting downstream needs are potentially severe if declining water supplies observed during droughts and dry seasons cannot be shared fairly (Lankford, 2004b). The third insight regards the hidden costs for water control by 'omission'; that selective competition for donor and government resources by water sectors obfuscates a more purposive holistic approach to water architectures. While IWRM and RBM should have provided an integrated approach, an examination of IWRM (Biswas, 2004; Neef, 2009) shows that it has been remarkably silent about water infrastructure.

Engineering Cultures

The third part of the hydromentality framework in Figure 16.1 is concerned with engineering and other stakeholder knowledges that feed into choices over design. These knowledges consider solutions that appear sensible but might not meet the scalar, variability, and connectivity dimensions of water control. Because of this lack, water debates are more likely to be captured by popular yet unproven fashions outlined in Figures 16.1 and 16.3 (e.g., small scaleness, boldness, and emphasis on storage or demand management).

There can be much written here about the interplay between water engineering, existing technology, natural environmental stresses, and long-standing educational and cultural factors (e.g., Trawick, 2001). Zwarteveen (2008) argues that engineers' subjectivities, epistemic communities, and reference points 'impact on their methods and conceptualisations'. Due to space constraints, I shall argue that an architectural theory and practice of water arises from two interrelated weaknesses related to the educating and socialising of water engineering. The first is that the engineer's task of creating architectures rarely puts people at the centre of the design process in ways that communicate the scarcity of water and its bifurcating nature. Instead, I believe, process is dominated by professionalised training and norms (Chambers, 1992; Lankford, 2004b; Zwarteveen, 2008) or is skewed by the terms of commission (c.f. the role of public–private partnerships in Morocco described by Houdret, 2012). This means that during the steps of the design process, the engineer insufficiently sees the system through the system's eyes (a phenomenological approach) or through the eyes of communities expecting to voice but not fully understand their concerns over water shortages.

The second weakness arises from a poorly articulated engineering objective. This should be the result of the process chosen (above), yet aim to create operable, transparent, structured, hybrid architectures that lean towards high levels of manageability of the complex problem of water sharing and distribution. In this second aim, one 'ergonomic' factor (Chapanis, 1996) should be kept uppermost: Designs usually fail to consider 'normal' human actions such as a desire to set gates and valves at their widest or highest setting, which skews water division towards privileged groups normally at the top end of the system.

The objectives in the previous two paragraphs are not solely the engineer's responsibility. Society's failure to hold engineering to account can be blamed on a number of factors—not least a lack of recognition of the abstractness of designing physical artefacts but also a dismissing of 'technical' matters, leading over time to a diminution of the engineering profession.[8] In addition are sources of critical thinking over how society experiments with water and then critically examines outcomes to rephrase new approaches and objectives. Four influential communities are identified that could enliven and shape a debate on water control but which, I speculate, are largely absent or ineffectual (Table 16.1).

Table 16.1 Critical communities influencing engineering learning and knowledges

Community	Beliefs / objectives	Units and ideas
Political ecologists	An observation of the changes in narrative associated with society's experimentation with water apportionment	Tradition, modernity, 'scale-ness', cultures of learning, terms of the debates
Water users, representatives, and social scientists	Engagement with negotiation and impacts of water control by various actors, with a particular regard for communities and individuals treated (in)equitably by water technology experimentation	Community, access, equity, water poverty; water user associations.
Other water engineers	Bringing differently trained engineers into the debate might lead to reform of professional norms within sectoral protocols	Design protocols, procedures, training, qualifications
Donors funding systems and reviews	The commissioning of projects that respond to perceived problems— for example, the millennium development goals of water and sanitation, with consequences for other investments in water	Social, environmental and economic priorities—costs and benefits; investment schedules and tools

Conclusions

A variety of factors imprint themselves on patterns of water transfer and distribution within a river catchment. This chapter has sought to highlight the role of infrastructure via four types of hydromentalities that interpret water control complexities, casting them as 'biases' and 'schools'. This was done purposively to nudge discussions on water infrastructure towards a more comprehensive approach—from a 'hydraulic mission' (Allen, 2001) towards a more inclusive 'hydraulic vision'. Four hydromentalities give different weights to demand, supply, and share management; to individual single structures; to schools of design-management interactions; and to competing water sectors. In turn, a moral economy and cultural environment of water science and engineering shapes these water control debates. The term given to this framework is water meta-control, leading to architectures of water structures hierarchically and spatially designed to deliver water security outcomes: the equitable, reasonable, timely, efficient, telegraphed water apportionment to competing users for productive and protective human/environmental needs in face of rising demands and limited variable supply over time and space. The term *telegraphed* offers water users opportunities to reflect on their space, technology, and context-mediated knowledge.

It is worth reiterating that the purpose of this framework is not to formulate design procedures or to include other climate- and water-facing criteria of design (Giordano, 2013), but instead to highlight water insecurities arising from inappropriate water infrastructure in a dynamic and/or water-scarce river basin, and to examine the social, economic, and epistemic factors that influence these architectural choices. If society and its professionals continue with fashion-driven methods to new and existing water infrastructure, then, in the guise of conventional engineering within but not across or between hydromentalities, we end up favouring certain sectors and politically powerful users with relative water gains—delivering to other groups, especially in times of scarcity, unwarranted levels of water insecurity.

Notes

1. For example, I have close-up experience of how the Commission for Africa (2005) ignored my advice for a technologically mixed approach to expand irrigation in Africa, only then to propose on page 251, 'Donors should support this . . . with an emphasis on small-scale irrigation'.
2. Flood infrastructure can influence water distribution during non-flood periods.
3. Zwarteveen (2008) refers to research by Oorthuizen on these connections.
4. Humans can adjust some of these natural endowments—a topic not covered here.
5. For the purposes of this article, my concept focusses on 'blue' water rather than soil (green) water, but does not preclude these.
6. The river basin is a well-understood unit of water management; however, water transfer structures and aquifers allow for water sharing between basins.

7. These tend to be seen as 'good' ways of managing water—though a debate around rigid schedules and structures of water in the face of shortages also exists.
8. Concerns regarding the relative decline and revival of engineering in the UK and elsewhere can be found in various publications, for example, The State of the Nation: Infrastructure 2010—Institution of Civil Engineers.

References

Agrawal, A. (2005) *Environmentality: Technologies of Government and the Making of Subjects,* Duke University Press Books.

Albinson, B. and C. J. Perry. (2002) *Fundamentals of Smallholder Irrigation: The Structured System Concept,* Research Report 58, IWMI, Colombo.

Allen, J.A. (2001) *The Middle East Water Question: Hydropolitics and the Global Economy,* I.B. Tauris, London, New York.

Biswas, A.K. (2004) 'Integrated water resources management: a reassessment', *Water International,* vol 29, no 2, pp248–256.

Bolding, A., Mollinga, P.P. and Van Straaten, K. (1995) 'Modules for modernisation: colonial irrigation in India and the technological dimension of agrarian change', *The Journal of Development Studies,* vol 31, no 6, pp805–844.

Boelens, R. (2008) 'Water rights arenas in the Andes: upscaling the defence networks to localize water control', *Water Alternatives,* vol 1, no 1, pp48–65.

Bos, M. G. (1987) 'Water management aspects of irrigation system design', in *Irrigation Design for Management Asian Regional Symposium, Volume II, Discussions and Special Lectures,* Kandy, Sri Lanka, 16–18 February, Wallingford, UK.

Chambers, R. (1998) *Managing canal irrigation: A practical analysis from south Asia,* Oxford and IBH Publishing, New Delhi.

Chapanis, A. (1996) *Human Factors in Systems Engineering,* Wiley Series in Systems Engineering, John Wiley and Sons, New York.

Commission for Africa. (2005) 'Our Common Interest', Report of the Commission for Africa. 11 March 2005, London.

Diemer, G. and Slabbers, J. (eds) (1992) *Irrigators and Engineers: Essays in Honour of Lucas Horst,* CABI, Wallingford.

Giordano, T. (2013) 'Adaptive planning for climate resilient long-lived infrastructures', *Utilities Policy,* vol 23, pp80–89.

Grobicki, A. (2009) *Water Security: Time to Talk across Sectors,* Water Front Articles, Stockholm International Water Institute, Sweden.

Grey, D. and Sadoff, C.W. (2007) 'Sink or swim? Water security for growth and development', *Water Policy,* vol 9, pp545–571.

Gupta, J. and van der Zaag, P. (2008) 'Interbasin water transfers and integrated water resources management: where engineering, science and politics interlock', *Physics and Chemistry of the Earth, Parts A/B/C,* vol 33, pp28–40.

Horst, L. (1983) *Irrigation Systems—Alternative Design Concepts,* Irrigation Management Network Paper 7c. ODI/IIMI Irrigation Management Network, Overseas Development Institute, London.

Houdret, A. (2012) 'The water connection: irrigation and politics in southern Morocco', *Water Alternatives,* vol 5, no 2, pp 284–303.

IUCN (2009) *Environment as Infrastructure—Resilience to Climate Change Impacts on Water through Investments in Nature,* IUCN, Switzerland.

Jowitt, P. W. (2009) 'Water infrastructure, the UN MDGs and sustainable development', *Desalinisation,* vol 248, pp 510–516.

Lankford, B.A. (1992) 'The use of measured water flows in furrow irrigation management—a case study in Swaziland', *Irrigation and Drainage Systems,* vol 6, pp113–128.

Lankford, B.A. (2004a) 'Irrigation improvement projects in Tanzania; scale impacts and policy implications', *Water Policy* vol 6, no 2, pp89–102.

Lankford, B.A. (2004b) 'Resource-centred thinking in river basins: should we revoke the crop water approach to irrigation planning?', *Agricultural Water Management*, vol 68, no 1, pp33–46.

Lankford, B.A. (2011) 'Responding to water scarcity—beyond the volumetric', in L. Mehta (ed) *The Limits to Scarcity: Contesting the Politics of Allocation*, Earthscan, London.

Lankford, B.A. and Gowing, J. (1997) 'Providing a water delivery service through design management interactions and system management', in M. Kay, T. Franks, and L. Smith (eds) *Water: Economics, Management and Demand*, E & FN Spon, London.

Lankford, B.A. and Mwaruvanda, W. (2007) A legal-infrastructural framework for catchment apportionment, in B. Van Koppen, M. Giordano, and J. Butterwort (eds) *Community-based Water Law and Water Resource Management Reform in Developing Countries*, Comprehensive Assessment of Water Management in Agriculture Series, CABI Publishing, Wallingford.

Lautze, J. and Giordano, M. (2007) 'Demanding supply management and supplying demand management', *The Journal of Environment and Development*, vol 16, no 3, pp290–306.

Molle, F. and Floch, P. (2008) 'Megaprojects and social and environmental changes: the case of the Thai "water grid"', *Ambio*, vol 37, no 3, pp199–204.

Molle, F. and Mamanpoush, A. (2012) 'Scale, governance and the management of river basins: a case study from Central Iran', *Geoforum*, vol 43, no 2, pp285–294

Neef, A. (2009) 'Transforming rural water governance: towards deliberative and polycentric models?' *Water Alternatives*, vol 2, no 1, pp180–200.

Pahl-Wostl, C. (2007) 'Transitions towards adaptive management of water facing climate and global change', *Water Resources Management*, vol 21, pp49–62.

Pittock, J. and Lankford, B.A. (2010) 'Environmental water requirements: demand management in an era of water scarcity', *Journal of Integrative Environmental Sciences*, vol 7, no 1, pp1–19.

Plusquellec, H., Burt, C. and Wolter, H.W. (1994) *Modern Water Control in Irrigation. Concepts, Issues, and Applications*, World Bank Technical Paper No. 246, Irrigation and Drainage Series, World Bank, Washington DC.

Pradhan, R. and Pradhan, U. (2000) 'Negotiating access and water rights: disputes over rights to an irrigation water source in Nepal', in B. Bruns and R. Meinzen-Dick (eds), *Negotiating Water Rights*, International Food Policy Research Institute, ITDG Publishing, London.

Renault, D., Facon, T. and Wahaj, R. (2007) *Modernizing Irrigation Management— The MASSCOTE Approach*, Food and Agriculture Organization (FAO), Rome

Rogers, P. (1997) 'Engineering design and uncertainties related to climate change', *Climatic Change*, vol 37, no 1, pp229–242.

Schuurmans, J., Schuurmans, W. and van Leeuwen, J. (1992) 'Improved real time control of water deliveries in International Conference on Advances' in J. Feyen, E. Mwendera and M. Badji (eds) *Planning, Design and Management of Irrigation Systems as Related to Sustainable Land Use: Proceedings of an International Conference*, Center for Irrigation Engineering, Katholieke Universiteit, Leuven, Belgium

Scott, A. (1996) 'Appropriate technology: is it ready—and relevant—for the millennium', *Appropriate Technology*, vol 23, no 3, pp1–4.

SEI (2011) *Understanding the Nexus*, Background Paper for the Bonn 2011 Nexus Conference, Stockholm Environment Institute.

Stakhiv, E.Z. (1998) 'Policy implications of climate change impacts on water resources management', *Water Policy*, vol 1, pp159–175.

Tortajada, C. (2006) 'Water management in Singapore', *International Journal of Water Resources Development*, vol 22, pp227–240.

Trawick, P. (2001) 'The moral economy of water: equity and antiquity in the Andean commons', *American Anthropologist. New Series*, vol 103, no 2, pp361–379.

Wisser, D., Frolking, S., Douglas, E. M., Fekete, B. M., Schumann, A. H. and Vörösmarty, C. J. (2010) 'The significance of local water resources captured in small reservoirs for crop production—A global-scale analysis', *Journal of Hydrology*, vol 384, no 3–4, pp264–275.

Zeitoun, M. (2011) 'The global web of national water security', *Global Policy*, vol 2, no 3, pp286–296.

Zwarteveen, M. (2008) 'Men, masculinities and water powers in irrigation', *Water Alternatives*, vol 1, no 1, pp111–130.

17 The Strategic Dimensions of Water
From National Security to Sustainable Security

Benjamin Zala

This chapter discusses the enduring preoccupation in policy circles with strategic questions related to water. In particular, the focus here is on the security implications of increased competition over water resources in defence and security analyses, despite the tendency in the academic literature towards an overly sceptical view of the potential for large-scale conflict over water. The chapter discusses the strategic dimensions of water with a focus on the role of population increases and climate change in framing the way water is treated in national security threat assessments, particularly in western countries. It examines the ways in which the academic literature on water security has diverged from the analysis of defence and security policymakers due to issues such as differing approaches to warfare, international institutions, and rationalist approaches to security and cooperation. This is followed by a comparison of two different responses to water security concerns, one based on a traditional 'national security' approach—or what is referred to as the *control paradigm*—and the other on what has become known as a *sustainable security framework*. It is argued that the latter approach provides the most appropriate principles for basing a holistic and long-term policy response to water, conflict, and cooperation in the years to come.

Water as a Strategic Issue

The strategic value of finite resources has long been recognised by military planners and strategists (Proença and Duarte, 2005, p648). Given the central importance of water to human survival, it is perhaps not surprising that this particular resource has received a great deal of attention within a strategic context (Amery and Wolf, 2000; Toset et al., 2000). Following the increased focus in the International Relations (IR) and security studies literature on the connections between the natural environment and conflict from the early 1990s onwards (Gleick, 1993; Dupont, 2008; Gleditsch, 2012), much of the current policy discourse around current and future global security threats emphasises water as a strategic resource linked to issues of stability, peace, and conflict. Whilst a significant body of literature now exists that convincingly makes the case as to why the overly pessimistic and deterministic 'water wars' thesis should be treated with some scepticism (Lowi, 1995; Rogers, 1996; Selby, 2005); it is, however, important to investigate the assumptions that underpin the arguments for why water should be viewed as a strategic issue if policymakers are to be guided towards outcomes that emphasise cooperation rather than conflict.

The earliest recorded case of an armed dispute over water dates back almost 5,000 years to ancient Mesopotamia between the Tigris and Euphrates rivers, where a conflict erupted between the cities of Umma and Lagash over the diversion of irrigation ditches. It is also the first recorded case of water being used as a military tool, as the supply of water to the city of Girsu was later cut off in a strategic move (Hatami and Gleick, 1994). The history of water-related conflicts includes interstate tensions, communal rivalry and civil conflict, insurgency, and even in more recent times, terrorism. Recent examples of the latter include the reported Al-Qaida involvement in the two weeks of tribal fighting in the Kurram region of Pakistan, near the Afghanistan border, over irrigation water and Laskar-e-Taiba's repeated threats against Indian dams.

There is a strong tradition in geopolitical and strategic thought that views water as a matter of national strategic interest, or as Peter Burgess has put it, such thinking regards water as 'a simple and finite object of political action. They presuppose water to be a good that, like other goods, can be taken up into a predetermined calculus of strategic advantage and disadvantage' (2008, p68).

Such thinking clearly underpinned the decision by the former Libyan leader Colonel Gadhafi's forces to attempt to cut off the water supply to Tripoli in their retreat in August and September of 2011 by, according to UN reports, vandalising the pumps that provide three-quarters of Tripoli's supply of water from the so-called Great Man-Made River (Miles, 2011). It is also evident in a recent editorial in the Pakistani newspaper *Nawa-i-Waqt* that argued, 'Pakistan should convey to India that a war is possible on the issue of water and this time war will be a nuclear one' (quoted in *The Economist*, 2011, p25), as well as the 'Jal, jungle, zameen' (water, forest, land) battle cry used by Naxalite rebels in India (Parenti, 2011, p135).

Of course, for all the evidence of previous disputes over water, there is a wealth of evidence for cooperative arrangements being used to help increase information flows, reduce transaction costs, and increase levels of trust between potentially conflictual parties over shared water sources (Tir and Stinnett, 2012). Indeed, the conflict between Umma and Lagash mentioned above has been referred to by Aaron Wolf as the last and only 'water war' in the full sense (Wolf, 1999, p263).

However, it should be noted that there is some debate over what should be thought of as genuine cooperation over water resources. Ana Elisa Cascão and Mark Zeitoun, for example, have pointed to certain so-called forms of cooperation that are in fact based on coercion or temporary submissiveness (Cascão and Zeitoun 2010, p29); they prefer the term *interaction,* which can take both conflictual or cooperative forms (with both being ever present). Jan Selby similarly opts the for the term *policy coordination*, arguing that 'situations of policy coordination without mutual adjustments or joint gains should instead be considered instances of 'domination' (2013, p1; see also Zeitoun and Allan, 2008).

Despite the trend in recent years to emphasise the likelihood of cooperation rather than conflict (in the broadest definition of both terms) over transboundary water resources (Barnaby 2009; Katz, 2011), a brief survey of some of the most prominent studies and policy papers from the national threat analysis of a number of western states demonstrates an enduring preoccupation with water and conflict. This is reflected not only in official assessment and policy documents but also in the work of a number of influential think tanks and research centres in the defence and security realm. Both in terms of scarcity-induced conflict and the use of water as a weapon, the projections in this discourse for future stresses on water supplies produce an image of increasing complexity, risk, and opportunity for conflict (Kundzewicz and Kowalczak, 2009; Parthemore and Rogers, 2010).

Two factors are particularly important in the framing of water issues in recent national threat analysis: projected increases in the global population and the physical impacts of climate change.

According to the World Economic Forum (WEF), over the next decade alone, the world population is expected to rise from the current 6.83 billion to approximately 7.7 billion, with most of the growth in emerging economies (WEF, 2011, p49). The United Nations projects a 50% increase in demand for food by 2030 (United Nations, 2012), and the International Food Policy Research Institute (IFRI) expects a 30% increase in demand for water, with other estimates rising to over 40% (WEF, 2011, p29).

Looking further ahead to 2040, the UK Ministry of Defence (MoD) is working on the assumption of a 'likely' population increase to 8.8 billion (DCDC, 2010a, p24). The fourth edition of the UK MoD's 'Global Strategic Trends—Out to 2040' report notes, 'Global population growth will be unevenly distributed with much of the growth likely to occur in regions, such as sub-Saharan Africa and the Middle East, that already suffer from stresses to food and water supplies. . . . When shortages are threatened, the adoption of export restrictions for food, and disputes over water flows, are likely to increase, affecting global supply, aggravating shortages and eroding trust' (DCDC, 2010a, pp110–11).

Added to population growth, and particularly the ways in which population is expected to be distributed, is what the Center for Strategic and International Studies has called the 'real game changer' of the role of climate change in exacerbating existing water-related tensions (Verrastro et al., 2010). The International Institute for Strategic Studies has argued, 'Changes in water resources will be the most visible impact of climate change on human society', and that 'when a country determines that its water resources are at risk, there are clear reasons for it to use all manner of statecraft in order to protect those resources, including war' (IISS, 2011, pp15–16).

Such analysis draws upon the growing consensus in the scientific community on the implications of the likely changes in the planet's climate over the coming decades (Foresight, 2011). A widely discussed article published by an international group of scientists in the journal *Nature* in 2010, which argued that nearly 80% of the world's population is exposed to high levels of threat to water security (Vörösmarty et al., 2010), was, according to one of the authors, 'a snapshot of the world about five or 10 years ago, because that's the data that's coming on line now' (quoted in Black, 2010). Dr. Peter McIntyre went on to tell the BBC at the time that 'people should be even more worried if you start to account for climate change and population growth' (Black, 2010).

Again, the UK MoD's threat analysis is illuminating in relation to the way that the combination of population pressures and climate change, over roughly the same time period (Zala and Rogers, 2011, p30), is framing the way in which it treats resource scarcity in a security context: 'population increases, resource scarcity and the adverse effects of climate change, are likely to combine, increasing the likelihood of instability and of disagreement between states, and providing triggers that can ignite conflict' (DCDC, 2010a, p73). This is not only confined to scenario planning and military analysis but, for example in the UK, is also now reflected in the most important national policy declarations, such as the National Security Strategy, which warns of the combination of 'drivers' such as climate change and resource scarcity that increase 'the likelihood of conflict, instability and state failure' (HM Government, 2010, p16).

Of course one of the most difficult aspects of factoring climate change into scenarios and plans related to water security and strategic threats is the high levels of uncertainty about the exact location, timing, and effect of the physical impacts of a changed global climate. This is often exacerbated in studies that rely on historical data alone to discuss future policy options (Tir and Stinnett, 2012, p217). Yet uncertainty is not a new variable for defence planners or military strategists, and the limitations of climate modelling do not, in principle, pose a greater problem than information gaps in other areas of security analysis (Zala and Rogers, 2011, p30). As Steven Jermy has noted on this issue, 'from a defence planning point of view, what matters is not to be able to say what will definitely happen, but rather to consider what could plausibly happen' (Jermy, 2011, p144).

Under conditions of potentially greater water scarcity at a global level, not only are the material conditions of unequal access to water a potential trigger of conflict and instability, but just as, and sometimes even more important, are perceptions of unequal access to water. This problem is particularly acute in regions in current or post-conflict situations in which issues of water security can become enmeshed in larger disputes and grievances. As Tony Allan has put it, approaches to water that are made in such circumstances 'confirm the truth that communities use evidence and make decisions on the basis of perception' (2000, pxv). This not only applies to questions of water access and distribution but also to the structural pressures on water availability discussed above, including climate change and population growth. A warming global climate and increasing population pressures do not have to actually make resources more scarce in quantitative terms to make communities and individuals feel insecure if a perception of scarcity exists.

The links between misperception, escalation, and conflict have been extensively explored in the IR literature on the causes of conflict and war (Jervis, 1976; Bueno de Mesquita et al., 1997; Booth and Wheeler, 2008). Even the aspects of this literature that depend on solely rationalist explanations of conflict allow for significant factors such as unequal information, differing perceptions of relative power, and commitment problems in crisis bargaining (Fearon, 1995). Further research by Fearon and others (Fearon, 1994; Filson and Werner, 2007) has also added domestic factors such as audience costs in shaping responses to heightened tension and crisis.

Water Security Research and National Threat Assessments

One reason for the divergence between the academic literature and the national threat analysis produced by defence, military, and intelligence analysts relates to changes in the character of modern warfare. The vast majority of the literature that has rightly pointed to the flaws in the water wars thesis has taken warfare to mean traditional interstate war (Wolf, 1998; Barnaby, 2009). Interstate warfare has declined steeply over the second half of the 20th century and this trend has continued into the 21st century. Since the late 1990s, defence and security policymakers around the world have become much more attuned to the changing nature of warfare and in particular to the increasing importance of asymmetric warfare (Terriff et al., 2008). More recently, this has evolved into a preoccupation with the concept of hybrid warfare—a dangerous and complex combination of insurgency, civil conflict, terrorism, pervasive criminality, and widespread civil disorder (Hoffman, 2009; DCDC, 2010b, p13). This has not only led to a major shift in the way that military planners and defence policymakers have approached issues such as training, procurement, and doctrine, but has also greatly affected the way threat assessments are being undertaken by national governments, regional organizations, and private sector analysts. This lowering of the threshold in terms of the kind of disputes, conflicts, and other actions short of interstate warfare that are of concern to defence and security policymakers greatly affects the way that water is treated in questions of strategy and defence (US National Intelligence Council, 2008, px). Concerns over the links between water and intrastate conflict and political violence now have a much stronger priority in this kind of analysis than was previously the case (Witsenburg and Adano, 2009; Sirin, 2011). As Geoffrey Dabelko has put it in relation to a 2012 U.S. National Intelligence Council report on water and security, 'In essence, the intelligence community recognizes that water doesn't need to "bleed to lead"; instead, there are a number of security concerns that are not connected explicitly to organized violence' (2012).

A second reason may be that even in relation to interstate conflict, a tendency in the academic literature on water security to follow the liberal institutionalist position—attributing more weight to international institutions as an intervening variable in international conflict—is common. Jaroslav Tir and Douglas Stinnett have argued, 'Forecasts that do not account for the important conflict management potential of international institutions will produce overly pessimistic scenarios' (2012, p212). Yet the other side of this argument is that scholarly research characterised by an uncritical treatment of the power of international institutions in mitigating security dilemmas (for an overview, see Young, 2001), and an over-reliance on empirical data from previous historical periods (in which population and climate pressures had far less impact), will produce overly optimistic results. Such research often takes the development of effective institutions and their survival under changed structural pressures as a given. As one study points out, 'institutions for regulating the use of internationally scarce resources sometimes fail to develop, and when they do, they are not always sufficiently resilient to deal with changing political and resource environments' (Giordano et al., 2005, p47). This research also ignores the underlying power dynamics of those institutions (Zeitoun and Warner, 2006, p454) and the potential for these dynamics to not only block cooperation but to even create conflict. It could be argued that those tasked with assessing long-term risks in policymaking circles simply do not have the luxury of relying on such assumptions in their forecasts or policy prescriptions.

A third reason may be that the vast bulk of the literature on transboundary water cooperation relies on a set of rationalist assumptions better suited to international political economy (IPE) than to national threat analysis (Warner, 2003). There is little doubt that concepts such as 'virtual water' have done much to illuminate the complex political economy of water use—particularly when compared to the often overly simplistic way that it is treated in much of the IR and security studies literature—but it has yet to be adopted in the realm of security and defence analysis in the same way it has in the water and IPE community. As Zeitoun et al. have pointed out, while the global trade in virtual water has always been politically feasible due to its invisibility, it only occurs when it is economically feasible (2010, p240). For example, if national water security is reliant on virtual water trade, sustainable, long-term solutions are therefore hostage to the same barriers, asymmetries, and inequalities that exist across the global trade in almost all commodities (Mori, 2003, p120). This matters in security terms because the poorest and most fragile states of the international system (who also suffer from relative economic disadvantage globally) are in the worst position to effectively solve their water scarcity problems 'outside of the river basin' (Allan and Mirumachi, 2010, p16). While an 'economically diversified economy can thus trade its way to water security' (Earle et al., 2010, p6), those analysing what Christian Parenti refers to as the 'new geography of violence' (2011) are less concerned with security threats emanating from relatively prosperous countries with diverse economies. Parenti's new geography of violence is particularly centred on what he calls the 'Tropic of Chaos'. This is a 'belt of economically and politically battered post-colonial states girding the planet's mid-latitudes' (2011, p9), in which one finds a 'catastrophic convergence of poverty, violence, and climate change' (p5). This is where security analysts are increasingly turning their attention to the 'networks of desperation' (Beebe and Kaldor, 2010, p37; DCDC, 2010a, p22; Rogers, 2010, pp80–87) that threaten the relative global stability seen since the end of the Cold War. In such countries, a diverse economy cannot be taken for granted. The challenges to economic diversification are many and include a lack of investment and trade, macroeconomic stability, a competitive exchange rate, good governance, and an absence of conflict (OECD-UN, 2010, p9).

All of these reasons only go part of the way towards explaining the enduring attention given to the strategic dimensions of water in defence and security analysis in ways that emphasise the potential for conflict (and all can and should be challenged). Yet understanding where they fit, particularly in current policy discussions, can provide an avenue for a more fruitful engagement between the water security scholarly community and the defence and security policy communities.

The Standard Response: National Security

Challenging the determinism of the overly pessimistic defence and security assessments on water security requires an understanding of the larger frameworks that are used as a basis for policymaking, as well as possible alternatives to the dominant paradigms. If we look to the current dominant approaches to global insecurity (for example, threats such as terrorism, organised crime, or the proliferation of nuclear weapons)—that which we might call a standard 'national security' approach—two major characteristics stand out.

The first is a major tendency towards reacting to the symptoms of insecurity. Despite the increased use of the language of conflict prevention in Western policy statements, on the whole, efforts at gearing our defence and security policies towards mitigating security threats are still minimal and the overwhelming majority of time, focus, and money is spent on crisis reaction. One of the best examples of this was the gap between the rhetoric of prevention in the threat analysis of the UK's updated National Security Strategy of 2010 and the actual policy responses contained in the white paper resulting from the simultaneous Strategic Defence and Security Review (Oxford Research Group, 2011).

The second characteristic of the current approach to national security is the preference for attempting to 'control' the symptoms of global insecurity. This is what is often referred to as the 'control paradigm' (McGwire, 2001) or what Paul Rogers refers to as 'lidism': the analogy here is to the symptoms of global insecurity such as conflict and instability as a pot of boiling water on a stove top and government approaches under the control paradigm are akin to putting a lid on the boiling pot rather than turning off the heat (Gill, 2004; Rogers, 2008).

The most likely result of a standard national security approach to water security based around the control paradigm is two-fold: one is the securitisation of the issue (placing it in discursive terms into the modality of threat/danger), and the second is an overly militarised response. While the former is not necessarily a problem per se, and in fact may assist in bringing greater and higher levels of attention to important water-related issues, the latter almost guarantees the silencing of cooperative policy options and gives preference to the zero-sum nature of realist strategic thinking. If the defence and security actors who are taking water security seriously are more likely to be persuaded of the importance of concepts such as virtual water for desecuritising water scarcity and proposals for multilateral management and dispute settlement mechanisms, a different way of thinking about security will need to be applied to the issue of water.

An Alternative: Sustainable Security

If we are then to seek out a more desirable alternative to the standard approach to global insecurity that could instead be employed in relation to the strategic dimensions of water, where should we turn? It is argued here that what has become known as a 'sustainable security' approach is more likely to encourage cooperative outcomes rather than falling into the traps of escalation, misperception, and miscalculation, and even all-out conflict. A sustainable security approach does not guarantee harmony in policy coordination on water issues, but rather decreases the likelihood for conflict by focusing specifically on addressing the causes of insecurity at the source.

The concept of sustainable security is a relatively new one that is gaining increasing ground with governments, NGOs, and analysts. While being a framework for approaching global security issues in the broadest sense, its application goes beyond the realm of traditional security studies and also includes issues such as development (Khagram et al., 2003), macro-socioeconomic and environmental trends (Zala and Rogers, 2011), food security (Fullbrook, 2010), and water management and use (Bowen, 2009).

A sustainable security approach prioritises the resolution of the interconnected and underlying drivers of insecurity and conflict, with an emphasis on preventive rather than reactive strategies. The central premise of the approach is that it is impossible to successfully control all the consequences of insecurity, and so efforts should instead be focused on resolving the causes—often referred to as 'underlying drivers' (Phillips et al., 2006; Stephenne et al., 2009). Four standout characteristics of this approach are worth highlighting for their applicability to the issue of water security.

The first is the shift in focus from the symptoms to the underlying drivers of global insecurity. A sustainable security approach prioritises the analysis of long-term global trends with a particular focus on trends that have the capacity to cause large-scale instability and loss of life of a magnitude unmatched by other security threats such as climate change, increasing competition over resources, and increasing socioeconomic divides both within and between states. The latter trend is of increasing concern to security analysts. A rising sense of resentment, anger, and violence amongst the marginalised majority of the world's population has been tracked and analysed for a number of years. One account has taken note of 'the more visible indicators of a deeper resentment among marginalised communities that are more educated, more literate and have far greater access to communications technology than ever before' (Rogers, 2010, p180). This is what the WEF, in its 2012 Global Risks Report, has referred to as the 'seeds of dystopia'—a state of affairs in which a large swathe of the global population experience everyday lives 'full of hardship and devoid of hope' (WEF, 2012). The potential for violent backlash, the WEF contends, threatens political stability across the international system as a whole by overwhelming 'the systems that underpin our prosperity and safety' (WEF, 2012). Rogers has pointed out that the 'alienated and often violent responses of recent years, and the even greater violence with which they have been repressed, have all evolved during a period of overall economic growth' (2010, p180). It is perhaps not surprising then that during a time of global financial crisis, and for much of the world relative stagnation, the security implications of this global trend are being increasingly analysed in connection with forms of instability and political violence (Richards, 2011, p16).

The second is an attempt to address these drivers of insecurity at the source rather than when they are deeply manifest and expressed in forms of violence. So, for example, if it is accepted that a warmer global climate will both exacerbate and amplify existing conflicts as well as create the conditions for new tensions, then a sustainable security approach, while being attuned to the need to make climate adaptation measures 'conflict sensitive', puts fast and effective mitigation strategies at the heart of national, regional, and international security policies.

This approach also emphasises the interconnections between the underlying drivers of global insecurity. For example, water security is, therefore, to be treated as being intimately connected to competition over other resources, as well as trends such as climate change and the security implications of a socioeconomically and politically marginalised majority world. This particular characteristic means that the key to operationalising a sustainable security approach is making a 'whole-of-government' approach to security more than just rhetoric and aspiration.

Finally, a sustainable security approach places a great emphasis on long-term and equitable solutions in addressing the factors that drive insecurity. This is a by-product both of looking at trends such as marginalisation as fuelling conflict, and of the emphasis on the interconnections between drivers.

This focus on long-term and equitable solutions can be thought of as an updated notion of conflict prevention that goes beyond simply examining locally specific conditions and historical grievances. As Lawrence Freedman has suggested, 'An updated notion of prevention, by contrast, might encourage recognition that the world in which we live is one in which the best results are likely to come from a readiness to engage difficult problems over an extended period of time' (Freedman, 2003, p114).

While many barriers exist to institutionalising this kind of approach to water security in mainstream defence and security policy circles, not least due to the dominance of the control paradigm, the ability of analysts, commentators, and NGOs to bridge the divide between academic analysis and policy threat assessments described above will go some way in promoting this approach. Research that emphasises the interconnected drivers of insecurity that influence the ways in which societies respond to water use, distribution, and availability, and carefully linkage of these to policy prescriptions (for one of the best recent examples attuned to the deficiencies in a traditional national security approach, see Zeitoun, 2011), offer a promising way forward. The following discussion highlights a number of avenues for future development.

Priorities for a Sustainable Security Approach to Water

There is no doubt that local-level water management, dispute resolution, and conflict prevention measures lie at the heart of a more sustainable approach to water scarcity and conflict. Particular attention must be given to addressing local-level perceptions of inequality and injustice, particularly in conflict and post-conflict situations. In such circumstances, there is no alternative to the laborious and sometimes painful task of sustained dialogue (Grey et al., 2009, p16) and attempts at trust-building, even amongst parties already in conflict (Booth and Wheeler, 2008). Focusing efforts at the local level allows for an approach to addressing the underlying causes of insecurity led from within communities where the effects of power asymmetries are most sharply felt. A holistic approach to water management and conservation, as well as issues of distribution and access at the local level, can be informed by concepts such as virtual water, as well as dealing directly with perceptions of inequality and long-term grievances (for example, for an IPE approach to water insecurity that addresses issues of marginalisation directly, see Baruah, 2012).

Yet, regional and global approaches are equally important. It is imperative that major powers, as well as regional and international organisations, coordinate to create genuinely global approaches to addressing the links between resource competition, climate change, and socioeconomic and political marginalisation in order to mitigate conflicts over water at the global level. As the 2011 edition of the WEF's Global Risks report puts it: 'It is at the local level that most opportunities can be found for improving resource efficiency and managing trade-offs between energy, water and food production. However, at the global and regional levels there are few initiatives to raise awareness, share leading practices and motivate consumers in an integrated approach' (2011, p47). Efforts at the regional and global level, attuned to power asymmetries and existing local conflicts, can be a powerful way of addressing perceptions of marginalisation and insecurity.

Nationally as well, greater efforts need to be made on making the connections between what are often divided or siloed into separate policy areas. For example, in the U.S. Quadrennial Diplomacy and Development Review released in 2010, the first of its kind, particular emphasis was given to the interconnections between security, diplomacy, and aid. Yet, issues such as water security slipped through the gaps. According to analysis by the Center for Strategic and International Studies' Global Water Policy Project: 'Overall, there was little reference to water in the report's discussion of three major activities—the Global Health Initiative, the Global Climate Change Initiative, and Feed the Future—yet it is clear that water quality, supply, and availability will be integral to the success of each of those endeavors' (Bliss, 2010). A deeper engagement between the impressively interdisciplinary water security scholarly community (for a dissenting view on this point, see Trottier, 2008) and defence and security policy community based on a greater understanding of some of the issues discussed above should assist in this task. A key first task in this will be the creation of avenues for regular dialogue between academics, NGOs, and others working on water security issues and defence and security policymakers and analysts who are increasingly factoring water issues into their security assessments and horizon scanning exercises. Such avenues can provide an opportunity for highlighting both the differences in each scholarly/policy community's respective approaches (and the reasons why, including the issues discussed above), as well as the commonalities that are not always readily apparent.

Conclusion

While the rationalist explanations as to why interstate war over water is unlikely point to the fact that 'Shared interests along a waterway seem to consistently outweigh the conflict-inducing characteristics of water' (Jarvis and Wolf, 2010, p138), the puzzle of wars is, as James Fearon has put it, that they 'are costly but nonetheless wars occur' (1995, p379), and therefore they cannot be simply explained away by simple cost–benefit analysis. Wars are planned, ordered, and fought by people who 'are subject to biases and pathologies that lead them to neglect the costs of war or to misunderstand how their actions will produce it' (Fearon, 1995, p379). When these considerations are added to the increasing importance of intrastate conflict for global security and the external variables of climate change and an increasingly crowded but socioeconomically divided world, the reasons why fears of 'water wars' prevail in both specialist literature (Schwarz et al., 2000), and particularly in defence and security policy discourse, become clearer. While this does not mean that those who have criticised the environmental determinism of neo-Malthusian pessimists have been wrong to do so, it does mean that a more nuanced position that takes the strategic dimensions of water seriously is an important middle ground. If the push towards an overly militarised response to increasing pressures on global water supplies is to be effectively countered, a much deeper engagement between the advocates of cooperative approaches to water management and those involved in threat analysis and national security planning will be necessary.

For a genuinely sustainable approach to water security, a major effort is needed in not only creating new approaches to security policy that put addressing the interconnected and underlying drivers of global insecurity at the heart of national, regional, and international policy initiatives today, but also to increase capacity for communicating complexity and risk to publics around the world faced with a worsening global economic environment and the reversion to self-interest and short-term outlooks that this often creates. This is now the central challenge for governments and civil society alike. A greater engagement between the academic water security community and their counterparts in the defence and security community on the most sustainable ways of addressing the strategic dimensions of water will be a useful first step.

References

Allan, T. (2000) 'Foreword' in H. Amery and A.T. Wolf (eds) *Water in the Middle East: A Geography of Peace*, The University of Texas Press, Austin.

Allan, T. and Mirumachi, N. (2010) 'Why negotiate? Asymmetric endowments, asymetric power and the invisible nexus of water, trade and power that brings apparent water security' in A. Earle, A. Jägerskog and J. Öjendal (eds) *Transboundary Water Management: Principles and Practice*, Earthscan, London and Washington, DC.

Amery, H. and Wolf, A.T. (2000) 'Water, geography and peace in the Middle East: an introduction', in H. Amery and A.T. Wolf (eds) *Water in the Middle East: A Geography of Peace*, The University of Texas Press, Austin.

Barnaby, W. (2009) 'Do nations go to war over water?', *Nature*, no 458, pp282–283.

Baruah, S. (2012) 'Whose river is it anyway? Political economy of hydropower in the Eastern Himalayas', *Economic and Political Weekly*, vol. XLVII, no. 29, pp41–52.

Beebe, S.D. and Kaldor, M. (2010) *The Ultimate Weapon is No Weapon: Human Security and the New Rules of War and Peace*, Public Affairs, New York.

Black, R. (2010) 'Water map shows billions at risk of "water insecurity"', BBC News, 29 September, www.bbc.co.uk/news/science-environment-11435522, accessed 3 November 2011.

Bliss, K. (2010) 'Water Issues in the QDDR', Center for Strategic and International Studies, 16 December, http://csis.org/publication/water-issues-qddr, accessed 3 November 2011.

Booth, K. and Wheeler, N.J. (2008) *The Security Dilemma: Fear Cooperation and Trust in World Politics*, Palgrave Macmillan, Houndmills.

Bowen, R.W. (2009) 'Water engineering for the promotion of peace', *Desalination and Water Treatment*, vol 1, no 1–3, pp1–6.

Bueno de Mesquita, B., Morrow, J.D., and Zorick, E.R. (1997) 'Capabilities, perception, and escalation', *American Political Science Review*, vol 19, no 1, pp15–27.

Burgess, P. (2008) 'Non-military security challenges' in C. A. Snyder (ed) *Contemporary Security and Strategy* (Second edition), Palgrave Macmillan, Houndmills.

Cascão, A. E. and Zeitoun, M. (2010) 'Power, Hegemony and Critical Hydropolitics' in A. Earle, A. Jägerskog and J. Öjendal (eds) Transboundary Water Management: Principles and Practice, Earthscan, London and Washington, DC.

Dabelko, G.D. (2012) 'Four Takeaways from the Global Water Security Intelligence Assessment', Woodrow Wilson International Center for Scholars, Environmental Change and Security Program, 27 March, www.wilsoncenter.org/article/four-takeaways-the-global-water-security-intelligence-assessment.

DCDC (2010a) *Global Strategic Trends—Out to 2040* (Fourth edition), Development, Concepts and Doctrine Centre (DCDC), British Ministry of Defence, Shrivenham.

DCDC (2010b) *Future Character of Conflict*, Development, Concepts and Doctrine Centre (DCDC), British Ministry of Defence, Shrivenham.

Dupont, A. (2008) 'The strategic implications of climate change', *Survival*, vol 50, no 3, pp29–54.

Earle, A., Jägerskog, A. and Öjendal, J. (2010) 'Introduction: setting the scene for transboundary water management' in A. Earle, A. Jägerskog and J. Öjendal (eds) *Transboundary Water Management: Principles and Practice*, Earthscan, London and Washington, DC.

Economist (2011) 'Unquenchable thirst', *The Economist*, 19 November, pp25–27.

Fearon, J.D. (1994) 'Domestic political audiences and the escalation of international disputes', *American Political Science Review*, vol 88, no 3, pp577–592.

Fearon, J.D. (1995) 'Rationalist explanations for war', *International Organization*, vol 49, no 3, pp379–414.

Filson, D. and Werner, S. (2007) 'Sensitivity to costs of fighting versus sensitivity to losing the conflict: implications for war onset, duration, and outcomes', *Journal of Conflict Resolution*, vol 51, no 5, pp691–714.

Foresight International Dimensions of Climate Change (2011) *Final Project Report*, The Government Office for Science, London.

Freedman, L. (2003) 'Prevention, not pre-emption', *The Washington Quarterly*, vol 26, no 2, pp105–114.

Fullbrook, D. (2010) 'Development in Lao PDR: the food security paradox', *Working Paper Series: Mekong Region*, No. 1, Swiss Agency for Development and Cooperation, Berne.

Gill, P. (2004) 'Securing the globe: intelligence and the post-9/11 shift from "liddism" to "drainism"', *Intelligence and National Security*, vol 19, no 3, pp467–489.

Giordano, M.F., Giordano, M.A., and Wolf, A.T. (2005) 'International resource conflict and mitigation', *Journal of Peace Research*, vol 42, no 1, pp47–65.

Gleditsch, N.P. (ed) (2012) 'Special Issue: Climate Change and Conflict', *Journal of Peace Research*, vol 49, no 1.

Gleick, P.H. (1993) 'Water and conflict: fresh water resources and international security', *International Security*, vol 18, no 1, pp79–112.

Grey, D., Sadoff, C. and Connors, G. (2009) 'Effective cooperation on transboundary waters: a practical perspective', in A. Jägerskog and M. Zeitoun (eds) *Getting Transboundary Water Right: Theory and Practice for Effective Cooperation*, Report No. 25, Stockholm International Water Institute, Stockholm.

Hatami, H. and Gleick, P.H. (1994) 'Conflicts over water in the myths, legends, and ancient history of the Middle East', *Environment*, vol 36, no 3, pp10–16.

HM Government (2010) *A Strong Britain in an Age of Uncertainty: The National Security Strategy*, Cm 7953, The Stationery Office, London.

Hoffman, F.G. (2009) 'Hybrid Threats: Reconceptualizing the Evolving Character of Modern Conflict', *Strategic Forum*, no. 240, Institute for National Strategic Studies, National Defense University, Washington, DC.

IISS (2011) *The IISS Transatlantic Dialogue on Climate Change and Security: Report to the European Commission*, International Institute for Strategic Studies, London.

Jarvis, T. and Wolf, A. (2010) 'Managing water negotiations and conflicts in concept and in practice', in A. Earle, A. Jägerskog and J. Öjendal (eds) *Transboundary Water Management: Principles and Practice*, Earthscan, London and Washington, DC.

Jermy, S. (2011) *Strategy for Action: Using Force Wisely in the 21st Century*, Knightstone Publishing, London.

Jervis, R. (1976) *Perception and Misperception in International Politics*, Princeton University Press, Princeton, NJ.

Katz, D. (2011) 'Hydro-political hyperbole: examining incentives for overemphasizing the risks of water wars', *Global Environmental Politics*, vol 11, no 1, pp12–35.

Khagram S., Clark, W.C. and Raad, D.F. (2003) 'From the environment and human security to sustainable security and development', *Journal of Human Development*, vol 4, no 2, pp289–313.

Kundzewicz, Z.W. and Kowalczak, P. (2009) 'Correspondence: the potential for water conflict is on the increase', *Nature*, no 459, p31.

Lowi, M.L. (1995) 'Rivers of conflict, rivers of peace', *Journal of International Affairs*, vol 49, no 1, pp123–44.

McGwire, M. (2001) 'The paradigm that lost its way', *International Affairs*, vol 77, no 4, pp777–803.

Miles, T. (2011) 'Gaddafi forces cut off Tripoli water supply', *Reuters*, 30 August.

Mori, K. (2003) 'Virtual water trade in global governance', in A.Y. Hoekstrapp, *Virtual Water Trade: Proceedings of the International Expert Meeting on Virtual Water Trade*, Value of Water Research Report Series No. 12, IHE Delft, Delft.

OECD and United Nations OSAA (2010) *Economic Diversification in Africa: A Review of Selected Countries*, United Nations Office of the Special Adviser on Africa and the NEPAD-OECD Africa Investment Initiative, www.nepad.org/system/files/OSAA-NEPAD_Study_Final5Oct.pdf, accessed 18 January 2012.

Oxford Research Group (2011) 'Written evidence to the house of commons defence select committee inquiry into the strategic defence and security review and the national security strategy', February, www.publications.parliament.uk/pa/cm201012/cmselect/cmdfence/761/761vw.pdf, accessed 3 November 2011.

Parenti, C. (2011) *Tropic of Chaos: Climate Change and the New Geography of Violence*, Nation Books, New York.

Parthemore, C. and Rogers, W. (2010) *Sustaining Security: How Natural Resources Influence National Security*, Center for a New American Security, Washington, DC.

Phillips, D., Daoudy, M., McCaffrey, S., Öjendal, J. and Turton, A. (2006) *Transboundary Water Cooperation as a Tool for Conflict Prevention and Broader Benefit Sharing*, Global Development Studies No. 4, Stockholm, Ministry of Foreign Affairs, Sweden.

Proença D., Jr. and Duarte E.E. (2005) 'The concept of logistics derived from Clausewitz: all that is required so that the fighting force can be taken as a given', *The Journal of Strategic Studies*, vol 28, no 4, pp645–677.

Richards, D. (2011) 'A soldier's perspective on countering insurgency', in D. Richards and G. Mills (eds) *Victory Among People: Lessons from Countering Insurgency and Stabilising Fragile States*, Royal United Services Institute, London.

Rogers, K.S. (1996) 'Pre-empting violent conflict: learning from environmental cooperation', in N.P. Gleditsch (ed) *Conflict and the Environment*, Kluwer Academic, Dordrecht.

Rogers, P. (2008) *Global Security and the War on Terror: Elite Power and the Illusion of Control*, Routledge, London and New York.

Rogers, P. (2010), *Losing Control: Global Security in the Twenty-First Century* (Third edition), Pluto Press, London and New York.

Schwartz, D., Deligiannis, T. and Homer-Dixon, T. (2000) 'The environment and violent conflict: a response to Gleditsch's critique and suggestions for future research', *Environmental Change & Security Project Report 6*, Summer, pp77–93.

Selby, J. (2005) 'The geopolitics of water in the Middle East: fantasies and realities', *Third World Quarterly*, vol 26, no 2, pp329–349.

Selby, J. (2013) 'Cooperation, domination and colonisation: the Israeli-Palestinian Joint Water Committee', *Water Alternatives*, vol 6, no 1, pp1–24.

Sirin, C. (2011) 'Scarcity-induced domestic conflict: examining the interactive effects of environmental scarcity and "ethnic" population pressures', *Civil Wars*, vol 13, no 2, pp122–140.

Stephenne, N., Burnley, C. and Ehrlich, D. (2009) 'Analyzing spatial drivers in quantitative conflict studies: the potential and challenges of geographic information systems', *International Studies Review*, vol 11, no 3, pp502–522.

Terriff, T., Karp, A. and Karp, R. (2008) *Global Insurgency and the Future of Armed Conflict: Debating Fourth-Generation Warfare*, Routledge, London and New York.

Tir, J. and Stinnett, D.M. (2012) 'Weathering climate change: can institutions mitigate international water conflict?', *Journal of Peace Research*, vol 49, no 1, pp211–225.

Toset, H.P.W., Gleditsch, N.P. and Hegre, H. (2000) 'Shared rivers and interstate conflict', *Political Geography*, vol 19, no 6, pp971–996.

Trottier, J. (2008) 'Water crises: political construction or physical reality', *Contemporary Politics*, vol 14, no 2, pp197–214.

United Nations (2012) *Resilient People, Resilient Planet: A Future Worth Choosing*, Report of the High-level Panel on Global Sustainability, United Nations, New York.

US National Intelligence Council (2008) *Global Trends 2025: A Transformed World*, Office of the Director of National Intelligence, United States of America, Washington, DC.

Verrastro, F.A., Ladislaw, S.O., Frank, M., Hyland, L. and Schlesinger, J.R. (2010) *The Geopolitics of Energy: Emerging Trends, Changing Landscapes, Uncertain Times*, Center for Strategic and International Studies, Washington, DC.

Vörösmarty, C.J., McIntyre, P.B., Gessner, M.O., Dudgeon, D., Prusevich, A., Green, P., Glidden, S., Bunn, S.E., Sullivan, C.A., Reidy Liermann, C. and Davies, P.M. (2010) 'Global threats to human water security and river biodiversity', *Nature*, vol 467, no 7315, pp555–561.

Warner, J. (2003) 'Virtual water—virtual benefits? Scarcity, distribution, security and conflict reconsidered', in A.Y. Hoekstrapp, *Virtual Water Trade: Proceedings of the International Expert Meeting on Virtual Water Trade*, Value of Water Research Report Series No. 12, IHE Delft, Delft.

Witsenberg, K.M. and Adano, W.R. (2009) 'Of rain and raids: violent livestock raiding in Northern Kenya', *Civil Wars*, vol 11, no 4, pp514–538.

Wolf, A.T. (1998) 'Conflict and cooperation along international waterways', *Water Policy*, vol 1, no 2, pp251–265.

Wolf, A.T. (1999) '"Water wars" and water reality: conflict and cooperation along international waterways' in S. Lonergan (ed) *Environmental Change, Adaptation, and Human Security*, Kluwer Academic, Dordrecht.

World Economic Forum (WEF) (2011) *Global Risks 2011: Sixth Edition*, World Economic Forum, Geneva.

World Economic Forum (WEF) (2012) *Global Risks 2012: Seventh Edition*, World Economic Forum, Geneva.

Young, O.R. (2001) 'Effectiveness of international environmental regimes: existing knowledge, cutting-edge themes, and research strategies', *Proceeds of the National Academy of Sciences of the United States of America*, vol. 108, no. 50, pp19853–19860.

Zala, B. and Rogers, P. (2011) 'The other global security challenges: socioeconomic and environmental realities after the war on terror', *The RUSI Journal,* vol 156, no 4, pp26–31.

Zeitoun, M. (2011) 'The global web of national water security', *Global Policy,* vol 2, no 3, pp286–296.

Zeitoun, M. and Allan, J.A. (2008) 'Applying hegemony and power theory to transboundary water analysis', *Water Policy,* vol 10, supplement no 2, pp3–12.

Zeitoun, M., Allan, J.A. and Mohieldeen, Y. (2010) 'Virtual water "flows" of the Nile Basin, 1998–2004: a first approximation and implications for water security', *Global Environmental Change,* vol 20, no 2, pp229–242.

Zeitoun, M. and Warner, J. (2006) 'Hydro-hegemony—a framework for analysis of trans-boundary water conflicts', *Water Policy,* vol 8, no 5, pp435–460.

18 Dances with Wolves
Four Flood Security Frames

Jeroen Warner

Introduction

In a humorous Dutch song of some decades ago (Drs P., 1974), a family on the way to the Russian city of Omsk riding in a three-horse open sleigh (troika) through the Siberian forest is suddenly faced with a pack of hungry wolves. They try to keep the wolves at bay by singing songs and throwing food, but the wolves soon catch up with the troika carrying the family. The desperate parents then decide to throw one child after the other overboard to mollify the insatiable wolves. In the end, the mother is also thrown out, and the father is left, but he is so giddy that the goal is within reach that he loses his balance and falls off the troika, too. Omsk, the song goes, is a very pretty city, but just that little bit too far away.

Like the unlucky family in the song, states are supposed to defend their own people; they are also prepared to sacrifice everything, even their own citizens, for their own safety. The role of the state is to keep the dangerous Other out. When the Other enters the gates, sovereignty means that everything should be done to restore order and show the Other its place—at any cost. Current debates on disaster governance ask questions about the division of roles in the story of the wolves, as will be illustrated by the case studies discussed later on. Are there ever really wolves on the prowl, or are we imagining them? Is singing songs and sacrificing kids the only thing we can do about it? Hasn't the fear of wolves obscured a proper view of what risks the sleigh riders are faced with? Is Omsk all that safe a destination?

A now-popular European, constructivist-leaning approach in security studies (Buzan, Waever, and de Wilde, 1998) argues that many 'threats' are in the eye of the beholder, and may be invoked because of the exceptional powers that calling a crisis legitimises. The state of exception derogates normal procedures, rights, and rules, and converts the status of some actors—whether 'enemy' or 'victim'—to one devoid of prevalent civil rights. Maybe our guardians are wolves in sheep's clothing? The concept of 'human security' and critical approaches to security suggest we're all on our own—in fear of the elusive predator, but also unsure of our protector.

The wolves in this chapter will particularly be read as 'water wolves'. In Dutch folklore and metaphor, a flood, the river coming out of its cage, is often represented by a wolf (Rooijendijk, 2009). We will, however, see that the wolf may partly be human-made. Floods claim more victims than any other disaster, including civil war. They have claimed hundreds of thousands of lives in single events in China and the Netherlands. Yet flooding brings many services—among others, irrigation, flushing, spawning ground for fish, drought protection, and replenishing environmental flows. The past decades have seen a kind of 'desecuritisation' of floods, as calls for 'living with the floods' are increasingly heard. On the other hand, the pendulum also swings back after floods (Hartmann, 2010).

Flood security, then, is a compromise between the threats and opportunities rivers, lakes, and rainfall bring. River regulation infrastructure and flood response are supposed to prevent disastrous floods without destroying the benefits floods bring. Engineers create often innovative infrastructural designs and increasingly devise participatory processes to discuss their plans with affected stakeholders. Such projects tend to get politicised despite technical and environmental soundness.

This chapter argues that unease over security distribution is a recurring factor, and that different frames exist on what the problem to be tackled actually is: the flood or the project, the government or the people. The framing of the enemy also influences who is seen as the hero and therefore confers legitimacy on that actor. The problem frame thus influences the mode of governance selected for dealing with it. The different role divisions, or governance alternatives, very roughly speaking, are state-led, market-led, and community-led. With the help of the metaphor of the 'water wolf', different images (frames) of floods and threats are presented in the second section. These frames may change in the course of a case (Warner, 2008). Casting a two-by-two typology of flood governance, applied to two examples from the Netherlands, the analysis puts into perspective a linear change in Dutch flood governance (third section), which is supposed to be in transition from flood control to risk management, from resistance to resilience (van den Brugge et al., 2005). The analysis is based on my doctoral research drawing on documentary analysis and interviews with multiple public, private-sector, and civil-society stakeholders (Warner, 2011).

Four Flood Frames

People rely on mental frames to make sense out of the bewildering mass of facts they are faced with every day. Although frames limit choices, people need frames as a sort of mental crutch—they determine what counts as a fact (Hoppe, 1999, p207; Wesselink and Warner, 2010). In this context, I am inspired by Van Eeten's (1997) narratological analysis of conflict on river management in the Netherlands. Van Eeten shows how flood risk stories (fairy tales) take the same narrative form; the discourse coalitions mirror each other, even though the discursive coalitions find themselves diametrically opposed

Table 18.1 Four flood frames related to disaster narratives

Flood frame	Disaster narrative	Relationship with flood	Security is about	Security Referent	Since	Ideology
1. Flood control	Hazard / structural	(Securitised) War with water	Reducing probability	Dikes, infrastructure	1920s	Statist
2. Risk management	Risk / behavioural	(Desecuritised) Peace with water	Reducing probability × Impact	Floodplain / polder	1950s	Liberal
3. Political ecology	Vulnerability	Hegemony	Reducing Probability × Impact × Vulnerability	Community	1980s	Critical
4. Post-normal risk management	Complexity	Mutuality of humans and nature	System	?	2000s	Liberal

to each other—the hero of one story is the villain in the other, the former's problem is the latter's solution. These mirroring frames especially come to show when flood projects become controversial (Warner, 2011). 'Contradictory certainties' (Thompson and Warburton, 1985) about what caused the problem gives rise to contradictory storylines that bring apparent certainty in the face of uncertainty (van Eeten, 1997). These storylines present different protagonists, enemies, and protégés: the wolf, the sleigh driver, and the kids in the story.

The four flood frames are summarised in Table 18.1, shown with their related disaster narratives as discussed here. The frames build on a trichotomy of a state-led, market-led, and community-led frames identified in an analysis of the Bolivian 'water war' of 2000 (Warner, 2004) and water conflict on the Nile (Warner, 2012), but are applied to floods. The four flood frames I selected (first column, Table 18.1) roughly match the 'flood waves' in flood management first identified by Green and Warner (1999), though other categorisations are certainly possible. They are then applied, by way of illustration, to four case studies.

The frames are also classed within the Omsk sleigh/wolves metaphor in Table 18.2. Frames and narratives are never fixed; they can be contested and politicised. Due to the antagonism and polarisation inherent to it, politicisation tends to take the form of hardened positions. However, if there is no a priori consensus on values, depoliticising real social tensions over an issue will only defer politicised confrontation. The politicisation of floods means the contestation of a dominant security frame. Hajer (1995, p59) defines *politics* as the struggle for discursive hegemony in which actors struggle to secure support for their definition of reality. This struggle, as we shall see, includes the question 'what/who is the wolf, and who shall do something about it?'

Table 18.2 Role distribution within each of the flood frames: wolf (villain), driver (hero), kids (protégé)

Flood frame	Paradigm (master frame)	Who is the enemy?	Who's in the driver's seat?	Who is to be protected?
1. Flood control	Hazards paradigm	Nature	State and its experts	The state's subjects
2. Risk management	Risk management paradigm	Nature and human behaviour	Risk managers	The general public
3. Political ecology	Vulnerability paradigm	The greedy elite enabling floods	The community	The community
4. Post-normal risk management	Complexity paradigm	Technoculture, risk society	Extended peer community	Modern society

Frame 1: Flood Control: Containing the Water Wolf

Etymologically, the word *disaster* means that the stars are misaligned, a cosmology that continues to inform many flood cultures. The earthquake of Lisbon in 1755, however, signified the start of a 'modern' approach to hazards. Rather than an act of God, disaster came to be treated as preventable. Controlling nature became the object of study and intervention, from a belief that modernity and development would deliver people from backwardness, pestilence, and disaster. The wolf might not be exterminated, but at least could be kept at bay. It became the mission of the 'enlightened West' to lift the 'unenlightened South' out of its disaster-prone backwardness (Bankoff, 2001). The secular state replaced God, as it were, as self-appointed sovereign protector against the 'outside' forces of risk and disaster.

But there is a risk that disasters overwhelm state systems. Sovereignty needs nature as its Other (Burke, 2007). Ophir (2010) notes that a catastrophe does not need to be declared to disrupt the sovereign power. Nature overthrows government by disabling it. This legitimises the accumulation of the means of power to resist nature. When floods are 'securitised', flood management becomes a national security concern, a threat to order, preceding every other concern. A society may be structurally subject to the state of exception. Egypt, Burma, and Singapore have made the state of exception permanent. This means information is classified and public participation curbed, with a highly securitised underlying system of control and surveillance of a pre-existing structure. But in the 'free West', a country like the Netherlands is on constant alert too, due to the resonance of a history of destructive floods. The military may be mobilised, and/or engineers are given a mandate to take decisions outside the political domain. Sovereignty is about the creation of boundaries within which authority and law can be exercised, and in so doing creates the conditions of possibility for political life (Burke, 2007).

Sovereignty ultimately means the state or its representatives can declare the state of exception, in which they shore up normal political rights and procedures. Under the Etatist 'Hobbesian' hazards contract (after English philosopher Thomas Hobbes), the government protects against external as well as internal enemies to prevent a return to the 'state of nature'. Enemies of the state can be placed outside the political order. For the political philosopher Carl Schmitt, influential on 'securitisation' theory, this power, how it is exercised, and how 'obedience is generated, represented and legitimised' (Ophir, 2007, p123) is the essence of the political. While disaster management is rarely that absolutist, the declaration of a state of emergency can legitimise extraordinary extra-legal measures for the sake of stability and survival. However, while 'securitised', nature (the 'wolf' in this frame) would not be tamed; it kept invading spaces believed to be safe. A wider approach started to take hold of disaster managers, focusing on the entire 'risk cycle' running from prevention, mitigation, and response to rehabilitation.

Frame 2: Risk Management: Making Space for the Wolf, Leaving Space for People

After engineers in the United States and elsewhere made a prolonged effort to keep floods away from people, White (1978) noted that floods had not become any less devastating. Human geographers started to realise there is no such thing as natural hazards—they are always mediated by social and political behaviour. Dissatisfied with the performance of infrastructural flood defences, an American liberal school of human geographers (White, 1978; Burton, Kates, and White, 1978) focused on human behaviour and choice: containing the risk rather than the flood. Rather than 'Hobbesian' technocratic, top-down protection, it prefers social responsibility (a 'Lockeian' hydrosocial contract, after John Locke, see Meissner and Turton, 2003; Pelling and Dill 2010). White's solution was to try and keep people away from floods with incentives such as institutional regulation (zoning), economic incentives (subsidies), and the availability of insurance. These influence the preferences within the range of alternatives individuals have at their disposal, to make up for people's 'bounded rationality' (Lindblom, 1959; see also Johnson et al., 2005). Levees became the alternative of last resort for 'buying down risk' (Masse et al., 2007). This retreat of infrastructure has been accompanied by state retreat, a liberalisation of land use.

Proposals to 'make space for the river' have been popular in Western Europe, combining a desire to renaturalise rivers with safety, regional economic development, leisure, and, in some cases, urban regeneration. This, however, opened a different can of worms. Flood security traditionally served food security; the 'space made', however, was more often than not agrarian land. Farmers therefore protested these projects. In places, space was found by allocating land to be inundated to save more economically important or more densely populated areas—controlled flooding. As we shall see, this sparked fierce controversy in flood-traumatised areas such as Zeeland (Southwest Netherlands).

The 'safety chain' approach to floods, used in the Netherlands and increasingly elsewhere in Europe, is a close cousin to the Disaster Risk Cycle. The premise is that reducing risk in each chain of the link will lead to a cumulative reduction in the risk as a whole, so that 'disasters' (where the challenge overwhelms coping capacity) would be a thing of the past. An underlying assumption is that disaster risk is manageable if done rationally. This led Australia's government, for example, to abolish designating drought as a disaster, and Ethiopia to present famine as a discomfort (Warner, forthcoming). If drought is a recurring hazard, Australians can see a drought coming on and can prepare for it by hoarding, getting insurance, and other risk transfer tools.

If floods are no longer considered an emergency issue by the key players (i.e., we make peace with water), we would expect a situation where 'closures' are opened up: The state is no longer automatically the lead security actor, rights and freedoms cannot be shored up for the common good, acceptance of sacrifices is not self-evident. The catch is that nonsecurity decision making also erodes the 'blank cheque' that a securitised status almost ensures—tough decisions need to be made on where to invest one's last euro, dollar, or pound. If the goal is no longer 100% risk reduction, a residual risk remains. This means that decisions made necessarily involve a degree of 'triage': who shall live in a flood, and who shall perish.

When the 'wolves' are already there, it is obviously too late. By making space for the river to enter the domain 'behind the dikes' but also making space for people in floodplains, authorities deliberately take a risk. They seek to tame that risk by arrangements such as collective or private insurance, building local defences and bunds, developing evacuation plans, launching awareness campaigns, and other Disaster Risk Reduction strategies. These, however, depend on the compliance of those expected to participate, so that they are accompanied by behavioural control techniques.

A liberal, economic rationale, as promoted in a desecuritised frame, promotes protection where the assets are, rather than making every life count irrespective of cost. As a consequence, poor areas get poor protection, as they live on cheap land close to rivers and volcanoes. Moreover, the nonstructural, 'behaviorist' approach led to a neglect of infrastructure, as revealed when Hurricane Katrina broke through New Orleans's meagre flood defence system.[1] This made some experts (Jongejan, Jonkman, and Vrijling, 2012) conclude that every last penny should be invested in hard defences. A third school, however, which may be termed 'political ecology', lamented something else: the fact that not all lives count the same (frame 3).

Frame 3: Political Ecology

A critical perspective that arose in the 1980s sees vulnerable people threatened by the political economy. People are 'at risk' (Blaikie et al., 1994) due to structural causes of security differentiation. Political ecology (e.g., Bryant and Bailey, 1997) instructs us that the relationship with nature always implies a relationship with society. Since the work of Maskrey and Romero (1983), it is now internationally accepted that natural hazards are, in fact, human hazards. It observes that some groups are far more vulnerable, living in unsafe conditions due to root socioeconomic causes. For them, disasters and their management highlight social inequity and open up windows for social change.

This frame faults disaster studies with being obsessed with control. Its sympathies lie with a Rousseauian,[2] community-centred, anti-hegemonic perspective on the wars with water, as they identify with local actors and initiatives rather than the state. In this tale, the 'wolves' are the greedy capitalist global elite. The marginalisation of the powerless goes hand in hand with the public 'facilitation' of the more powerful to grab a hold of prized resources (Collins, 2008).

A similar critique of risk management, coming from security studies, takes a Foucauldean tack. Risk policies are about anticipating rather than responding. Recent times have seen a fixation on risk management, ensuring a smooth landing when things inevitably go wrong. The modern state not only makes sure it can act in the event of a wolf, but it also prepares for the possibility of wolves, even in areas where wolves have never been sighted, and calculates how many limbs can be sacrificed to ward off mortal danger. Much more unobtrusive social controls, such as insurance charts, decide where we can live. In Scotland and Canada, the private sector resorts to fiscal instruments, such as withholding insurance cover from people who build in dangerous places (Crichton, 2005), and thus 'tame' people's behaviour on the basis of risk profiling. This makes human behaviour the cause of disaster risk: 'Wolves R Us'. While the liberal 'peace with water' drive often involves a degree of public participation, civil-rights-oriented analysts and activists worry about the permanent and largely invisible 'war on risk' evoked by risk management technologies unobtrusively influencing human behaviour, and decisions being made by computers.

We end up with a more subtle form of structural peacetime 'securitisation' of risk, a modality of social control that may be termed the 'managerialisation of risk' (Aradau, 2001) or 'riskisation' (Corry, 2012), which discusses how the object of governance is moving from threat to risk, from actuality to potentiality, from an identified danger that can be meaningfully thwarted/personalised/contained by 'walls' to a possible scenario and a policy proposal preventing risk from materialising in real harm (Corry, 2012).

Disaster Risk Reduction (DRR) risks the expansion of the sphere of exceptionality to the whole cycle, as social behaviour is scrutinised and influenced—openly or unobtrusively—by a providential state (Ophir, 2007), or private insurance in the post-providential state, to minimise loss to these actors. This normalises security practices that may clash with civil rights and liberties as well as accountability and transparency: obedience as realised through hidden controls based on forensic charts.

Frame 4: Holistic Post-Normal Risk Management

While the third frame takes an explicitly political approach, and therefore can have explanatory value in understanding risk conflicts, the all-out confrontation of politicised conflict may bring an impasse, while a technocratic handling takes the risk out of the political debate. De Marchi and Ravetz (1999) have called attention to a different approach to risk: the type in which both facts are uncertain and values are divergent. It engages with the fundamentals of modernity in that scientific authority is now increasingly distrusted.

'Wicked' (intractable) problems, where both values and facts are disputed or uncertain, and decisions need to be taken urgently, cannot be solved by a political process only. They may require a mix of politics and technocracy (post-normal science). For such problems, Hisschemöller and Hoppe (1998) recommend undertaking a signalling and social learning process first. While De Marchi and Ravetz (1999) see a natural progression from classical hazards to be dealt with by applied science, to complex risks engendered by complex socioeconomic interplay, it can be argued that relatively straightforward hazards can also translate into complex problems due to complex nonlinear social dynamics. At the edge of chaos, improvisation is as important as planning.

In a post-normal science approach, uncertainty is not banished but managed, and values are not presupposed but are made explicit (Funtowicz and Ravetz, 1993; Healy 1999). In the same vein, Renn (2008) calls attention to systemic risk, where 'butterfly effects' can transfer risks. Post-normal science, especially popular at the turn of the millennium, promises a way out of Beck's problem of technoculture producing its own problems by extending democracy into the scientific realm. De Marchi and Ravetz (1999) suggest a modality in which an extended peer community becomes involved in deliberations over a risk.[3] Engel and Engel (2012) note that when on the edge of chaos, we have to 'plan to improvise'.

In this perspective, it is not nature or humanity, but the interplay or conjunction between the two—their mutuality—that creates disaster risk in nonlinear ways. Hilhorst (2003) calls this approach the 'complexity paradigm' in disaster studies, where experts, managers, and locals find ways to act in concert, without assuming they will necessarily reach consensus; the opposite is just as likely due to their different ways of knowing and acting.

Conceptual Framework

Considering the shifting distribution of roles (Table 18.2) together with the varying degree of agreement over facts and values, as introduced in the previous discussion, we can derive the dynamic two-by-two matrix of Table 18.3 (after Warner, 2008). As Table 18.1 shows, the dominant flood frame may shift over time, and we may surmise flood frames move through the cells of Table 18.3 over time. The following section applies the framework to two cases of flood (in)security in the Netherlands. Thereafter it

Table 18.3 Conceptual framework, based on 'participatory security governance' matrix (axes based on Hisschemòller and Hoppe, 1998, after Warner, 2008). Dominant frames are expected to shift over time between the cells.

		[high] Agreement on values [low]	
		Threat not open to dispute	*Threat open to dispute*
Agreement on facts	*High*	Securitisation (foreclosing debate by speech act)	Open conflict, internalising antagonisms (power of argument)
	Low	Routinisation/managerialism (foreclosing debate through risk management)	Dialogue (power of better argument)

will be discussed whether these changes have also been accompanied by a change of attribution of the 'wolf' to be contained, based on the typology introduced in Table 18.2.

Two Cases of River Flood Frames in The Netherlands

The West Netherlands are famously below sea level, yet the Netherlands did not always lie so low. Peat extraction eroded the basis for safety. Wherever this created holes, these tended to fill up as the 'water wolf' came knocking, fast eroding river banks and lake shores. The 'water wolf' is an animalisation of the erosive power of rivers and sea. Lake Haarlemmermeer, for example, in the West Netherlands, devoured three villages in 1591 and 1611. The management of floods was in the hands of thousands of competing water boards. Until in 1798, the French occupying authorities imposed a French-style centralised agency that has remained in force until today. After the 1953 floods, the agency became a powerful 'state within a state', charged with closing the gaps in the Dutch coastline and reinforcing the river dikes.

Until the mid-1990s, the Netherlands considered itself a bastion of flood safety—in terms of our tale, a 'mega-Omsk' by the sea. The 'providential state' built dikes paid for out of special taxes so that the good citizens of the Low Countries could all sleep safely. In the 1990s, however, a 'desecuritisation' took place, a lowering of protection standards, 'greening' of flood policy, as well as the decentralisation and downsizing of the public works department and room for stakeholder participation in decision making.

The case studies below, however, suggest a greater range of modalities for (non-)participation than 'wartime' or 'peacetime' decision making, with different degrees of actor involvement in considering policy alternatives. Tables 18.4 and 18.5, compiled and systematised by the author (see also Warner, 2008), apply the modalities found to the present case study; numbers trace the dispute's development through time. It will be shown that the frames held by the various authors reflect dissent on who is being protected by whom, and under what kinds of conditions.

Table 18.4 Conceptual frame applied to the Border Meuse. Numbers indicate chronological development of six phases.

		[high] Agreement on values [low]	
		Threat not open to dispute	*Threat open to dispute*
Agreement on facts	High	[Securitisation] 1995–1997 Emergency law (DGR) 1995–2005 Ban on building in floodplain 2004 Limburg incorporated in national security framework: dikes 1, 5	[Open conflict] 2001–2003 Protest, lawsuits 3
	Low	[Routinisation/Managerialism] 2 1997–2001 Space for the river, discussion behind closed doors	[Dialogue] 4, 6 2001 Participation process led by provincial authorities

Table 18.5 Conceptual frame applied to the Ooij polder. Numbers indicate chronological development of four phases.

		[high] Agreement on values [low]	
		Threat not open to dispute	*Threat open to dispute*
Agreement on facts	High	[Securitisation] Seriously considered, not implemented 1	[Open conflict] 2002–2005 Mayors side with platform, mobilisation of politicians 4
	Low	[Routinisation/managerialism] 3 2000–2002 Consultation with intermediate organizations and local authorities, not citizens	[Dialogue] 2 Considered not implemented: joint learning with local stakeholders

Broadening and Deepening the River Meuse

The success of the Delta coastal defence works had also proved a weakness. Dutch engineers are known to cite a variety on the Lord's Prayer: give us our daily bread, and a flood every 10 years. After all, a flood keeps the momentum up: We need the wolves to come around now and then, so we can build traps for them. We need the actualisation of risk—disaster—to enhance a social support base for building expensive defence works. While the power of the Dutch Public Works department has been 'naturalised', based on its role of protector against floods, due to the structural 'securitisation' of the water in the Netherlands (flood frame 1, assigned the top left corner of Table 18.4 as phase 1), its sovereignty over the river and waterways needs to be reaffirmed, reasserted, and justified.

While the low-lying West has remained dry, minor (high-incidence, low-consequence) flooding in the Southern province of Limburg, which is above sea level, is not unheard of, and local people are used to it—they tend to take a 'desecuritised' perspective. The scale of two consecutive high-water events on the river Meuse in 1993 and 1995 served as a wake-up call for Limburg that bigger floods are possible. Its main effect, however, was at the national level. Dramatic pictures at Borgharen, where the Meuse enters Holland, incited national politicians to make bold promises from the Water Minister: 'safety by 2015'. The event propelled flood-security measures to the top of the agenda. But as the Meuse leaves room for timely escape, civil engineers in her department doubted that the flood issue is really a security concern and therefore eligible for urgent, special treatment, as the province of Limburg claimed, and preferred to enforce the ban on building in the floodplain (flood zoning, i.e., frame 2).

The Grensmaas Plan, a plan to broaden and renaturalise the Common Meuse, promoted as a nature development scheme in the 1980s, was given a boost by the need to sweeten a voluntary agreement concluded in 1990 between gravel kings and regional authorities to dig up 35 million tones. The river scheme sought to strike a happy balance between flood defence, natural values, and sand and gravel extraction, the latter paying for the costs of making the river greener and safer ('security-plus'). Stakeholders, however, continued to see the project in a different light: a blank cheque lifeline for the regional gravel-digging industry by another name ('gravel-plus'), harming the environment without eliminating flood risk—destruction rather than protection (frame 3).

The province of Limburg was already plagued by political scandals over construction and gravel extraction. The Grensmaas project seemed moribund when the floods of 1995 came to the rescue. While the original renaturation project focused on the Common Meuse area (Border or Common Meuse), the plan developed into something much bigger after the flood of 1995. This high-water crisis event sparked special legislation in which everything seemed possible: special powers, unlimited resources, informal cooperation with citizens. Emergency funds were raised in close informal cooperation with citizens, land was acquired, and environmental and participation rules were disregarded under an open-ended budget.

There is a strong impression the flood opened the door to solutions lying in wait for a suitable problem. It also opened a window for the central-level Public Works Department, disregarded before the floods, to take the reins and act as a hegemon. It convened a consortium with public, private, and civil-society actors. After the Special Law expired by 1997, however, the process went back to normal. The national government, however, took over the leadership in the project, combining it with other interventions, to make the next stretch of the Meuse safer and greener. The province had expected the Minister to provide extra funding for the Meuse works as a 'security project', but was refused, while 'budget neutrality' remained a precondition for provincial actors. This way, the consortium painted itself into a corner. The Environmental Impact Assessment for the Sandy Meuse was rejected by the EIA Commission, necessitating several rounds of replanning behind closed doors (phase 2). After over a year's silence, a new version reduced environmental amenities and increased hard structures and gravel quarrying. The environmentalist consortium partner, Natuurmonumenten, almost pulled out in protest.

Provincial and local politicians started to question the motives for the project, seeing it as a project to legitimise gravel digging. Using 'flood safety' as a lever for multipurpose projects combining nature conservation, gravel excavation, property development, and so forth leads stakeholders to question the motives for flood protection projects (phase 3). After strong protests, Limburg's provincial government decided to take over in 2001 and decided to do things differently; in close consultation with stakeholder groups, it came up with a new plan within six months, taking into account people's own problem analyses (phase 4). Scattered local protest groups had united into the Bewoners Overleg Maasvalle umbrella, which acted as a constructive sparring partner for Limburg. After this fresh restart, the project got into all manner of crises and lawsuits over European environmental and antitrust legislation and Belgian objections, as costs spiralled again and various partners threatened to pull out. Topsoil depots were to be manufactured from unsellable Meuse topsoil, as a useful alternative to dumping them in gravel pits, but it was found to contain chemical pollutants.

While flood protection is a powerful securitiser, the regional 'shadow of the past' left space for discrediting 'space for the river' and Grensmaas, questioning its motives and criticising the process of selecting alternatives. Old misgivings about Limburg's 'colonised' (hegemonised) status versus Holland resurfaced. Moreover, local citizens did not see why the water should be tamed, as they were used to frequent minor flooding. Dikes built around population centres create a 'bath tub', turning a high-incidence event into a high-consequence event; as a local speaker phrased it, 'from a damp rug every year to a dead Limburg citizen every 10 years'. This way, displacing risk, the Limburg authorities are courting disaster, aiding and abetting the risk of fatalities. The deal made after the security window ran out in 1997 lumped several river interventions together into the Maaswerken project, but exposed these officials to the problem frame of opportunistic gain-seeking, as it was the province who had brokered the gravel compromise. This perspective moved the focus back from risk management to flood defence (frame 1, phase 5): keeping the wolf out, rather than taming and living with the animal. And while after the high water of 2003 both politicians and activists called for speedy flood works, that same year saw petitions against the Grensmaas project as its modality.

In 2005, however, the final contract was signed starting the project, while the regional authorities and water boards had successfully managed to include the River Maas in the national flood security framework, making the central government also financially responsible to furnish extra money for dike upkeep. Meanwhile, a more participatory process was started for a long-term vision of flood security on the River Meuse to deal with higher climate-induced peak flows. While the participants indeed represented an extended peer group, reminiscent of post-normal science (frame 4, phase 6), the complexity of the system and the rather complicated hydrological models and a simulation tool, the Blokkendoos, implied an expert-dominated process (Wesselink, 2009).

Sacrificing the 'Kids'? Controlled Flooding of the Ooij Polder

In 1995, the Dutch government preventatively evacuated 200,000 people and declared a state of emergency after a fairly insignificant flood. The event, however, enabled extraordinary measures such as emergency dikes and the threat of forced expropriation (phase 1 of Table 18.5). Yet after the dust had settled, authorities were taken to account for actions taken in the emergency phase. This means that in liberal democracies, authorities cannot act with impunity.

In the Netherlands, fear of increasing flood peaks, possibly because of climate change, led central government to earmark some sparsely populated areas for controlled flooding, to keep more densely inhabited areas with more economic assets safe in case of a major flood. In 2000, senior policy-makers in the national government came up with a plan to give three such areas 'controlled flooding' status. Should a really big flood peak on the River Rhine and hit the Netherlands, these areas would be flooded first, to save downstream areas. Controlled flooding as flood management is practiced in different countries, but what is special about the Netherlands is that every space is inhabited and planned for, so that reserving space for emergencies is bound to bring conflict.

The emergency flood storage posed a problem to decision makers: Are we to treat this in a closed, security mode (frame 1, phase 1) or in a more open modality? Seeking the front pages to boost her political profile, Vice-Minister De Vries decided to forego the option of conducting security policy behind closed doors. The ministerial flood managers tended to the former approach, but the Vice-Minister was advised by her communication advisers to enter into a dialogue with stakeholders (frame 4, phase 2). What the Vice-Minister had in mind was indeed a form of deliberative democracy.

The stakeholder dialogue for joint learning she appeared to favour (phase 2), however, was swiftly prevented by her department and replaced by a commission that allowed a limited, controlled form of societal consultation to 'sell' the policy (phase 3). The exclusion of the local stakeholders led to civic protest and indeed skilful politicisation of the issue (phase 4). The opposition pictured the Vice-Minister as 'the enemy', conducted a 'knowledge guerrilla' unearthing an apparently classified document, undercut the assumptions of the department's frame, and successfully counterposed a frame in which controlled flooding was the problem rather than the solution.

The compromise was a mix of both: behind-closed-doors expert deliberation looking at alternatives, plus a highly Dutch form of consultation of nonpublic stakeholders (civil society and private sector and local-level authority leaders), which bypassed the policy-affected local population. The road chosen by the State Advisory Commission is most reminiscent of the 'managerial' approach (frame 2). Civil-society organisations were consulted in controlled focus groups without veto power, but affected citizens were not. One of the polders slated for controlled flooding, the Ooij, is inhabited by some 15,000 citizens. When they heard about the designation of their polder in 2002, they declared 'war' on the responsible Junior Minister when she proved unresponsive to their worries. More's the pity, as interviews suggest the residents of the Ooij were willing to discuss alternatives. The area had historically been inundated each year. This custom had gone out of practice for decades, while the last high-water event had been over five years ago, and both the analysis of the problem and the values leading to a solution turned out to be contested, not only with local stakeholders but also within the expert community.

Moreover, there was great uncertainty over the facts: Did new climate scenarios justify a higher protection standard and thus more intrusive measures? Even though the Ooij citizens know they are first in line to be flooded by forces of nature, they were not prepared to be flooded without consultation by the forces of government. They refused running the risk of being sacrificed by their supposed protectors (frame 3). Moreover, the designation of the area for controlled flooding had an economic impact: It would discourage investment and make house values go down (Roth, Warner and Winnubst, 2006). Excluded stakeholders then decided on a counterattack, politicising the issue such that it became a parliamentary debate. Given the basic willingness of local stakeholders, their positions were not immutable and a multistakeholder process of 'joint learning' might have opened alternatives rather than the politicisation that ensued. In this sense, the road taken was a missed opportunity for hammering out a mutually acceptable deal with all key stakeholders.

Discussion and Conclusion: Taming Floods, Taming People in Flood Security

While the Copenhagen school on security has widened the conceptualisation of 'security', its leaders have tended to remain wedded to the Westphalian state as the ultimate initiator and provider of security. But others can and do also speak security. A crisis, normally the domain of state representatives, can successfully be called by other public speakers' crisis counter-narratives. Natural disasters are human disasters, mediated by land use and people management and planning. While we cannot influence the incidence of earthquakes much, floods are different in that we influence flood risk by building in flood plains and eroding natural barriers. Humans have quite some bearing on both the incidence and consequences of floods through land-use planning. In that sense, Gilbert White's (1978) axiom claiming that floods are 'acts of God' but flood losses are largely 'acts of man' wants qualification: Even the probability of an event is largely socially constructed so that both sides of the risk equation (risk = probability × loss) are man-made.

The asymptotic road to Omsk, the oasis that we never really reach, is symbolic of risk management: We can never arrive at total safety. 'Safe-fail' risk management, currently the dominant approach in flood management, is reflected in the concept of Disaster Risk Reduction (frame 2). It does not seek to kill or chase the wolf, but to domesticate it. This, however, means both that someone has to shoulder the residual risk, and that to reduce the sum total of risk, people need to be tamed. Treating water as a friend, giving it space, and using 'green' technology for natural embankments as practiced in the UK and the Netherlands means to accept certain self-organising, 'chaotic' aspects of the river.

This does not necessarily sit well with local stakeholders. Local protesters claim that multipurpose 'security-plus' projects are safe enough and, in fact, increase the sense of insecurity. This became clear when the environmentalists initially advocated the 'ecological flooding' of polders in the Rhine Basin. A flood is a great opportunity for environmental restoration. Yet when the Ooij polder was slated for controlled flood storage, Gelderland's provincial environmentalist alliance advocated uncontrolled flooding. But the local chapter changed their minds about this 'let it flood' alternative when protests against the uncontrolled flooding mounted. In the Ooij case, both the analysis of the problem and the values leading to a solution turned out to be contested, not only with local stakeholders but also within the expert community. Yet a nonparticipatory mode was opted for, triggering politicised citizen action, where people saw economic gain erode their safety prospects. As the flood safety argument has special resonance in the Netherlands, many river projects on the Dutch rivers Meuse, Rhine, and Ijssel are sold as 'security-plus' projects, while local stakeholders suspect security to be used as a lever for developments for the rich and their elite hobby, environmentalism (e.g., Warner and van Buuren, 2011).

It not only proves difficult for central government to be less controlling, however, but also for Dutch citizens to break an ingrained culture of relying on government for its security. The mutuality or complexity paradigm accepts that up to a degree we all (not just the big bad state or transnational companies) are part of the creation of disaster, in complex and unexpected ways, calling for an extended participatory approach to disaster risk management, as was contemplated in the Ooij case and tried in the Meuse case. Yet as Wesselink and Hoppe (2011) aver, the post-normal approach is often simply a synonym for 'dialogue', failing to contribute to solving problems—maybe deliberately so, as it is in actors' interests to frame a problem type a certain way (Hoppe, 2010), even to frame it as irresolvable (Wesselink and Warner, 2010). In that sense, post-normal science may not necessarily do better than a confrontational, politicised approach (see Warner, 2007).

The illustrative Dutch examples show that the road taken in flood control is not linear, from securitised to desecuritised, or from government to governance, but can rather take more complex, iterative itineraries, with different actors finding themselves in the picture. While it may take a flood or a shock like 9/11 to return to the securitised mode (and the comeback of social engineering), different, opportunistic agendas may also play a part. In the Maas case, for example, both the securitised and post-normal desecuritised phases were revisited within a decade. Hartmann (2010) suggests that the fading memory of flood disaster promotes more recklessness in colonising the flood plain.

The analysis illustrates that the role distribution on the story may change before the tale is through. Before boarding the Omsk-bound troika, we are well advised to check for wolves in sheep's clothing, and vice versa.

Let us, finally, imagine an alternative scenario to the troika tale, in which the travellers safely reached the city. We can imagine Omsk as a city surrounded by thick walls, a moat, cannons, and a defence force against wolves, bandits, and foreign invaders, with a stock of food and goods in case of a siege of the city, to survive scarcity of vital goods. Whoever rules Omsk has the authority over these defence forces and infrastructure. This authority includes the power of decision on what happens in a state of siege in the name of keeping nature out (total fire suppression, total flood control). However, if the city is never attacked, citizens of Omsk will be less willing to respect the authority of its rulers and pay taxes for the upkeep of defence forces and emergency stock, unless the rulers provide other peacetime desirables. It may pay politically to invoke the spectre of the wolf now and then (aided by real or imagined actual events) to legitimise the defence infrastructure. If the wolves were not already there, states (and communities) would have to invent them.[4]

Notes

1. Interestingly, as in other recent disasters such as L'Aquila in Italy, there was an elaborate plan for a flood scenario. New Orleans could have been well prepared, but the information was simply not used.
2. After French philosopher Jean-Jacques Rousseau.
3. Much of this literature is likewise Habermasian (after the Frankfurt scholar Jürgen Habermas), a rationality starting from the assumption that it is possible to overcome institutional and political constraints to arrive at consensus.
4. After Voltaire: if God didn't exist, He would need to be invented.

References

Aradau, C. (2001) 'Beyond good and evil: ethics and securitization/desecuritization techniques', *Rubikon*, December.
Bankoff, G. (2001) 'Rendering the world unsafe: vulnerability as Western discourse', *Disasters*, vol 25 no 1, pp19–35.
Blaikie, P., Cannon, T., Davis, I. and Wisner, B. (1994) *At Risk: Natural Hazards, People's Vulnerability and Disasters*, Routledge, London.

Bryant, R.L. and Bailey, S. (1997) *Third World Political Ecology*, Routledge, London.

Burke, A. (2007) 'What Security Makes Possible. Some Thoughts on Critical Security Studies', Working paper, Australian National University, Department of International Relations, Canberra.

Burton I., Kates, R.W. and White, G.F. (1978) *The Environment as Hazard*, Oxford University Press, New York.

Buzan, B., Waever, O. and de Wilde, J. (1998) *Security. A New Framework*, Harvester Wheatsheaf, Hampstead.

Collins, T.W. (2008) 'The political ecology of hazard vulnerability: marginalization, facilitation and the production risk to urban wildfires in Arizona's White Mountains', *Journal of Political Ecology* vol 15, www.csulb.edu/rodrigue/geog696/ debriefing/3human/tabag_collins.pdf.

Corry, O. (2012) 'Securitisation and "riskification": second-order security and the politics of climate change', *Millennium—Journal of International Studies*, vol 40, no 2, pp235–258.

Crichton, M. (2005) 'Flood Risk and Insurance in England and Wales: Are There Lessons to be Learnt from Scotland?', Technical Paper Number 1, University College London, Benfield Greig Hazard Research Centre, Department of Earth Sciences, London.

De Marchi, B. and Ravetz, J. (1999) 'Risk management and governance—a post normal science approach', *Futures*, vol 31, no 7, pp743–757.

Drs P. (1974) Dodenrit (Troika hier, troika daar). Polydor 7".

Engel, K.E. and Engel, P.G.H. (2012) 'Building resilient communities. Where disaster management and facilitating innovation meet', in A. Wals and P.B. Corocan (eds) *Learning for Sustainability in Times of Accelerating Change*, Wageningen Academic Publishers, Wageningen, Netherlands.

Funtowicz, S.O. and Ravetz. J. (1993) 'Science for the post-normal age', *Futures*, vol 25, no 7, pp733–755.

Green, C. and Warner, J. (1999) 'Emerging Models of Flood Hazard Management'. Paper presented at the Stockholm Water Symposium, August.

Hajer, M.J. (1995) *The Politics of Environmental Discourse: Ecological Modernization and the Policy Process*, Oxford University Press, Oxford.

Hartmann, T. (2010) 'Reframing polyrational floodplains, Land policy for large areas of temporary flood storage', *Nature and Culture*, vol 5, no 1, pp15–30.

Healy, S. (1999) 'Extended peer communities and the ascendance of post-normal politics', *Futures*, vol 31, no 7, pp655–669.

Hilhorst, D. (2003) 'Responding to disasters: diversity of bureaucrats, technocrats and local people', *Journal of Mass Emergencies and Disasters*, vol 21, no 3, pp37–55.

Hisschemoller, M. and Hoppe, R. (1998) 'Weerbarstige beleidscontroverses: een pleidooi voor probleemstructurering in beleidsontwerp en –analyse', in R. Hoppe and A. Pieterse (eds) *Bouwstenen voor een argumentatieve beleidsanalyse*, Elsevier, Den Haag.

Hoppe, R. (1999) 'Policy analysis, science, and politics: from "speaking truth to power" to "making sense together' *Science and Public Policy*, volume 26, number 3, June 1999, pp 201–210.

Immink, I. (2007) *Voorbij de risiconorm. Nieuwe relaties tussen ruimte, water en risico*, PhD dissertation, Wageningen University.

Johnson, C.L., Tunstall, S.M. and Penning-Rowsell, E.C. (2005) 'Floods as catalysts for policy change: historical lessons from England and Wales', *International Journal of Water Resource Development*, vol 21, no 4, pp561–575.

Jongejan, R.B., Jonkman, S.N. and Vrijling, J.K. (2012) 'The safety chain: a delusive concept', *Safety Science*, vol 50, no 5, pp1299–1303.

Lindblom, C. (1959) 'The science of muddling through', *Public Administration Review*, vol 19, no 2, pp79–88.

Maskrey, A. and Romero, G. (1983) *Como entender los desastres naturales*, PREDES, Lima.

Masse, T., O'Neil, S. and Rollins, J. (2007) *The Department of Homeland Security's Risk Assessment Methodology: Evolution, Issues, and Options for Congress*, US Congressional Research Service, RL33858, Washington, DC.

Meissner, R. and Turton, A.R. (2003) The hydrosocial contract theory and the Lesotho Highlands Water Project. *Water Policy*, vol. 5 no 2 pp115–126.

Ophir, A. (2007) 'The two-state solution: providence and catastrophe', *Journal of Homeland Security and Emergency Management*, vol 4, no 1, pp1–44.

Ophir, A. (2010) 'The politics of catastrophization', Research Architecture: A Laboratory for Critical Spatial Practice, http://roundtable.kein.org/node/1094.

Pelling, M. and Dill, C. (2010) 'Disaster politics: tipping points for change in the adaptation of socio-political regimes', *Progress in Human Geography*, vol 34, no 1, pp21–37.

Renn, O. (2008) *Risk Governance. Coping with Uncertainty in a Complex World*, Earthscan, London.

Röling, N. and Woodhill, J. (2001) 'From Paradigms To Practice: Foundations, Principles and Elements for Dialogue on Water, Food and Environment', Background Document for National and Basin Dialogue Design Workshop, Bonn, December 2001, Secretariat for Global Dialogue on Water, Food and Environment.

Rooijendijk, C. (2009) *Waterwolven. Een geschiedenis van stormvloeden, dijkenbouwers en droogmakers*. Amsterdam: Atlas.

Roth, D., Warner, J. and Winnubst, M. (2006) 'Een noodverband tegen hoogwater. Waterkennis, beleid en politiek rond noodoverloopgebieden', Boundaries of Space Series, Wageningen University, Wageningen.

Thompson, M. and Warburton, M. (1985) 'Uncertainty on a Himalayan scale', *Mountain Research and Development*, vol 5, no 2, pp115–135.

van den Brugge, R., Rotmans, R. and Loorbach, D. (2005) 'The transition in Dutch water management', *Reg Environ Change*, vol 5, no 4, pp164–176.

Van Eeten, M.J.G. (1997) 'Sprookjes in Rivierenland: Beleidsverhalen over wateroverlast en dijkversterking', *Beleid en Maatschappij* vol 14, no 1, pp32–43.

Warner, J (2004) 'Water, Wine, Vinegar, Blood. On politics, participation, violence and conflict over the hydrosocial contract (with special reference to the Water War of 2000 in Cochabamba)' in: World Water Council, Proceedings of a seminar on 'Water and Politics', 26–27 February 2004, Marseille.

Warner, J. (ed.) (2007) *Multi-Stakeholder Platforms for Integrated Water Management*, Ashgate, Aldershot.

Warner, J. (2008) 'Emergency river storage in the Ooij Polder—a bridge too far?', *International Journal of Water Resources Development*, vol 24, no 4, pp567–82.

Warner, J. (2011) *Flood Planning. The Politics of water security*, IB Tauris, London

Warner, J. (2012) 'Three lenses on water war, peace and hegemonic struggle on the Nile', *International Journal of Sustainable Society*, vol 4, no 1/2, pp173–193.

Warner, J. (2013) 'The politics of 'disasterisation': changing contracts?', in D. Hilhorst (ed) *Disaster, Conflict and Society in Crises. Everyday Politics of Crisis Response*, Routledge, London.

Warner, J. and van Buuren. M.W. (2011) 'Implementing space for the river: narratives of success and failure in Kampen, the Netherlands', *International Review of Administrative Sciences*, vol 72, no 3, pp395–415.

Wesselink, A. (2009) 'Hydrology and hydraulics expertise in participatory processes for climate change adaptation in the Dutch Meuse', *Water Science & Technology*, vol 60, no 3, pp583–595.

Wesselink, A. and Hoppe. R. (2011) 'If post-normal science is the solution, what is the problem? The politics of activist environmental science', *Science Technology Human Values*, vol 36, pp389.

Wesselink, A. and Warner, J. (2010) 'Reframing floods: proposals and politics', *Nature and Culture*, vol 5, no 1, pp1–14.

White, G.F. (1978) *Natural Hazards: Local, National, Global*, Oxford University Press, Oxford.

19 Household Water Security and the Human Right to Water and Sanitation

Jonathan Chenoweth, Rosalind Malcolm, Steve Pedley, and Thoko Kaime

Introduction

There are varying degrees of access to water and sanitation, and thus, in many developing country cities, it is difficult to determine whether individual households achieve water security at the household level. In such cities, it is the poor who are most likely to depend on poor quality water and sanitation services and, thus, are most likely to suffer from water insecurity at the household level. Here we explore how the human rights to water and sanitation are addressed in the national water laws of two countries (Ethiopia and Kenya) where water and sanitation systems fail to achieve universal coverage and thus household water insecurity is a significant problem.

According to the Joint Monitoring Programme of the World Health Organisation (WHO) and United Nations Children's Fund (UNICEF), 89% of the world's population in 2010 had access to an improved water source and, of these, 54% had access to a piped water supply on their premises; 11%, or 780 million people globally, however, lacked access to improved water supplies. Furthermore, 63% of the global population were served by improved sanitation systems (WHO/UNICEF Joint Monitoring Programme for Water Supply and Sanitation, 2012), meaning that far more people lack access to improved sanitation than water. Looking regionally, while developed countries had achieved near-universal coverage of improved water and sanitation services, in sub-Saharan Africa only 61% of people had access to improved water supplies, and a mere 16% to piped water, while only 30% had access to improved sanitation (WHO/UNICEF Joint Monitoring Programme for Water Supply and Sanitation, 2012). Although the world appears to have met the Millennium Development Goal of halving by 2015 the proportion of people lacking access to safe drinking water globally, it is likely to fail to achieve the equivalent sanitation target globally (United Nations, 2011a). It will almost certainly fail to meet both the water supply and sanitation targets in sub-Saharan Africa. Thus, despite considerable progress being made at meeting water and sanitation needs, many households are insecure in terms of basic access to water and sanitation.

This chapter focuses on water security at the household level. Grey and Sadoff (2007, p548) give a general definition of *water security* as 'the availability of an acceptable quantity and quality of water for health, livelihoods, ecosystems and production, coupled with an acceptable level of water-related risks to people, environments and economies'. Applying the concept of water security at the household scale would suggest that water security means ensuring households have a sufficient quantity of water of sufficient quality to maintain the health of household members. Household water security is closely tied to sanitation provision, since achieving household access to acceptable quality water for health and livelihoods generally requires sanitation provision also; a lack of sanitation in a community makes the provision of safe water supply in that same community difficult to achieve.

Water security at the household level is tied to the concept of human rights since human rights deal with the maintenance of individual health and well-being, and now also include access to water and sanitation. The concept of a human right denotes an entitlement that is fundamental and inalienable. It is universal to all—everyone is entitled to human rights simply by virtue of being human. A human right can only be disregarded in law when it conflicts with other, very limited, duties on a state, such as when there is a state of emergency. The modern concept of human rights goes beyond classical ideas of justice or political or cultural legitimacy and thus elevates the status of household water security to a right that states have a duty to protect regardless of other factors (United Nations Human Rights Council, 2010).

This chapter begins by addressing the issue of household water security as a human right in international law, expressed as a right to water and sanitation. Then, via the case studies of Kisumu, Kenya, and Addis Ababa, Ethiopia, it explores how these human rights are addressed in the national water laws of two countries where water and sanitation systems fail to achieve universal coverage. The chapter then goes on to examine how households in these case study countries meet their water and sanitation needs despite the poor quality of services received. The chapter concludes by asking how the legal systems in the case study countries can practically facilitate better water security at the household level for the poor.

Kisumu and Addis Ababa were chosen as case studies because they show how the human right to water is being dealt with in two low-income developing countries where available resources for investment in the water sector are limited, but where both countries have explicitly acknowledged that there is a right to water and sanitation either in their constitution or in key national legislation. Kisumu is a city of approximately half a million people on Lake Victoria, while Addis Ababa is a city of 3 million built on hilly terrain. Both Kenya and Ethiopia are classified as having a low level of human development (United Nations Development Programme, 2010), while according to the World Bank, Kenya and Ethiopia are classed as low-income countries (World Bank, 2011).

The Right to Water and Sanitation

While a range of human rights have been formally identified through an extensive body of international agreements and resolutions, the most important of which is the 1948 Universal Declaration of Human Rights, none of these early classical expressions of human rights (with the possible exception of the later African Charter on Human and Peoples' Rights) have contained explicit statements of rights pertaining to the environment in general, or water and sanitation in particular. The African Charter, developed and agreed by the African states, which are members of the African Union, declares in Article 24 that 'all peoples shall have the right to a general satisfactory environment favourable to their development' (African Commission on Human and People's Rights, 1981). By contrast, the 1948 Universal Declaration on Human Rights states that '[e]veryone has the right to a standard of living adequate for the health and well-being of himself and of his family, including food, clothing, housing and medical care and necessary social services' (United Nations, 2011b). The right to an adequate standard of living has been further affirmed by subsequent United Nations declarations and treaties. The most notable of these is the 1966 International Covenant on Economic, Social, and Cultural Rights. Article 11 of this convention provides for a right to an adequate standard of living, while Article 12 establishes a right to health. So, it can be argued that the right to water and sanitation are implied within the content of these documents, since access to both are key determinants of health and well-being. Additionally, the majority of states that ratified the International Covenant on Economic, Social and Cultural Rights have subsequently reaffirmed in political declarations that the right to an adequate standard of living necessarily also includes water and sanitation.

The first formal direct recognition of a right to water came in 1977 at the United Nations Water Conference when it was stated in the Conference's Action Plan that all people 'have the right to have access to drinking water in quantities and of a quality equal to their basic needs' (UN-Water Decade Programme on Advocacy and Communication, 2011). The United Nations General Assembly in its resolution A/Res/54/175, The Right to Development, in 2000 affirmed that the right to clean water was a fundamental human right, and in General Comment No 15 (2002) formulated this right under Articles 11 and 12 of the International Covenant on Economic, Social and Cultural Rights. But it was on July 28, 2010, that the United Nations General Assembly took the important step of directly recognising that the human rights to water and sanitation were fundamental human rights in its resolution 64/292, The Human Right to Water and Sanitation (United Nations General Assembly, 2010). Two months later, the United Nations Human Rights Council adopted resolution 64/292, The Human Right to Water and Sanitation, and clarified the foundation for recognition of the right and the associated legal obligations that relate to this right (United Nations Human Rights Council, 2010). Thus, it is clear that the human rights to water and sanitation are fully accepted under international law as basic human rights.

While virtually all governments agree in principle that the universal right to water and sanitation exists, there is less consensus about what the fulfilment of this right means in practice. Tying the concept of the right to water and sanitation to the principle of water security, and the Grey and Sadoff (2007, p548) definition of water security noted previously, would suggest that ensuring that all people have access to 20 litres per capita per day is insufficient. (This is the amount which is specified by the WHO and UNICEF in their global assessment of water supply as the minimum amount required for a person to be deemed to have access to improved water supply; (WHO/UNICEF Joint Monitoring Programme for Water Supply and Sanitation, 2011.) Such an amount would not allow the fulfilment of other basic human rights, such as the right to work and the right to sufficient food, and it is doubtful that 20 litres per capita per day would allow health requirements to be fully met. Chenoweth (2008) argues that estimates of minimum basic water requirements must consider minimum requirements for domestic household use for activities such as drinking, cooking, and washing, as well as minimum water requirements for water-efficient economic activities to provide employment. He suggests on this basis that 85 litres per capita per day are sufficient for basic domestic water requirements at the household level, and 120 litres per capita per day if economic activities are included also. Achieving household water security and fulfilling the human right to water and sanitation means achieving more than the basis access level assessed by the WHO and UNICEF.

Case Studies of National Law on Water, Sanitation, and Human Rights

Given that virtually all governments accept in principle that there is a human right to water and sanitation, it is not surprising that this right is specified legally in some countries, either in the constitution or via key national legislation. The South African constitution, for example, states in Section 27.1 (b) that everyone has the right to have access to sufficient food and water (Republic of South Africa, 2009). The South African Water Services Act (1997) expands on the constitution by stating in section 3 that, 'Everyone has a right of access to basic water supply and basic sanitation' and that 'Every water services institution must take reasonable measures to realise these rights' (Republic of South Africa, 1997). Thus, a duty to ensure the realisation of the human right to water and sanitation in South Africa is placed on the water services institutions of the country.

Kisumu, Kenya

Like South Africa, Kenya has also acknowledged the right to water and sanitation. The 2010 constitution of Kenya states, in Article 43, that '[e]very person has the right . . . to reasonable standards of sanitation . . . [and] to clean and safe water in adequate quantities' (Government of the Republic of

Kenya, 2010). However, the current legal regime for water resources management in Kenya dates back to reforms that began with the publication of the National Policy on Water Resources Management and Development as Sessional Paper No. 1 of 1999 (Ministry for Water Development, 1999). The significant reform proposed in this water policy paper was changing the focus of government away from being a direct provider and developer of water resources to that of regulator and policymaker. This policy paper led to the creation of the Water Act 2002, which has been in force since March 18, 2003.

Section 49 of the Kenyan Water Act 2002 requires the Minister for Water to publish a National Water Services Strategy, with one of the objectives of the strategy '(a) to institute arrangements to ensure that all times there is in every area of Kenya a person capable of providing water supply and (b) to design a programme to bring about progressive extension of sewerage to every centre of population in Kenya' (Government of Kenya, 2002). The National Water Services Strategy is required to document areas underserved with water and sanitation services and to contain details of plans for extension of services to these areas, including timeframes and investment plans.

The National Water Services Strategy (2007–2015) produced by the Kenyan Ministry of Water and Irrigation states that the overall goal of the strategy is 'to ensure sustainable access to safe water and basic sanitation for all Kenyans' (Ministry of Water and Irrigation, 2007, p13). The strategy states that sustainable access to safe water and sanitation is a human right, with water and sanitation provision for the poor to be enabled by social tariffs ensuring at least 20 litres per person per day. In terms of specific targets, the strategy sets the following as goals to be achieved by 2015: to increase access to safe drinking water from 60 to 80% of the population of urban areas and from 40 to 75% in rural areas, and to increase access to waterborne sewage services from 30 to 40% in urban areas, and 5 to 10% in rural areas (Ministry of Water and Irrigation, 2007). Access can be via communal access points as well as private connections.

One of the major reforms of this act was the separation of water resources management and water services supply, with the establishment of Water Services Regulatory Boards, which were charged with regulating the supply of water, and Water Services Boards, which own the supply system within their specified jurisdictions (Government of Kenya, 2002). According to Section 55 of the Water Act 2002, a Water Services Board may exercise its powers and functions under licence via one or more Water Service Providers. Thus, delegation to municipal authorities and commercially orientated autonomous bodies as part of a strategy for increasing access is permitted under the Kenyan Water Act 2002 and is also encouraged by the 2007 National Water Services Strategy.

In the city of Kisumu, Kenya, the Kenyan government has established the Lake Victoria South Water Services Board, which is a public corporation established to develop water supply facilities in the Lake Victoria South

Basin and is a licensee of the Water Act 2002. In line with this Act, the Lake Victoria South Water Services Board has contracted the Kisumu Water and Sewerage Company (KWASCO) to operate the water and sewerage system and, thus, provide water and sewerage services to residents. It was estimated that in 2008–2009, KWASCO had achieved 29% coverage with its water supply network, up from 26% in 2006–2007, and 6% coverage for sanitation, up from 5% in 2006–2007 (Water Services Regulatory Board, 2008, 2010).

KWASCO is estimated to provide water services to 153,083 people out of a population of 525,313 via 14,084 connections; thus, an average of 10.9 people are being served by each KWASCO water connection, receiving an estimated 42 litres per capita per day (excluding unaccounted for water) (Water Services Regulatory Board, 2010). Accurately estimating water coverage is problematic, as surveys of households in Kisumu show that most households without a direct water connection to KWASCO's network get their water from a variety of sources (Okotto et al., 2010).

Thus, the majority of the population in the city of half a million rely on some form of intermediate or independent water provider, such as standpipe operators reselling water from KWASCO under licence, resellers operating without a resellers licence, or private well and borehole operators who have dug a well or sunk a borehole on their land and sell the water. Standpipe operators, unlicensed resellers, and well and borehole operators all sell directly to consumers, as do as mobile water vendors who deliver water directly to households either by handcart or tanker truck. In addition, there is one small mini-utility on the edge of Kisumu which is operating under licence for the Lake Victoria South Water Services Board. This mini-utility is a small privately owned water treatment works which supplies local residents and tanker trucks with treated drinking water.

With the exception of the licensed standpipe operators and the mini-utility, these independent and intermediate water providers are operating illegally under Kenyan law. The Water Act 2002 does not expressly deal with independent and intermediate water sellers, but prohibits the supply of water in excess of specified quantities without a licence. The Water Act 2002 in Section 56 (Requirement for License) states:

> No person shall, within the limits of supply of a licensee (a) provide water services to more than twenty households; or (b) supply—(i) more than twenty-five thousand litres of water a day for domestic purposes; or (ii) more than one hundred thousand litres of water a day for any purpose, except under the authority of a license. A person who provides water services in contravention of this section shall be guilty of an offence.

Given these limits, most well and borehole operators, as well as mobile water resellers, operate illegally. Inevitably due to the illegal nature of their operation, the quality or price of water supplied by intermediate and independent water vendors is not regulated and, thus, the safety of the water supplied is uncertain.

In the low-income and high-density suburbs of Kisumu, pit latrines are the predominant form of sanitation. The groundwater level in these suburbs rises and falls with the different seasons, but at no time is it particularly deep. As a consequence, the vertical separation between the bottom of the pit latrines and the groundwater is very small, even when the groundwater level is at its lowest. During the rainy season, the groundwater level rises by enough to inundate the latrines, which creates a pathway for the contents of the latrine to diffuse into the groundwater and then disperse over a wide area. Since the distance between the latrines and the private well and borehole used for water supply can be very small—often just a few metres—the latrines represent a significant contamination risk (Wright et al., 2011). But subsurface transport of contaminants is not the only risk to point water supplies in Kisumu; the poor quality of construction and lack of adequate protection around the wells makes surface contamination of the water inevitable. Surveys of the water quality confirmed these assumptions, recording levels of contamination by faecal indicator bacteria several orders of magnitude above the most relaxed water quality guideline value of the WHO (Okotto et al., 2010). The incidence of water-related diseases in these suburbs is thus high, and the population is particularly vulnerable to sudden and widespread outbreaks of waterborne disease. It is clear that water security at the household level is not being met for many residents in Kisumu.

Addis Ababa, Ethiopia

According to Article 90 of the Ethiopian Constitution, 'To the extent the country's resources permit, policies shall aim to provide all Ethiopians access to . . . clean water', while Article 92 states that the '[g]overnment shall endeavour to ensure that all Ethiopians live in a clean and healthy environment', and '[g]overnment and citizens shall have the duty to protect the environment' (Government of the Federal Democratic Republic of Ethiopia, 1994).

The Water Resources Management Policy reiterates the aim that every Ethiopian citizen should have access to sufficient water of acceptable quality to satisfy basic human needs; it also prioritises water for domestic water supply and sanitation above other water uses (Ministry of Water Resources, 2001). The overall objective of the policy is the provision of adequate, reliable, and clean water supply and sanitation services to the Ethiopian people, as well as the provision of water for economic uses. The Ethiopian Water Resources Management Proclamation 2000 also declares that water resources of Ethiopia are to be used for their highest social and economic value, with domestic use having priority over any other use (Government of the Federal Democratic Republic of Ethiopia, 2000).

Within Addis Ababa, the Addis Ababa Water and Sewerage Authority (AAWSA) is the department of the Addis Ababa city government responsible for water supply and sanitation within the city. It was re-established in 1995 with the purpose of supplying safe and adequate water and sanitation to the city (Government of the Federal Democratic Republic of Ethiopia, 1995). It has been estimated that 98% of the water consumed in Addis Ababa can be traced back to water provided by AAWSA; however, the proportion of households with their own connection to AAWSA's network is estimated to be around 39%. Seventy-four percent of households in Addis Ababa rely on a pit latrine, with only 17% having a flush toilet, and 7% defecating in the open (Central Statistical Authority, 2005).

In Addis Ababa, while 98% of water consumed in the city can be traced to AAWSA, a significant proportion of the city's population rely on intermediate water providers who resell water purchased from AAWSA (Ayalew et al., in press). These intermediate water resellers include city government–administered public standpipes and community group–owned and –operated public standpipes that operate under license from AAWSA. The most common forms of intermediate water providers, however, are neighbour resellers and unofficial water kiosks, who do not operate under a contractual relationship with AAWSA that permits reselling, and mobile water vendors who transport water from areas of the city with running water to areas without.

Discussion: Improving Household Water Security in Kisumu and Addis Ababa

While there are legal and institutional frameworks in place in Kisumu and Addis Ababa for achieving household water security, results on the ground are still falling a long way short of what is required. The services of the official water and sanitation authorities do not provide water security as defined by Grey and Sadoff (2007, p548), and the independent and intermediate water and sanitation providers operating in either city also do not currently allow the achievement of household water security. Even with these unofficial water and sanitation providers, many households are not able to access water of a sufficient quantity and quality to satisfy their needs. Despite this, the independent and intermediate water and sanitation providers are playing a significant role in bridging the gap between what the official (municipal systems) provide and what households require in terms of water and sanitation. They thus have a role to play in improving water security.

The water quality study conducted by Okotto et al. (2010) showed that mobile water resellers in either Kisumu or Addis Ababa do not cause significant deterioration in water quality, with any such deterioration occurring primarily during household storage in the case of water sourced from the municipal piped network. Water vendors effectively extend the coverage of the municipal water system in both cities into underserved and unserved areas, with the advantage of being able to compensate for the intermittent municipal supply.

In the case of Kisumu, while water drawn from boreholes and wells is significantly lower in quality than is municipal supplied water, and therefore not the preferred source for drinking, this well and borehole water plays an important role in bringing total average water consumption in the city up to an acceptable level, where it is used for nonconsumptive purposes. Thus, this water is important for household water security from a quantitative perspective. According to the household surveys, whereas average daily water consumption in Addis Ababa is only 14.6 litres per capita per day, in Kisumu it is 34 litres per capita per day (Okotto et al., 2010). At the same time, water quality from many of the private wells and boreholes in Kisumu is poor, and since this water is cheaper for mobile water resellers to purchase but has the same aesthetic qualities as the municipal water, households face uncertainty about the origin, quality, and safety of the water they do not purchase personally directly from a public standpipe. Regulation is therefore needed to ensure that water resellers follow best-practice guidelines for their operations, with routine water quality testing and random inspections introduced to improve consumer information about the quality of the water supplied and to prevent water resellers passing off the lower quality well water as the more expensive but higher quality municipal water.

Conclusions

The nature of a human right elevates concepts of water security to fundamental, universal, and inalienable rights which impose duties on governments. In terms of actually achieving household water security, however, the picture on the ground in Kisumu and Addis Ababa, as in many developing country cities, shows that household water security has not yet been achieved, and even assessing the extent of progress is far more complicated than the assessments relating to achieving the human right to water and sanitation generally suggest. There are varying degrees of access to water and sanitation, and thus it is difficult to determine whether individual households have achieved their human right to safe water and sanitation, and thus the extent to which household water security is being fulfilled across both cities.

Clearly, the water and sanitation services most residents receive in Kisumu and Addis Ababa are completely inadequate by developed country standards, and are in breach of any concept of human rights in practice understood and realised in such countries. There is no prospect of this changing in the near future. In the near future, water security at the household level can perhaps most effectively be advanced through better regulation of independent and intermediate water providers. These providers need to be formally recognised and regulated so uncertainty about the safety of the water they supply can be reduced, thus improving access to safe water even if this water is not supplied directly via the official piped water network. Such a relatively low-cost step would go a long way to improving water security at the household level in both cities, since so many households depend upon their services.

References

African Commission on Human and People's Rights (1981) *African Charter on Human and People's Rights*, African Commission on Human and People's Rights, Banjul.

Ayalew, M., Chenoweth, J., Malcolm, R., Mulugetta, Y., Okotto, L. and Pedley, S. (2014) 'Why regulate the independent water providers? A regulatory framework for privatized water vendors in Kenya and Ethiopia', *Journal of Environmental Law, vol* 25(1), in press.

Central Statistical Authority (2005) *Welfare Monitoring Survey 2004*, Federal Democratic Government of Ethiopia, Addis Ababa.

Chenoweth, J. (2008) 'Minimum water requirement for social and economic development', *Desalination*, vol 229, pp245–256.

Government of Kenya (2002) *Water Act 2002*, Government of Kenya, Nairobi.

Government of the Federal Democratic Republic of Ethiopia (1994) *Constitution of The Federal Democratic Republic of Ethiopia*, International Constitutional Law Project, www.verfassungsvergleich.de/, accessed 28 October 2011.

Government of the Federal Democratic Republic of Ethiopia (1995) 'Addis Ababa Water and Sewerage Authority Re-establishment Proclamation', *Proclamation No. 10/1995*, Federal Negarit Gazeta, Addis Ababa.

Government of the Federal Democratic Republic of Ethiopia (2000) 'Ethiopian Water Resources Management Proclamation', *Proclamation No. 197/2000*, Federal Negarit Gazeta, Addis Ababa.

Government of the Republic of Kenya (2010) *The Constitution of Kenya*, International Constitutional Law Project, www.verfassungsvergleich.de/, accessed 28 October 2011.

Grey, D. and Sadoff, C. W. (2007) 'Sink or swim? Water security for growth and development', *Water Policy*, vol 9, pp545–571.

Ministry for Water Development (1999) 'Sessional paper No. 1 of 1999 on National Policy on Water Resources Management and Development', Government of Kenya, Nairobi.

Ministry of Water and Irrigation (2007) The National Water Services Strategy 2007–2015', Ministry of Water and Irrigation, Nairobi.

Ministry of Water Resources (2001) 'Ethiopian Water Resources Management Policy', Government of the Federal Democratic Republic of Ethiopia, Addis Ababa.

Okotto, L., Ayalew, M. M., Chenoweth, J., Malcolm, R., Pedley, S. and Mulugetta, Y. (2010) 'The establishment of legal frameworks for independent and small-scale water providers', Report presented at case study workshops in Addis Ababa, Ethiopia, 22 May 2009 and Kisumu, Kenya, 27 May 2009, *Surrey Law Working Papers*, School of Law, University of Surrey, Guildford.

Republic of South Africa (1997) *Water Services Act, 1997*, South African Government Information, Pretoria.

Republic of South Africa. (2009) *Constitution of the Republic of South Africa, 1996*, South African Government Information, Pretoria, www.info.gov.za/documents/constitution/index.htm, accessed 26 October 2011.

UN-Water Decade Programme on Advocacy and Communication (2011) *The Human Right to Water*, United Nations Office to Support the International Decade for Action 'Water for Life' 2005–2015, Zaragoza.

United Nations (2011a) *The Millennium Development Goals Report 2011*, United Nations Secretariat, New York.

United Nations (2011b) *The Universal Declaration of Human Rights*, United Nations, New York, www.un.org/en/documents/udhr/, accessed 26 October 2011

United Nations Development Programme (2010) *Human Development Report 2010*, Oxford University Press, New York.

United Nations General Assembly (2010) 'Resolution adopted by the General Assembly: 64/292 The human right to water and sanitation' United Nations General Assembly, New York.

United Nations Human Rights Council (2010) 'Resolution adopted by the Human Rights Council: 15/9 Human rights and access to safe drinking water and sanitation', United Nations Human Rights Council, New York.

Water Services Regulatory Board (2008) 'Impact: A performance report of Kenya's water services sub-sector', Water Services Regulatory Board, Nairobi.

Water Services Regulatory Board (2010) 'Impact: a performance report of Kenya's water services sub-sector', Water Services Regulatory Board, Nairobi.

WHO/UNICEF Joint Monitoring Programme for Water Supply and Sanitation (2011) *Drinking Water: Equity, Safety and Sustainability*, WHO/UNICEF, Washington, DC.

WHO/UNICEF Joint Monitoring Programme for Water Supply and Sanitation (2012) *Progress on Drinking Water and Sanitation: 2012 Update*, WHO / UNICEF, Washington, DC.

World Bank (2011) *Data: Countries and Economies*, World Bank, Washington, DC, http://data.worldbank.org/country, accessed 26 October 2011.

Wright, J. A., Okotto, J., Yang, H., Pedley, S. and Gundry, S. (2011) 'Understanding the inter-relationship between millennium development goal targets: a spatial analysis of pit latrine density and groundwater source contamination', *American Association of Geographers Annual Meeting*, American Association of Geographers, Seattle.

Part IV

Conclusion

20 Food-Water Security

Beyond Water Resources and the Water Sector

J. A. (Tony) Allan

Introduction

The purpose of this chapter is to highlight those who manage food-water and who, therefore, actually deliver sustainable water security. It will emphasise the contributions of agents in the food supply chain, mainly farmers, who are the de facto managers of food-water. The term *food-water* is introduced to focus attention on the 90% of water needed by an individual for their food consumption.[1] The other 10%—water needed for domestic and industrial uses—will be called *non-food-water*. Food-water can be sourced from green—that is, root-zone—water, accounting for about 70% of food production, or from blue surface or ground waters that account for about 30% of food production. Non-food-water has to be blue water sourced.

It is shown that societies and their governments and markets have made it impossible for farmers—society's water managers—to prioritise effectively our collective water security. The global significance of water in food supply chains is gaining currency, but for the most part the significance of food-water is economically and politically invisible (Allan, 2010, 2011; Larson, 2011; Segal and MacMillan, 2009; Sojamo 2010). Most importantly, the significance of food-water is invisible to food consumers.

Current national and global political economies of food production steer farmers towards water managing practices that do not ensure food-water security or towards the increasingly necessary stewardship of ecosystem services. This danger is especially relevant to water ecosystem services. Farmers are major agents in the food supply, as they manage 90% of food-water. They know their contribution to food security, but even the farmers are unaware that they are the world's strategic water managers. In addition, other agents in the food supply chain, the major food commodity traders, the major food retailers, as well as all the consumers of water-intensive food commodities, are equally ignorant of the ways that consumer food choices and behaviour, and the way the ag industries help to meet them, affect water security. Food traders and food manufacturers convey, preserve, and add value to food commodities. They have little impact on water security, as they themselves use negligible volumes of water compared with farmers. But they are very prominent and could be very influential in influencing the way water is managed and ecosystem services are protected. These corporations are potentially the main messengers. They could highlight the importance of the key players—the farmers—in improving water productivity. They could also highlight the importance of the *sustainable* intensification of water use, plus water ecosystem stewardship.

Policymakers in public policymaking and regulatory agencies have also not yet grasped the vital contributions made to water security by the agents in the food supply chains in private sector markets. They have neither identified the importance of the delivery of high returns to water on farms, nor are they aware of the equally vital role that farmers play in the stewardship of water and in underpinning the ecosystem services of water.

The chapter will first, very briefly, identify the underlying hydro and economic fundamentals of water security. It will review the nature of the endowments of green water and blue water. Secondly, it will examine the forces driving and shaping demand—demography, socioeconomic development, climate change, and cheap food policies. Third, the roles of farmers, traders, food-processing corporates, supermarkets, and consumers in the food supply value chain will be highlighted. It will be shown that these players beyond the water sector determine whether political economies are food and water secure.

Important Underlying Fundamentals and Assumptions: Food-Water and Non-Food-Water

> [S]tatistics on the productivity of green and blue water use are much more important than raw numbers on water endowments per head of population.
>
> Poverty determines water poverty: water poverty does not determine poverty.
>
> The big truth we have surely discovered is this: a country's water endowment is less relevant to its water security than the strength of its economy. Having water is not what counts; how an economy uses its water is the crucial factor. (Allan, 2011b, p293)

Wherever you irrigate, you will always run out of blue water.

The purpose of this section is to draw attention to the political economy processes beyond the normal concerns of water sector professionals and scientists that have already, and will in the future, determine water security. The analysis will focus on food-water—on the massive volumes of water needed by society (about 90%)—that is managed by farming communities who feed society. It will also highlight the importance of the stewardship of water. Good water stewardship is also mainly achieved by farmers because it is farmers who use the massive volumes of water in producing the food that they sell into food supply chains.

The analysis will not consider non-food-water—that is the blue water used for domestic and industrial services. Very few communities in the world lack the small volumes of blue water needed for domestic uses. These uses account for about 5% of water needed by a community. Another 5% or so is required for their industrial uses. Communities are not non-food-water insecure because they lack water and access to it. If communities do not have these water services—which require proportionately very small volumes of water—it is because they do not have diverse economies and enhanced human capital. They also lack financial capital and investing skills, as well as effective engineering, commercial, and regulatory capacities. One of the few economies that faces the extreme predicament of having only 5% of the water needed, Singapore, has become a very diverse and advanced OECD economy. It has demonstrated that extreme food-water and non-food-water poverty does not determine poverty. Poverty and the associated absence of the suite of capacities to remedy poverty determine non-food-water poverty.

Food-water security and non-food-water security depend partly on local green and blue water endowments. Local and global water security, in turn, mainly depends on the extent to which these endowments are allocated and managed effectively. In practice, these local endowments are usually inadequate to underpin local food-water security, even when managed very effectively. It has become normal for most economies in the world to operate in food-water deficit. Increasing water demands driven by rising populations frequently exceed local water endowments. Nor can farmers yet reverse the serious erosion of local water ecosystem services. About 160 out of 210 economies worldwide are net food—and, therefore, embedded water—'importers'. They rely on the global systems. Being a food-water deficit economy is certainly normal.

Water resources in water surplus economies have—for millennia—been remedying the water deficits of water-scarce regions. The remedy is delivered by internal and international trade. Embedded water can be 'imported'[2] in food commodities water, together with notional ecosystem services of water. The direction of such trade is not determined by water endowments, but by capacities associated with socioeconomic advancement and the ability to optimise opportunity costs. Outcomes are the result of how farmers in individual economies combine their water endowments with other inputs (Reimer, 2012, p138).

Farmers are the key players in achieving food-water security. They have, for example, worked water productivity miracles—especially in the past 200 years. Since 1800, they increased green water productivity in wheat production by 10 times on the rain-fed farms of temperate northwest Europe. Clearly, statistics on the productivity of green and blue water use are much more important than the almost meaningless raw numbers on water endowments per head of population. In northwest Europe, the area cultivated and the water used after two centuries by farmers was much the same. Farmers did not find more water. They multiplied the tonnage of the crops produced by 10 times with the same green water (Allan, 2011a, p121).

Hydrological data are not very helpful in defining water security. Heroic efforts have been made for decades to quantify the water resources of nations and river basins (FAO Aquastat, 2012). These datasets, unfortunately, do not help with the analysis of water security, for two reasons. First, they focus too narrowly on the apparent supply and demand for blue water. Secondly, they do not capture green/root-zone water's inputs to food production or, most importantly, on the returns to green and blue water on farms. Water and food scientists and managers have started to consider the wider hydrology of food production that includes root-zone water (FAO Aquastat, 2012; IWMI 2007, p7). But they have not acknowledged their role in keeping in place for a century or more misleading assumptions and datasets unhelpful both for science and for policymaking. The point is emphasised here because the partial approaches of hydrologists have delayed by decades the adoption of an understanding of food-water security based on the actuality of the political economy of food-water.

Indicators such as volumes of water available per head are not irrelevant, but they are certainly not sufficient to underpin a proper conceptualisation of water resource security. The past two centuries have witnessed increases in the productivity of water on the farms of both emerging and industrialised economies that have been food-water securing. At the same time, it has also become clear that where blue water has been a significant input, it has always been over-allocated. The use of the word *always*[3] is very uncomfortable for the politicians and engineers unwittingly complicit in the over-allocation of blue water during the hydraulic mission era. The voice given by society to the water environment came too late to save the rivers that do the heavy lifting of irrigated crop production. These include the Nile, the Tigris Euphrates, the Jordan, the Aral Sea rivers, the Indus, the Colorado, the Red River, and the Murray-Darling River. They all flow from areas of substantial rainfall into semi-arid regions or deserts, where the temptation to avail of the 30°C temperatures at which most food crops prosper proved too irresistible. It was not until the 1980s that governments of neoliberal industrialised economies of California, Israel, and Australia began to listen to the precautionary messages of the NGOs and grapple with the extremely difficult challenge of reducing allocations of natural water to agriculture so that water could be put back into the environment (Gilmont, 2013).

Farmers are the key players, as they have the capacities to sustainably intensify, or not, local water endowments. At this point in the history of irrigated agriculture, they have been asked already, in neoliberal industrialised economies, to change their approach by becoming stewards of blue water. Irrigators worldwide will in due course become good stewards of blue water. The local economies in which farmers operate are also very important, for other reasons. Where economies are diversified, strong farmers can participate in global food commodity trade. International trading systems have, for millennia, proved to be able to meet the needs of the water and food scarce, both locally and internationally. It will be concluded that a version of this approach underpinned by principles of sustainable intensification will deliver future water and food security, provided the further intensification of natural resource use is well stewarded.

Society revisits the issue of natural resource security at least once a century. Malthus raised the spectre of the natural resource gap in 1800, when the global population was about 1 billion. By 1950, the jury was still out on whether we were natural resource secure. Farmers had managed to keep abreast of rising global food demand to meet the food needs, by then about 2.5 billion. By 1990, the population had more than doubled again. Farmers were still keeping pace, albeit at a significant cost by then to the blue water environment. In the early decades of the 21st century, we are revisiting the question of water security again.

The aim of this scene setting has been to highlight a wider suite of underlying assumptions—relating to farming, economics, and politics—than those that are the usual focus of the analysis water security—namely, natural endowments and increasing demand for these natural resources.

Farmers and the Food Supply Chain

Since the dawn of Neolithic farming over 13,000 years ago, food security and water security have almost wholly depended on how effectively farmers have used green water. This green, root-zone water, used for rain-fed crop and livestock production, at first provided sufficient food for subsistence farming families as well as for the small urban populations of the past.

The past two centuries of rapid urbanisation have been very different. They have witnessed two water resource and farm-related production and productivity miracles. First, in some temperate regions, such as northwest Europe, rain-fed yields of wheat per hectare have risen by as much as 10 times. Rainfall has not changed significantly. The tonnage of crops produced has not required 10 times the water. Very importantly, the impact of these increases in the productivity of rain-fed farming have not seriously damaged the ecosystem services of green water.

What is food-water security when it depends so much on the green water–managing practices of farmers rather than on the volumes of blue water mobilised by nature or hydraulic engineers?

The second farmer and food-water related production miracle has been the management of blue water. This additional food production is a consequence of investment since the mid-19th century in hydraulic structures that store and distribute blue water. For the first 100 years of this hydraulic mission, it was usually public investment alongside social and other investments by village communities. In the past five decades, while investment has been mainly public, there has also been a great deal of private investment, especially in on-farm investment in groundwater infrastructures.

This second farmer- and food-related productivity miracle has resulted in irrigated farming being associated with about 30% of the food produced worldwide. Nowhere are there sound data on the proportions of blue water associated with this irrigated crop and livestock production. We have not fathomed how to estimate accurately across river basins the proportions of blue water used in irrigated crop production. The proportions of blue and green in the total consumptive use for crop evapotranspiration vary greatly from year to year. Blue water applications can account for very varying proportions of the water input. It can vary from the 100% levels in Egypt where over one meter depth of blue water per year is applied, to the 10% levels of application—of only 150 mm depth—for the minor, but essential supplementary applications in southwest Michigan in the United States and in East Anglia and in Lincolnshire in the UK.

Irrigators use blue water. Blue water is the water diverted from rivers and pumped from groundwater aquifers. The only use to which green water can be put is for crop and livestock production and to provide some ecosystem services. Blue water, diverted by irrigators, by contrast, often competes with uses that bring very much higher economic returns. It competes with water for manufacturing and other services. Blue water is always reallocated from food production to higher value social uses such as domestic water services, as well as to a wide range of other mainly urban water services.

A major and very important difference between green and blue water is in their vulnerability to over-allocation. It is difficult to over-allocate green water. Green water in the soil profile cannot be raided in a drought year. Green water tends to look after itself. The impacts of farming on the root-zone water environment need not be serious and negative. Negative impacts can certainly be avoided and frequently can be reversed.

Blue surface and ground water are different. They cannot look after themselves. Both natural and engineered blue water are always drawn upon beyond natural recharge. If this does not occur in the first year of a drought, then certainly it will in subsequent drought years. Irrigators can draw on last year's blue water if there is water in the reservoir. More dangerously, they can draw on next year's groundwater recharge, the volume of which is unknown. They can even draw notionally on blue water that might come in the next three or more years, if there is equivalent storage to draw on.

As a consequence, wherever we irrigate, we always run out of blue water. The decision to irrigate always leads to a form of blue water over-allocation.

The costs of the over-allocation are borne by the politically voiceless water environment. Society has been slow to recognise that the blue water environment, wherever it has had demands for irrigation placed upon it, has ceased to provide blue water ecosystem services. Reversing the over-allocation is always very politically stressful, as the costs of such a remedy have to be borne via the impaired livelihoods of irrigators and of the communities that serve them. The irrigators may be rich and powerful or poor and vulnerable. Politicians and water policymakers rarely have the political and economic resources to confront the strength of the rich and powerful. Nor do they have the means to provide alternative livelihoods for impoverished irrigators on small farms.

The Forces Driving Demand That Test Food-Water Security and Impair Ecosystem Services: Demography, Socioeconomic Development, Climate Change, and Cheap Food

In this section, we have to grapple with unavoidable and very uncomfortable uncertainties associated with the error bars associated with the metrics of three key driving forces—demography, socioeconomic development, and climate change. The uncertainties are large and often bewilderingly unhelpful. They are discussed at some length in other chapters. The features of the assumptions briefly critiqued in this chapter have been identified via experience rather than being framed to serve a modelling process.

Demography and Food-Water Security

One of the main forces that drives increased water demand is population. A feature of the water footprint of an individual—unlike their energy footprint—is that it has not increased significantly since the beginning of farming 13,000 years ago. Their energy footprint has increased hundreds of times in just the past 200 years. Food is different. In rich OECD countries, the water footprint of an individual, 90% accounted for by food consumption, could at the most have doubled on average over the past 200 years. But it has generally only increased by 30%. The food-water footprint for those living in advanced industrialised economies is also not much different from that of a human living 50,000 years ago. Nor is it much different from that of someone living 500,000 years ago. By contrast, the individual energy footprint of OECD individuals could be 400 times as much as it was 200 years ago.

When resource security was being examined in the 1970s and 1980s, it was assumed that the world's population would increase to catastrophic levels vaguely towards over 20 billion (Ehrlich, 1975). At the Cairo UN Conference on Population in 1994, a dramatic reassessment of future populations was presented and mainly accepted. There was a low scenario levelling off in the second half of the 21st century at 8 billion, a medium scenario of 10 billion, and a high scenario of 12 billion. We have to recognise that the estimate of demographic trends has been subject to massive volatility. The current tweaking by pessimists and optimists is small beer compared to the challenges of past tendencies to overestimate.

Demography is a driving force associated with a big error bar. The current spread between a future increase of 1 billion, and more than 2 billion, is dangerous. It tempts the pessimist to adopt the high population scenario. All the analysts who want to get society and policymakers to focus on the issue of water security need to nudge the estimate of future population towards, and often beyond, 9 billion by 2050. They also tend to bring the levelling-off of population forward to 2050 rather than leaving it more comfortably in the second half of the century, as originally presented in Cairo in 1994. Presumably, these pessimists would have appreciated the estimates of four decades ago with numbers double the humble 9 billion.

Farmers have in practice kept pace with the food-water demand of a seven-fold increase in global population since 1800. They have worked the world's green water resources very hard and, in working the water resources hard, they have seriously impaired the sustainability of blue water resources wherever water has been used for irrigation. Farmers will be required to increase food production for the food-water needs of another 1 or 2 billion people and, at the same time, they are now being asked to steward blue water resources in unprecedented ways. They can certainly do the former. Society needs to help its water-managing farmers to be able to adopt the additional priority of protecting water ecosystem services.

Socioeconomic Development and Food-Water Security

Socioeconomic development is another important driving force. It is associated with increased water demand as well as with reducing food-water demand. Advanced economies increased their water consumption of non-food-water and food-water until the 1980s. They have experienced the more important feature of socioeconomic development with respect to food-water security, namely the capacity of advanced economies to bring about increased crop yields and the associated increased returns to water.

The 'developing' economies of Africa have for the most part, to date, lacked the stimulus of vigorous socioeconomic development that leads to the vast improvements in water productivity—especially in food-water productivity. These economies have underutilised green and blue water resources. Well-focused socioeconomic development could mobilise some blue water as well as very significantly improve the returns to both blue and green water. The crop and livestock output from food-water could be doubled. Even higher gains are feasible. It can be concluded that there will be enough African food-water to support the extra billion Africans who will be around in the second half of the century. It will be the development of existing water resources in diverse and strong socioeconomies that will be the most important factor.

Climate Change and Food-Water Security

To these major uncertainties about future population numbers, food consumption preferences, and the impacts of socioeconomic development must be added the uncertainties of climate change impacts on water resources. As with the other uncertainties, scenarios and assumptions are in currency. At one extreme, it is suggested that we shall be safely food-water secure with well-stewarded water ecosystems because there are sustainable intensifying options. At the other extreme, however, it is argued that we could face armed conflict over water because global food-water supplying systems can no longer cope.

Compared with the uncertainties associated with demography and socio-economic development, the negative impacts of climate change on crop productivity are relatively less significant. A further complication is that there will be some regions that are more negatively affected by climate change than others. Farmers in the Middle East, sub-Saharan Africa, and Australia will have to become even more effective than in the past to maintain food production. Farmers in the temperate latitudes, on the other hand, will in some regions enjoy more secure supplies of food-water.

The most important feature of current modelling of climate change and its impacts over the past two decades is the difference between the models available for the major river basins. Models are contradictory. Some predict lower flows. Others predict higher flows. The best that can be offered are *ensemble means* that result from an averaging of the available predictions. These ensemble means are at best uncomfortable and at worst very misleading (Gosling et al., 2010, 2011).

Cheap Food

Clearly there are many major uncertainties associated with the availability and demand for food-water. Farmers have to contend with these uncertainties in meeting the food demands generated in local and global food supply chains. For farmers, these uncertainties are all managerially and livelihood challenging and even life threatening. These are not favourable, appropriate, or secure circumstances for the precious managers of scarce food-water—the farmers—who ensure society's food security and who must be the guardians of water ecosystems services.

In this precarious world of food-water management where there are numerous major uncertainties, one feature has been enduringly certain for the past two centuries: Food commodity prices fall. This certainty has proved to be possibly even more challenging for farmers than the uncertainties related to demography, socioeconomic development, and climate change.

Falling food commodity prices have been interrupted by occasional global food price spikes. They were conflict induced in the 1940s. Oil price spikes induced those of the 1970s. Again in 2008 and 2011, there have been global commodity price frenzies, with associated food price spikes. Food prices are currently falling back. We shall have to live through another decade to see whether the shocks of 2008 and 2011 have reversed the long-term trend towards ever cheaper food prices. The jury is out.

The reason for the decline in food prices is partly because most farmers have enjoyed the availability of ever more effective inputs. They have been able to deploy capital increasingly effectively and have adopted improving agronomic practices. But there are two other important reasons for the fall in food commodity prices that are food-water related and, therefore, food-water security related. First, the cost of water as an input has been assumed to be free or nearly free. Such costs have certainly not figured in the accounts of the agents in the food supply chain, and certainly not of farmers who manage 90% of food-water. Secondly, the negative consequences of misusing and polluting water resources—especially in irrigated agriculture—have not been captured in the exchange value of food commodities.

The food supply chain is an extraordinary market phenomenon. Figure 20.1 shows that the agents in the whole supply chain—from farmers to the food commodity consumers—are all in the private sector. They produce, trade, manufacture, process, retail, and consume via private sector markets. This contrasts markedly with non-food-water services, which are almost exclusively organised and delivered in the public sector. The UK is an exception.

In the food supply chain, farmers and private corporations run everything from family farms to major transnational corporations that trade, manufacture, and retail food. They all use accounting rules that are blind, first to the cost of water inputs and second to the impact of food and fibre production on the ecosystem services of water, especially the associated misallocation and mismanagement of blue water in irrigation.

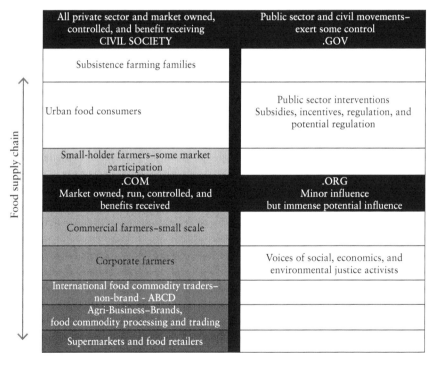

Figure 20.1 An analytical structure of the agents and social solidarities involved in the food supply chain, from food production by farmers to food consumption by individual consumers. The food-water in the private sector food supply chain—90% of the water needed by an individual or economy—is allocated and managed by farmers and then supplied by agents in the private sector markets using water-value blind rules.

Source: Allan, 2010

Who imposes the iron grip that keeps in place these dangerous cheap food assumptions that encourage unsustainable water managing practices? The first question must be, who has established the all-pervasive idea that food should be cheap and that food prices should fall? Second, who decreed that the accounting rules in the food supply chain should be blind to the costs of food-water inputs and of ecosystem impairment?

In the industrialised OECD economies, where cheap food policies are most deeply entrenched, the drive for cheap food has been established and kept in place by the recurringly dangerous alliance of state and market that is characteristic of all the strategic sectors of the neoliberal project. Big oil and the state. Big auto and the state. The military-industrial alliance. Big food and farming and the state is the state-market alliance that impacts food-water security.

There have always been large low-income communities that have needed cheap food if societies were to remain politically quiescent. The state and the private sector agents in the supply chain both have reasons to deliver cheap food to society. Consumers are served by markets and in democracies can vote to advance their interests. The cheap food monster seems to have developed two heads in the past three decades. The first reflects the persistent numbers of the low-waged and unemployed. These numbers have if anything proportionately increased in the past three decades because of long-term trends in the political economies of neoliberal states (Harvey, 2010). The second head has emerged because of the steadily rising numbers of the aging retired whose incomes have no chance of meeting the costs of 'properly priced' food.

The environmental and economic havoc brought about by the alliances of big oil and the state, of big auto and the state, of big finance and the state, and of the military-industrial alliance are well documented. Tinkering with any of these relationships has always been associated with very high politics and very contested national politics. The state-market pact on cheap food has a particularly long history. This, together with its unfathomable complexity, could possibly make big food and the state the most difficult of all the pacts to restructure and reform. The politics are tangled and impossible to penetrate. Meanwhile, the individual consumer/voter has a rare collective individual leverage as both a voter vis-à-vis the state and as a customer in food chain markets. The converged interests of food consumer, market, and state to deliver and access cheap food—in a system without a moral compass or safe accounting rules—will always be difficult to confront. Future engagement will also be very challenging for those wanting to steer farmers to allocate and manage water effectively and to become sound ecosystem stewards. This unholy pact of consumer, voter, state, and market is not an obvious target for those wanting to understand how to bring about food-water security. But it is argued that there must be engagement with this unholy pact if we are to engage effectively in a process that will bring about local and global food-water security.

Cheap food policies are installed and maintained in both rich and poor political economies. The unrelenting drive for cheap food in OECD economies is of great relevance to food-water security. The cheap food mantra has created a dysfunctional market for food in which water is not valued and water ecosystems are damaged by the over-allocation of blue water to irrigation. The farmers and other agents in the food supply chain compete and operate rationally according to the accounting rules they have been given. These rules are water-value blind and have created a juggernaut that is very destructive with respect to food-water security and has no capacity to change the rules that make it blind.

The grip of the cheap food doctrines of the OECD economies is globally significant. These high-yielding farm economies have been putting subsidised food staples into world markets for over half a century at prices that do not account for the costs of water. The extraordinary farm politics of the United States and Europe have for over half a century been exerting downward pressure on the world prices of staple food commodities (Paarlberg, 2011). These include wheat, corn, and soy, as well as the international prices of animal protein. The impacts have been perverse in the ways they have impacted the unprotected livelihoods of many African farmers.

The world prices of staple food commodities have for half a century not even reflected the accountable inputs of commodity production, nevermind the costs of food-water or of protecting the ecosystem services of blue water. For vulnerable farm livelihoods in African countries, the consequences have been deadly. Almost always when there has been a chance to increase prices of locally produced crops, and thereby enable local farm investment, the opportunity has been blocked by the intervention of the local national government. They judge that they have no option but to import the U.S.- and European-subsidised cheap staple commodities on the world market. Such cheap imports eke out the purchasing power of their increasingly numerous very-low-income urban populations. The outcome is that farm incomes spiral downwards. This deadly cycle of low or even declining farm investment for much of the past half century substantially explains the persistence of the low productivity of green and blue food-water achieved on African farms.

If African farmers are to feed an additional billion people by the second half of the 21st century, three things must happen. First, the perverse food politics of the OECD economies must change, or at least the reach of their subsidies via world markets must be curtailed. Second, African farmers must be provided with conditions that incentivise them to increase the productivity of their green and blue water while protecting water ecosystem services. Third, the agricultural research, regulatory, and marketing infrastructures have to be enhanced so that farmers can sustainably intensify their use of food-water.

Achieving Food-Water Security: Concluding Comments

The analysis has focused on the 90% of green and blue water allocated and managed by farmers. This water has been called food-water to capture the extent to which food-water and food security are inextricably linked. It has been argued that society—consumers and voters together with the state and the market—has entrusted its food-water security to farmers, without putting in place appropriate reporting and accounting systems. These systems are needed to highlight the value of water as an input as well as the ecosystem costs of mismanaging water. It has also been shown that food consumers and the many private sector agents in food supply chain markets—traders, manufacturers, supermarkets, and other retailers—are unaware of the impacts of their compliance with the dysfunctional price signals of existing food supply chain markets. They are oblivious to the dangers of operating the water-value blind accounting systems of the current food supply chain markets. At the same time, society is dangerously unaware of the impacts on their food-water security of wasteful practices in producing and storing food commodities, as well as being addicted to wasteful food choices and even more wasteful food disposal practices.

The analysis has also identified the role of the many perverse impacts of the existing national farm politics and international trade politics on farmers in OECD, emerging economies, and especially on those in poor developing economies. The widespread preoccupation of states and markets with ensuring the availability of cheap food has been highlighted to expose the risks that such policies have on food-water security. They impair the capacities of farmers—in OECD, emerging, and developing economies—to operate in ways that ensure food-water security.

The analysis also highlighted the very uncomfortable and unavoidable uncertainties of key drivers—population, socioeconomic development, and climate change. Climate change and climate change impact science currently provide contradictory predictions on future flows in river basins that make them very unhelpful in predicting future food-water security. The prediction history of demographic and socioeconomic development suggests that society will be able to allocate and manage food-water to provide future food and water security. On the basis of a scenario that requires an increase of 40% in food production, it is concluded that the world can be food-water secure when the population peaks in the second half of the century. Farmers will increase green and blue water productivity. At the same time, some communities will 'consume' more food and more embedded water. But this tightening in the relationship between supply and demand will be turned as farmers in regions that currently achieve low returns to green and blue water increase their water productivity. Others communities in advanced economies will 'consume' less embedded water and they will also change their food choices and their food-waste behaviours to reduce their water footprints.

But this optimistic prediction will only come about if the messages of this chapter are heard and adopted by key players. First, by food consumers. Second, by all the agents in food supply chains. Third, by those who determine food regimes and food politics nationally and internationally and food-trade politics at the international level. The main purpose of the chapter has been to highlight the importance of the role of farmers in allocating and managing water to both intensify, as well as sustainably steward, green and blue water.

Notes

1. The reader is asked to recognise that in using the term *food-water* they are being asked to assume that it include other crops such as those that provide services of natural and ornamental vegetation, as well as fibres and biofuels.
2. The 'movement' of embedded or virtual water must be clarified. Economists (Merrett, 2003; Wichelns, 2011) have pointed out that when a food commodity is traded, there is no actual movement of the water that was used to produce the commodity. It is recognized that the importing economy does enjoy a range of benefits from the import of the commodity. Food importing economies do not have to face the painful economic challenge of mobilizing the water to produce the commodity. More important, they do not have to cope with the very stressful politics of accessing the water to produce the commodities. The rapid increase in the international trade in food commodities of the past half century occurred during a period of falling prices. As a consequence—despite the elemental sensitivity of the process with respect to national security—because the 'trade' was economically invisible and politically silent, it brought widespread but utterly misunderstood water security. In this chapter, in order to highlight the difference between the *traded commodity* and the *embedded water associated with the commodity traded*, the following convention will be adopted. Where the import, export, movement, flow, or trade of embedded or virtual water is being discussed, then the process will be shown as virtual water 'trade', virtual water 'import', virtual water 'export', and so forth. The importing economy is 'exporting' a water footprint equivalent to the water embedded in the commodity imported. It also 'exports' the noninternalised environmental impacts associated with the production of the commodity.
3. The use of the word *always* is deliberately confronting. All the rivers with annual flows below 150 cubic kilometres per year are dysfunctional in terms of water ecosystems. Some others that are much bigger—such as the Indus—are also severely impaired. The Mekong seems destined for a similar fate. Only Bangladesh can claim to have a water-using regime that is not over-allocating. Bangladesh is at the end of a system supplied by three huge rivers, albeit experiencing simultaneous seasonal variation. Of these three rivers, only the Ganges experiences significant diversions for irrigation. The Brahamputra and the Meghna provide annual flows at close to natural historic level. The politicians and the hydraulic engineering community of Bangladesh have not yet over-allocated the low seasonal flows. Perhaps this will be a case where the stewardship message will influence allocative outcomes before the ecosystem services of water have been seriously impaired.

References

Allan, J.A. (2001) *The Middle East Water Question: Hydropolitics and the Global Economy*, IB Tauris, London.

Allan, J.A. (2010) 'Prioritising the processes beyond the water sector that will secure water for society—farmers in political, economic, and social contexts and fair international trade', in L. Martinez-Cortina, A. Garrido and E. Lopez-Gunn (eds) *Re-Thinking Water and Food Security*, Fourth Marcelino Botin Foundation Water Workshop, CRC Press and Taylor and Francis, Leiden and Oxford.

Allan, J.A. (2011a) 'The role of those who produce food and trade it in using and 'trading' embedded water: what are the impacts and who benefits?', in A.Y. Hoekstra, M.M. Aldaya and B. Avril (eds) *Proceedings of the ESF Strategic Workshop on Accounting for Water Scarcity and Pollution in the Rules of International Trade*, Amsterdam, 25–26 November 2010, Value of Water Research Series No 54., Water Footprint Network, University of Twente, Enschede.

Allan, J.A. (2011b) *Virtual water: tackling the threat to the planet's crucial resource*, IB Tauris, London.

Ehrlich, R. (1975) *The Population Bomb*, Rivercity Press, New York.

FAO Aquastat (2012) *Aquastat On-Line Data for the Nations of the World*, FAO, Rome.

Gilmont, M. (2013) 'Decoupling dependence on blue-water: reflexivity in the regulation and allocation of Israel's Water', *Water Policy*, in press.

Gosling, S.N., Bretherton, D., Haines, K. and Arnell, N.W. (2010) 'Global hydrology modelling and uncertainty: running multiple ensembles with a campus grid', *Philosophical Transactions of the Royal Society: A.*, vol 368, no 1926, p4005–4021.

Gosling, S.N., Taylor, R.G., Arnell. N.W. and Todd, M.C. (2011) 'A comparative analysis of projected impacts of climate change on river runoff from global and catchment-scale hydrological models', *Hydrology and Earth System Sciences*, vol 15, p279–294.

Harvey, D. (2010) *The Enigma of Capital*, Exmouth Press, London.

IWMI (2007) *Water for Food: Water for Life, The Comprehensive Assessment of Water Management in Agriculture*, Earthscan, London.

Larson, E.A. (2011) *Corporate Social Responsibility and Transboundary Flows of Virtual Water: Bringing Multinational Food Companies into a Political Economic Discussion of Water Security*, unpublished MSc dissertation, Department of Geography, King's College, London.

Merrett, S. (2003) 'Virtual water and Occam's razor', *Water International*, vol 28, no 1, pp103–115.

Parry, M.L., Rosenzweig, C., Iglesias, A., Livermore, M. and Fischer, G. (2004) 'Effects of climate change on global food production under SRES emissions and socio-economic scenarios', *Global Environmental Change*, vol 14, pp53–67.

Paarlberg, R. (2011) *Food Politics: What Everyone Needs to Know*, Oxford University Press, New York.

Reimer, J.J. (2012) 'On the economics of virtual water trade', *Ecological Economics*, vol 75, pp135–139.

Segal, R. and MacMillan, T. (2009) *Water Labels on Food: Issues and Recommendations*, Food Ethics Council, London.

Sojamo, S. (2010) *'Merchants' of Virtual Water: The 'ABCD' of Agribusiness TNCs and Global Water Security*, unpublished MSc dissertation, Department of Geography, King's College, London.

Wichelns, D. (2011) 'Assessing water footprints will not be helpful in improving water management or ensuring food security', *International Journal of Water Resources Development*, Vol 27.3, pp607–619.

21 A Synthesis Chapter

The *Incodys* Water Security Model

Bruce Lankford

Indebted to the book's authors and inspired by their arguments and ideas, this concluding chapter reflects on the idea of water security. My objective here is to make sense of the chapters in this book and literature elsewhere by proposing that water security and insecurity are not necessarily polar opposites; that water security need not be the opposite or 'converse' (Grey and Garrick, 2012) of water insecurity.[1] Rather, by establishing two gauges of water security (sufficiency of water security and equity of water security), *four* conditions of water security are proposed, hinted at in the title of this chapter (*incodys*). First, 'in' is shorthand for *insecurity;* second, 'co' is short for *collective security,* and third, 'dys', implying inequitable security, is short for *dys-security.* The fourth condition is *co-insecurity*, but this is not repeated in the title. The use of these terms also invokes the idea of *a-security,* implying the lack of an issue of water security, which is briefly discussed in the chapter.

Readers will therefore recognise the conceptual nature of this chapter; the ideas contained here are theoretical and intended to summarise this book's thinking on water security. The *incodys model* (as I term it) allows, amongst other matters, the consideration of how metrics and indicators might inform water security. Furthermore, and despite cautions levelled in Chapter 1 against single definitions of water security, I use this analysis to offer an explanation of water security: 'In creating four conditions of insecurity, co-insecurity, dys-security and co-security, placed in a two-axis graph of "sufficiency" and "equity", water security is viewed as a transitive space of water securities through which a community, individual, or system moves as a result of natural and human water-related drivers'.

However, despite this move towards synthesis and conciseness, I reiterate that water security is sufficiently complex to warrant multiple understandings; that, too, is one of water's defining qualities. It would be rash to claim that the model proposed here comprehensively reflects the many social, economic, and environmental dimensions of water and water management (a viewpoint at the centre of Zeitoun's web analysis). This sense of complexity is also captured in the UN-Water (2013) brief on water security. Thus, this chapter attempts to steer a path between a linear conception of a single axis of insecurity–security and a wholly complex multiple 'nth' dimensional abstract space of composite axes and indicators of water security. My 'middle path' solution is to formulate water security as a two-axis understanding of water security that can be graphed on a standard x–y axis graph. This graph, in turn, gives rise to four conditions of water security, introduced above and explained below. The other option of selecting three gauges of water security (for example sufficiency, equity, and productivity) is partly discussed in the chapter; however, this would require a three-dimensional cuboid representation of water securities and the possible formulation and naming of *eight* water security conditions.

Initial Insights: Water Security 'Opposites' and 'Transitions'

To lay the groundwork for the *incodys* model, I begin with three key distinctions. First, I argue that water security need not be 'the opposite' of water insecurity. In other words, there is not a single linear scale between water insecurity and water security. This point regarding multiplicities is also echoed by Boelens (Chapter 15), who discusses 'divergent water securities'.

Second, similar to the first point, there are places in the world where water security and insecurity are not an appropriate frame to view the water situation. This is where patterns of supply and demand are purely 'natural', with no or a very small influence or anthropogenic demand from society. This idea, described by the term *a-security,* further indicates the limitations of a linear scale reaching from insecurity to security.

Third, I am interested in water security actions and 'transitions' suggested by the term *water securitisation.* 'Transitions' implies a shift from one state to another state, for example, from insecurity to security. However, 'securitisation' is not necessarily the process of making human and ecological parties water secure; it is not the action that assures water security. This is because, following this book's authors (e.g., Leb and Wouters, Chapter 3; Zala, Chapter 17; Zeitoun, Chapter 2), I interpret securitisation in a highly strategic and controlling sense (both making water the object of a security concern, then pursuing forms of appropriation). For example, securitisation of water for a nation is seen by Leb and Wouters as highly political and at the possible cost of neighbours when sharing a transboundary watercourse. Hepworth and Orr (Chapter 14) also warn of private benefits accruing to corporates from securitisation. I, therefore, retain securitisation to mean the removal of water scarcities and vulnerabilities for few, usually powerful and hegemonic parties, leaving other, usually less powerful and more marginalised, parties more vulnerable.

Drawing from these three distinctions, one can clarify and reiterate seven points. One is that *water security* is a term that invokes a sense of being somewhere, being in a state, or facing a certain condition. For example, a community might be 'water insecure'.

Second, water security also invokes the sense of change and transition—either actual change (e.g., being in a drought) or the chance of forthcoming change (e.g., being in an environment prone to drought).

The third insight, closely related to the first two, is that water security invokes the idea of gradation—that there are grades of difference between a wholly insecure context and a wholly water secure context. Following Mason (Chapter 12), these grades of difference might be discernible if indicators are appropriately designed and measured.

Fourth, stepping back, society and its decision makers must discern under what conditions the idea of water security arises. In other words, in the rush to frame water via water security (drawing on Cook and Bakker's identification of the rising usage of the term), we forget to consider situations where water security is a nonissue and how this informs the idea of water security. The term *a-security*, alongside the *incodys* framework, helps make this distinction (see later).

Fifth, one might argue that the ideas of 'to be water insecure' or 'to be water secure' (and the grades in between) suggest a classification of different types of water *securities* (Boelens, Chapter 15). This classification is developed in the *incodys* model. Therefore, a community can find itself facing a given set of water conditions (e.g., at the risk of flooding, or having less water than required, or facing poor water quality) and, assuming this can be measured, the community's condition can be classified accordingly.

Sixth, from previous points emerges the idea of action or intervention, for example, in making an insecure community more secure. In other words, there is an intention to improve the conditions a community faces—to make things better. This 'doing' propels a given party along a trajectory and invokes an idea of transition(s) and of a governance of these transitions. Boelens (Chapter 15) expresses this as 'realizing water security', although 'doing' also covers situations when conditions become more insecure.

Lastly, from these observations and from a desire to steer a path between oversimplification and overcomplication, I consider two scales or gauges that seem to be of major significance in water security. These two scales, termed *sufficiency* and *equity,* create a two-dimensional water security '*incodys* space' (described later). Sufficiency implies sufficient quantity and quality of water and sufficient protection from excessive quantities of water. Thus, 'sufficiency' is inferred by Grey and Sadoff (2007) as "the availability of an acceptable quantity and quality of water for health, livelihoods, ecosystems and production" (p 545) and is considered by Cook and Bakker (Chapter 4) to occur when "[a human] has access to sufficient safe and affordable water". Sufficiency is also considered in terms of accessible water of a good quality being connected to the human right to water, recognising water's centrality to health, the home, and production (Chenoweth et al., Chapter 19; Zeitoun, Chapter 2). The idea of sufficiency/sufficient is also exposited by UN-Water (2013) as central to water security for humans (and the human right to water; see also Leb and Wouters, Chapter 3), and for the mutual connections between healthy ecosystems and societies. Control of floods or protection from floods is also part of suffi-ciency. Multipurpose infrastructure for flood protection (as well as drought provision and hydropower) is identified by Mirumachi (Chapter 11) and Garrick and Hope (Chapter 13), while Warner (Chapter 18) adds to this by correctly drawing out the political provenance of the notion of sufficient flood protection. Finally, sufficiency should recognise the multiple types of water (e.g., soil water, groundwater) that society draws on (Falkenmark, Chapter 5; Allan, Chapter 20).

Equity is the other major axis or gauge, and is implied when Zeitoun (Chapter 2) asks the question, 'Water security for whom?' Tickner and Acreman (Chapter 9) argue that social and economic security is not possible without environmental and ecological security. Hepworth and Orr conclude their analysis of the involvement of corporates in securitising water to meet their profit interests by suggesting a more sustainable notion of mutuality: "navigation between corporate securitisation and stewardship of the water commons, towards a more genuinely shared, wider water security" (Chapter 14). Garrick and Hope (Chapter 13) point not only to risk but also to the sharing of risk. The idea of equity is central to Boelen's (Chapter 15) entire chapter; in his first sentence, he links water scarcity and insecurity to the ways in which water and water services are distributed rather than to absolute availability of water. Confirming equity and equitability as strong themes running through Leb and Wouter's chapter, as key underpinnings of transboundary water law, the proposition of 'water justice' (Clement, Chapter 10) also picks up the distributive dimensions of water security. Equally, the 'working definition of water security' (UN-Water, 2013; p1) implies equity (but does not mention it directly) by referring to water's different interests and benefits: sustaining livelihoods, human well-being, socioeconomic development, and preserving ecosystems.

Water Security Hydro-Physical Outcomes: Six Measures

The *incodys* water security model, described in the next subsection, describes a transitive space made up of two axes of sufficiency and equity. However, to place a water community, system, or situation into this transitive space requires selected indicators to be measured and assessed. Table 21.1 and Figure 21.1 arrange six water security indicators within the two classes of sufficiency and equity. Each of the two axes and their three respective indicators are discussed below.

Table 21.1 suggests examples of metrics that might assist in defining indicators that can be measured or classified, and confirms the hypothesis that the *incodys* model primarily describes hydro-physical measures of water either as dimensional ratios (such as available water per capita), dimensionless ratios (e.g., measures relative to a physical standard), measures of difference (e.g., variations in supply or protection), or as categories (e.g., pass/fail) or as count data recast as percentages (e.g., the number of community households satisfied with their water provision).

Table 21.1 Definitions and measures of water sufficiency and equity/justice

Main axis	Indicator	Details	Example units
Sufficiency	Volumetric sufficiency	Exact or relative measures of per-capita water availability	Cubic metres per capita
	Water quality	Exact or relative measures of water quality (e.g., pollution, turbidity, salinity, oxygen), or measures of client satisfaction in urban areas	Classes (high, good, mod, poor, bad)
	Flood protection	Risk map of protection from flooding (distance-depth for example)	Depth, extent, risk of event (e.g., return period)
Equity/ justice	Water allocation/ equity	Bulk water allocation between sectors; measures might include difference from expected or variation measures such as the Christiansen coefficient, interquartile ratio, or coefficient of variation	Exact % ratios relative from expectation or history; measures of difference
	Dynamic apportionment	Scarcity allocation during drought or when levels of supply change dramatically and dynamically; low flow ratios and access measures during drought might apply	Percent coverage of minimum requirement; measures of difference
	Productivity/ efficiency	The specific production from units of water and/or land/ labour using carefully accounted denominators of withdrawal and/or consumption	Unitless efficiency; economic or biophysical productivity

As explained elsewhere, social, political, and economic measures could also be collated and analysed and would very much support the physical character of the *incodys* model. Furthermore, nonphysical measures could also be arranged under two criteria of sufficiency and equity. For example, taking water laws and agreements, one could analyse whether they were sufficient (in number and detail) and equitable (in terms of their intention to support or undermine the physical apportionment of water to different users and sectors).

Table 21.1's grouping within two themes agrees with Mason's advice: 'then the emphasis should rather be on reducing the variables of concern, and the complexity in how they are amalgamated' (Chapter 12). Critical to the coherence of the *incodys* approach are indicators that are closely grouped conceptually—in other words, that are commensurate to each other and to their meta-indicator of either sufficiency or equity. The two

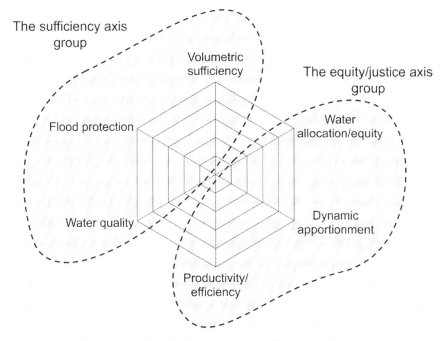

Figure 21.1 The hexagonal radar chart of water sufficiency and equity

significant weaknesses of the water poverty index (WPI) (Sullivan, 2002) stem from the incongruity of the five indicators selected (resources, capacity, use, access, environment) and the disparate constituents by which these five indicators are measured (Mason, Chapter 12; Molle and Mollinga, 2003). In searching for an acceptable level of commensurability (see also Cook and Bakker), the *incodys* model differentiates hydro-physical measures from other socioeconomic metrics and amalgamates the six subindicators into two themes of sufficiency and equity.

Furthermore, as expressed elsewhere in this chapter, the interpretation of these measures (both in terms of selection/design, but also their meaning) should be clearly acknowledged as being relative and subjective. The *incodys* model is very much intended to be a dialogue tool and decision aid; who gets to select the measures, to reflect on them, and to employ them in classifying a system should be transparent and highly processual.

The Sufficiency Axis Group

The sufficiency axis primarily speaks of biophysical outcomes related to infrastructure designed to remove constraints to insecurities generated by inadequate water quantity, poor water quality, and excessive flood risk. In other words, water systems on this axis grade from 'insufficiency' on the left to 'sufficiency' on the right. Clearly not all of these attributes (quantity, quality, and flood) might exist contemporaneously; indeed, water scarcity and floods might be mutually exclusive. (Recall whether and how these physical attributes are managed so that water and water benefits are 'shared out' between water users is covered by the equity/justice axis—see below).

- *Volumetric sufficiency.* The objective here is to assess whether water demand is adequately provided for by a combination of supplies of water, including surface water, groundwater, rainwater/green water (Falkenmark, Chapter 5), desalinization, and virtual water (Allan, Chapter 20). Volumetric sufficiency is linked by chapter authors to, for example, human security (Falkenmark, Chapter 5), food security and production (Clement, Chapter 10), and to sustainable cities (Earle, Chapter 7). Given the subjective, transient, and relative nature of demand, a locally derived measure of sufficiency is probably the most sensible route to take. Thus, for a selected system, water needs might be derived from its domestic, agricultural, and industrial functions, but validated against norms and standards taken from similar situations (or indeed with reference to international standards; Falkenmark, Chapter 5). Sufficiency measures of demand and supply can then be converted into a range of indicators such as a ratio/percentage (e.g., 125% would indicate a surfeit of 25% supply over demand) or as a ratio against an accepted standard (e.g., where 25% indicates a surfeit of 25% over a recognised scarcity limit).

- *Water quality.* Nearly all the contributors to this book argue either explicitly or in passing that water security includes water quality by invoking the idea of 'safe' water for human needs and/or sanitation (examples include Cook and Bakker, Chapter 4; Hepworth and Orr, Chapter 14; Zeitoun, Chapter 2; Chenoweth et al., Chapter 19). In addition, Mirumachi (Chapter 11) also considers water quality to be part of environmental needs, and Conway (Chapter 6) links water quality concerns to climate change. A number of pollutants and quality criteria might be considered in drawing up a single measure of water quality. Depending on location, typical concerns include: the presence of agricultural chemicals and poisonous compounds such as heavy metals and metalloids (e.g., arsenic in Bangladesh groundwater); low oxygen content; and waterborne pathogens that cause ill-health in humans and animals. Interventions to raise water quality include control and regulation and water treatment post and prior to abstraction or consumption. Scoring water quality might adopt the EU Water Framework approach of classes or a pass/fail system (Kallis and Butler, 2001; EU, 2012) or community definitions of acceptable standards and whether they are met or not (e.g., the number of hours spent fetching drinking water).

- *Flood protection/risk.* Authors identify risks associated with the damage wrought by flood (Warner, Chapter 18) and protection from floods (Earle, Chapter 7). For the purposes of the *incodys* model, a broad definition of flooding is taken, including heightened depths and destructive energies of water from stream flood events, persistent waterlogging, and combined freshwater and seawater level rises and surges. A variety of interventions and works grapple with the dangers presented by floods, including technologies such as inserting and/or raising bunds, providing river training, dredging and removing bottlenecks, adding safe storage areas such as floodplains, redesigning urban and industrial architecture, introducing improved emergency services, and effecting upstream land-use changes to the catchment. While these describe some of the management inputs, the outcomes measured for the purpose of assessing security from floods could utilise the Water Framework Directive model where classes of risk are mapped spatially, and from which, for example, some insurance companies derive their premiums.

Equity/Justice Axis Group

The equity/justice axis measures three different criteria that reflect the equitability and justice dimensions of water management. This axis promotes the idea of water sharing—either of water abundance or of water scarcity. Furthermore, this group of measures reminds water managers and decision makers that 'partial security' is not security, that those who gain at the expense of others might have their advantage questioned and removed.

- *Allocation.* This measure captures the idea of how bulk or annualised volumes are shared between different sectors or communities within a given basin, subcatchment, or country. Because allocation is a measure of sharing, and because uneven allocation must reflect a difference relative to practice or expectations, measures of allocation must be carefully defined. Coefficients of variation or of difference are likely to be incorporated into this type of indicator. Different emphases on blue and green water allocation (Falkenmark, Chapter 5) would reflect the types of systems being studied.
- *Dynamic apportionment.* Critical to the notion of equality is whether all users and sectors gain under two particularly challenging environments: one is during drought, and the other is during periods when systems are moving quickly between wetness and dryness. Dynamic apportionment is a suitable term that covers equitable distribution under such conditions. Both variability and drought are set to increase as a result of climate change (Conway, Chapter 6), and both provide opportunities for advantaged parties to gain in relative terms alongside others. As well as incorporating this idea in this book (Lankford, Chapter 16), three papers (Lankford, 2004; Lankford and Beale, 2007; Lankford and Mwaruvanda, 2007) elaborate upon how fast moving and stochastic nonequilibrium conditions allow upstream irrigators to unpredictably and disproportionally gain compared to others.
- *Productivity/efficiency.* Some might argue that productivity and efficiency are substantively a different type of descriptor of water systems, in effect creating a 'third' axis allowing systems to be described using the criteria of sufficiency, equity, and productivity. However, seeking purposively to keep with two axes and four water security states, I have joined productivity

and efficiency to equity/justice rather than to sufficiency. This decision flows from the argument that well-shared timely co-managed water is likely to be more productive for users between and within sectors than if only a select few obtain 'their secure supply'. Nevertheless, as Clement (Chapter 10) hints, the economic argument for water to flow to the most productive sector (e.g., industry rather than agriculture) counters the sharing principle of this equity/justice axis. However, Clement's observation on economic efficiency suggests why biophysical efficiency might be linked to equity rather than to sufficiency or stand alone as a separate indicator. The principle at work here is one of allocating water on the basis of a level playing field so that potential (i.e., higher) biophysical efficiency not actual (probably low) biophysical efficiency drives both the allocation decision and the amounts of water transfers involved. In this sense, allocation is then seen as equitable (in the legal sense) rather than purely on the basis of political priority and a misapprehension that agricultural sectors are, de facto, inefficient. To exemplify: Low-productivity irrigation comes under focus for allocation of water to another sector via raising its efficiency rather than deducting water from its net beneficial needs.[2] Indicators for productivity and efficiency can be generated from using either exact measures (e.g., crop per drop or dollar per drop) or relative measures set against local and achievable standards.

The Incodys Water Security Model

Using points made in the introduction and in the sections above, the *incodys* model of water security (Figures 21.2 and 21.3) utilises a two-dimensional field of water security constructed from 'sufficiency' and 'equity' based on six criteria given in Table 21.1. The *x*-axis is the sufficiency axis, while the *y*-axis depicts equity. Furthermore, it is the movement up the equity/justice axis (away from the *x*-axis) that invokes the idea of cooperation and collectiveness. Hence the naming of two conditions includes the prefix *co-* for those at the top of the graph where sharing and equity are expressed most clearly. This cooperative ethos is expressed by Zala (Chapter 17), Mirumachi (Chapter 11), and in words by Leb and Wouters: 'cooperation and not securitisation is at the heart of achieving effective water security'. One consequence from the two-axis model is the identification of four types of water securities, explored in the subsections below. A fifth condition, *a-security*, begins the discussion.

Water a-Security

Not graphed in either the radar charts of Figure 21.2 or the *incodys x–y* chart of Figure 21.3 is a condition of *a-security*. A-security is discussed first because it usefully distinguishes what the *incodys* model does and does not address. The *incodys* model primarily applies to situations where water demand and supply are in a state of balance/imbalance and where both demand and supply

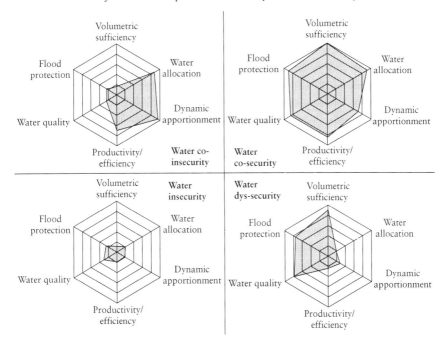

Figure 21.2 Four conditions of water security depicted in the radar charts

[A] = High levels of water insecurity
[B] = Some progress made

[C] = Certain factions water secure
[D] = Lack of supply; water insecurities shared
[E] = Collective water security improved

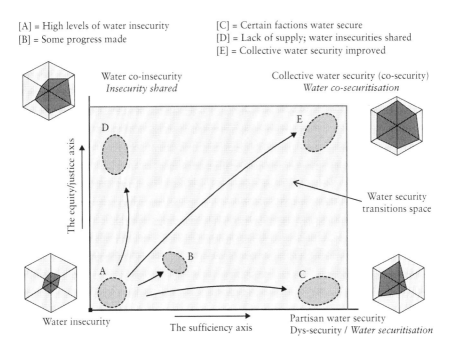

Figure 21.3 The *incodys* transitions space of water security

are significantly shaped by anthropogenic influences (or where ecological concerns are represented by humans).

On the other hand *a-security* describes conditions where extant and future anthropogenic and ecological demands are predictably and quantitatively small. Examples of this in humid zones might be found in central rainforests of Amazon and Congo regions, which experience annual rainfall amounts of approximately 2,500 mm/year or more and where few humans live. At the dry end of the spectrum, examples of human absence and where nature is attuned to aridity are found in tropical, sub-tropical, and artic deserts, such as the Sahara. While there are communities in these environments that face water insecurities (pointing to the need for an appropriate scale focus), it would be questionable to impute the term 'water insecure' to environments predictably very dry or very wet inhabited by very few humans. However, a-security is subjective and in flux; future balances between supply and demand in a climate-changing and population-changing world are unpredictable.

Placing a-security prior to the subsections on the four conditions of water security prefigures the discussion below on perspectives: that while water security and measures of water cannot be arrived at objectively, we nevertheless should treat critically claims regarding water insecurity. The identification of four *incodys* securities and use of measurable indicators serve discussions that would otherwise poorly disentangle the interests and perspectives at work.

Water Insecurity

Shown in the bottom left frame of Figure 21.2, water insecurity describes a condition of both insufficient and poorly shared water resources. This state arises because there is inadequate supply over demand, poor water quality with associated health impacts, and/or high risks of extensive flood damage. In terms of equity and justice, despite there being little excess water, some sectors gain at cost of others both during normal conditions and when droughts hit. Water productivity is low and uneven, which combined with reduced volumes of water, leads to low economic growth.

Water Co-Insecurity

Communities face water co-insecurity (top left of Figure 21.2) when scarcity, poor water quality, and/or the risk of flooding are shared more equitably between individuals, communities, and systems. Unfortunately, without good monitoring and communication, users experiencing water co-insecurity may blame their neighbours for over-consuming water; thus, in some situations, co-insecurity might 'feel' much like water insecurity. For example, without an emphasis on a sufficiency of good quality water, health impacts are still likely to be negative. Productivity, although likely to be more even and perhaps slightly higher than in an insecure situation, may not feed through to high production because of water supply constraints.

Water Dys-Security

In Figure 21.2, bottom right, water dys-security is marked by highly uneven, partisan, and factional water security. Thus, although at one scale there is sufficient supply to meet demand, this has been captured by a few players or sectors. This has occurred because of geographic sequencing (for example, where large areas of irrigation lie upstream of wetlands) or because of imbalances in water law (where environmental demands are poorly recognised). Another example is found in parts of India where via a combination of technology and energy pricing, richer farmers have sunk deeper boreholes, leaving shallow tubewells supplying poorer members of the community. The particular feature about water dys-security is that these imbalances are felt throughout the hydrological regime from abundance during normal to wet years and during downturns and dry periods. Although flood protection and good quality water are afforded to some, the overall sense is that that some sectors of the community or environment are poorly provided for. Imbalances in water productivity are also notable, with the effect that in total the benefits of water are not optimised.

Water Co-Security

The radar chart for water co-security (upper-right of Figure 21.2) views water and flood protection as sufficient for all parties and evenly shared between the parties involved. In particular, sectors and users are treated equitably when analysed using both bulk annual volumes and during periods of scarcity and rapid change. Furthermore, combinations of sufficient water (shared equally) with good quality clean water and flood protection also serve to boost water productivity, healthy human populations, and environmental goods.

Water Security Transitions

Figure 21.3 brings together the above states of water security into a transitions space. The first, 'water insecurity' is located in the bottom-left corner, characterised by insufficiency and inequity. In the bottom right-hand corner, water securitisation has delivered sufficiency to certain factions but leaves others insecure. In the top left of Figure 21.3, sharing delivers water co-insecurity more equitably yet parties still face forms of insufficiency. Finally, in co-security, towards the top-right of Figure 21.3, interventions improve sufficiency and sharing. Figure 21.3 shows a system hypothetically moving through this space from situation A (water insecurity) to C, D, and E, each relating to the four corners of the graph. Situation B offers an example whereby some progress is made, perhaps by the installation of a new water treatment works, but the community continues to argue that it remains short of sufficient quantities of water or that this better drinking water is more expensive and is not reaching all members of the community.

Discussion

A number of observations can now be made about the design of the *incodys* model and how it might be applied. Readers are reminded that this conceptual framework is at an early stage of development and needs additional work on metrics and indicators using case material.

Socioeconomic Water Security Outcomes

I reiterate here that in the search for commensurability, the *incodys* model seeks to portray hydro-physical (e.g., cubic metres of water consumed by sectors) rather than socioeconomic outcomes (e.g., number of water agreements in place). However, this is not to downplay the latter. With respect to socioeconomic water security outcomes, a number of authors in this book identify examples. Outcomes include general health and poverty indicators (Chenoweth et al., Chapter 19); urban, industrial, and economic activity associated with water (Earle, Chapter 7); and impacts on other resources such as energy, land, and food (Froggatt, Chapter 8). The lack of space precludes further exposition; however, it is important to mention that these metrics are highly complementary of hydro-physical measures of sufficiency and equity.

The Incodys Transitive Space

I have purposively not divided the *incodys* space into four exact quadrangles (or placed the axes intersection at the centre of the graph), as I believe this would mistakenly signal that exact and even thresholds apply to many, if not all, systems. Instead, I perceive the *incodys* model as being a dialogue tool to elicit and subsequently test subjective understandings of water security in relation to physical metrics. In other words, researchers using the model might discover that communities and individuals believe that insecurity occupies the majority of the space of Figure 21.3, leaving the notion of verifiable mutually agreed co-security a minor part of the field. Furthermore, the conceptual location of the fifth state of a-security is not placed in the *incodys* space because it does not arise through the criteria of sufficiency and equity of water security. In other words, because patterns of supply and demand are not significantly problematic (being almost entirely natural and without the presence of human concern), a-security sits outside of the *incodys* space.

Sufficiency and Equity Interactions

Early discussions of the model with Masters of Science (MSc) students on the UEA Water Security degree in March 2013 confirmed an array of interactions might exist between sufficiency and equity. One is that 'sufficient' water quality, quantity, and flood protection might automatically 'trickle down' into equity. Others viewed that these two dimensions need not be necessarily linked and that high levels of sufficiency might be marked by low levels of equity between parties. These views imply further work on whether the *incodys* model is interested in relative or absolute measures of water security. For example, a new large dam might generate greater benefits for more people (in absolute terms), but some members of society gain tremendously from this infrastructure compared to others.

Participation, Perceptions, and Metrics

Working with ideas of water security at different spatial and temporal scales will inevitably summon critical problems related to subjectivities and perceptions. As Zala (Chapter 17) writes: 'Particular attention must be given to addressing local-level perceptions of inequality and injustice, particularly in conflict and post-conflict situations'. Similarly, Garrick and Hope (Chapter 13) posit: 'Individual and social perceptions of risk are fundamental to decision making to manage water security risks and tradeoffs'.

For these reasons, the *incodys* model promotes the collection and analysis of metrics as paramount objectives to inform the deliberative and participatory process that then attempts to reconcile disparate users' perceptions. It is the lack of substantiated metrics that make many participatory workshops somewhat hollow affairs. It should be noted that participation itself is not part of the *incodys* set of physical metrics, but it may be recorded as a socioeconomic outcome.

Insecurity Reversals

Further work is required on how the model might depict a severe reversal in water insecurity—as might happen during a protracted drought.[3] While one answer to this phenomenon might be a continuation of the axes into negative territory using minus numbers or the adoption of a logarithmic scale, I am more inclined to think that the *incodys* space is not anchored to a particular baseline. In other words, severe water scarcity would see the graph rescaled. This fits points made elsewhere in the chapter that the model is primarily a metric-informed dialogue tool.

Mapping Water Security at the Global Scale

Finally, because water security is made up of two themes, it will be highly unlikely that the *incodys* model can be turned into an index that can be used as a global map of water security. I judge this as a major benefit and caution against attempts to do so. World maps that 'find' damp Northern Europe to be one colour, zero rainfall Sahara desert to be another, Greenland to be blank, and the 'hotspots' of the Indus and Nile (for example) to be another colour are entirely unconvincing, given that they miss the local detail of the multiple and transient aspects of water. The *incodys* model would, with its two axes, be difficult to transform into different colours. Thus, because of explicitly recognised problems of data, perspective, and temporal flux, a global map of *incodys* numbers or classes would be nonsensical.

Conclusions

Rather than rephrasing the overview offered in the opening chapter or identifying further work and spelling out gaps in this book (of which there will be many), this final chapter has drawn on authors' chapters to propose a water security synthesis. This chapter has applied a framework to distinguish between water security governance and water security. With regards to the latter, a two-axis formulation of water security, based on ideas of 'sufficiency' and 'equity', has been proposed. This, in turn, gives rise to an *incodys* transitive field comprising four states of water security: insecurity, co-insecurity, dys-security, and co-security. It has also grappled with the scope of water security considering that there are situations where, due to environmental circumstances and an absence of anthropogenic interests, no water security problem arises (so called *a-security*). Clearly, this conceptual treatment of water security has left open many questions of how to define and measure sufficiency and equity and to govern water security transitions.

Notes

1. I am more persuaded of the 'opposites' argument in the field of food security; that the 'opposite' of food security is food insecurity. This is because, in some respects, food is an endpoint, while water is both something received but also consumed, nonconsumed, distributed, and passed on. The collective use and distribution of water, and therefore equity, is central to successful water management.
2. The aim of this discussion is to link efficiency and productivity to equity. Without doubt this logic in the real world would be messy and complex because of the difficulties in separating beneficial, nonbeneficial, and recovered flows.
3. I am grateful to Jenny Fraser (UEA MSc Water Security) for her question on the issue of water security reversals.

References

EU (2012) 'The EU Water Framework Directive—integrated river basin management for Europe', http://ec.europa.eu/environment/water/water-framework/index_en.html, accessed 24 March 2013.

Grey, D. and Garrick, D. (2012) *Brief No.1: Water Security as a Defining 21st Century Challenge*, University of Oxford, Oxford.

Grey, D. and Sadoff, C. (2007) 'Sink or swim? Water security for growth and development' *Water Policy,* vol 9, no 6, pp545–571.

Kallis, G. and Butler, D. (2001) 'The EU water framework directive: measures and implications', *Water Policy,* vol 3, no 2, pp125–142.

Lankford, B. A. (2004) 'Resource-centred thinking in river basins: should we revoke the crop water approach to irrigation planning?', *Agricultural Water Management,* vol 68, no 1, pp33–46.

Lankford, B.A. and Beale, T. (2006) 'Equilibrium and non-equilibrium theories of sustainable water resources management: dynamic river basin and irrigation behaviour in Tanzania', *Global Environmental Change,* vol 17, no 2, pp168–180.

Lankford, B. A. and Mwaruvanda, W. (2007) 'A legal-infrastructural framework for catchment apportionment', in B. Van Koppen, M. Giordano, and J. Butterworth (eds) *Community-based Water Law and Water Resource Management Reform in Developing Countries*, Comprehensive Assessment of Water Management in Agriculture Series, CABI Publishing, Wallingford.

Molle, F. and Mollinga, P. (2003) 'Water poverty indicators: conceptual problems and policy issues', *Water Policy* vol 5, pp529–544.

Sullivan, C. (2002) 'Calculating a water poverty index', *World Development*, vol 30, no 7, pp1195–1211.

UN-Water (2013) *Water Security and the Global Water Agenda: A UN-Water Analytical Brief*, United Nations University, Institute for Water, Environment & Health (UNU-INWEH), Canada.

Index